HERITAGE ECOLOGIES

Heritage Ecologies presents an ecological understanding of heritage that furthers a concern for how its making and unmaking always involves a wide range of human and other-than-human actors.

Recognizing the entangled nature-cultures of heritage is essential in the Anthropocene era, where uncertainty and rapid environmental change force us to recast common conceptions of inheritance and to envision new strategies for preservation. Heritage sites are meant to be open and shared spaces, and a recurring argument in the cases presented here is that this openness inevitably also overrides our selections, orders and appreciations. Through a diverse range of case studies, the chapters collected in this book aim to explore the affects and memories engendered by diverse heritage ecologies where humans are neither the sole makers nor the only inheritors. The common call is that the experiential, perceptive and informational plenitude enabled through contributions of other-than-human actors is key to an ecological rethinking of heritage in the twenty-first century.

Heritage Ecologies is unique in bringing heritage studies into closer proximity with a wide variety of non-representational and object-oriented theories and is an important volume for students and researchers in archaeology and heritage studies.

Torgeir Rinke Bangstad is a researcher at the Department of Archaeology, History, Religious Studies and Theology at UiT – The Arctic University of Norway.

Þóra Pétursdóttir is Associate Professor of Archaeology at the University of Oslo, Norway.

Archaeological Orientations

Series Editors:
Gavin Lucas
University of Iceland, Iceland
Christopher Witmore
Texas Tech University, USA

An interdisciplinary series that engages our on-going, yet ever-changing, fascination with the archaeological, *Archaeological Orientations* investigates the myriad ways material pasts are entangled with communities, animals, ecologies and technologies, past, present or future. From urgent contemporary concerns, including politics, violence, sustainability, ecology, and technology, to long-standing topics of interest, including time, space, materiality, memory and agency, *Archaeological Orientations* promotes bold thinking and the taking of risks in pressing trans-disciplinary matters of concern.

Ruin Memories
Materialities, Aesthetics and the Archaeology of the Recent Past,
Edited by Bjørnar Olsen, Þóra Pétursdóttir

Elements of Architecture
Assembling Archaeology, Atmosphere and the Performance of Building Spaces
Edited by Mikkel Bille and Tim Flohr Sorensen

Reclaiming Archaeology
Beyond the Tropes of Modernity
Edited by Alfredo González-Ruibal

Interrogating Human Origins
Decolonisation and the Deep Human Past
Edited by Martin Porr and Jacqueline M. Matthews

Heritage Ecologies
Edited by Torgeir Rinke Bangstad and Þóra Pétursdóttir

For more information on this series please visit: www.routledge.com/Archaeological-Orientations/book-series/ARCHOR

HERITAGE ECOLOGIES

Edited by Torgeir Rinke Bangstad
and Þóra Pétursdóttir

Routledge
Taylor & Francis Group
LONDON AND NEW YORK

First published 2022
by Routledge
2 Park Square, Milton Park, Abingdon, Oxon OX14 4RN

and by Routledge
605 Third Avenue, New York, NY 10158

Routledge is an imprint of the Taylor & Francis Group, an informa business

British Library Cataloguing-in-Publication Data
A catalogue record for this book is available from the British Library

Library of Congress Cataloging-in-Publication Data
Names: Bangstad, Torgeir Rinke, editor. | Þóra Pétursdóttir, 1978– editor.
Title: Heritage ecologies / [edited by] Torgeir Rinke Bangstad & Þóra Pétursdóttir.
Description: London ; New York, NY : Routledge, Taylor & Francis Group, 2021. | Series: Archaeological orientations | Includes bibliographical references and index.
Identifiers: LCCN 2021014270 (print) | LCCN 2021014271 (ebook) | ISBN 9781138294646 (hardback) | ISBN 9781138294608 (paperback) | ISBN 9781315101019 (ebook)
Subjects: LCSH: Cultural property. | Human ecology—History—Methodology. | Collective memory.
Classification: LCC CC135 .H4565 2021 (print) | LCC CC135 (ebook) | DDC 363.6/9—dc23
LC record available at https://lccn.loc.gov/2021014270
LC ebook record available at https://lccn.loc.gov/2021014271

ISBN: 978-1-138-29464-6 (hbk)
ISBN: 978-1-138-29460-8 (pbk)
ISBN: 978-1-315-10101-9 (ebk)

Typeset in Bembo
by Apex CoVantage, LLC

For Product Safety Concerns and Information please contact our EU representative:
GPSR@taylorandfrancis.com
Taylor & Francis Verlag GmbH, Kaufingerstraße 24, 80331 München, Germany.

CONTENTS

List of figures *viii*
List of contributors *xii*
Acknowledgments *xvii*

PART I
Introduction **1**

1 An ecological approach to heritage 3
Torgeir Rinke Bangstad and Þóra Pétursdóttir

PART II
Anthropocene **29**

2 Legacies: rethinking the futures of heritage and waste in the Anthropocene 31
Rodney Harrison

3 Scars: living with ambiguous pasts 49
Anna Storm

4 Wilderness heritage: for an ontology of the Anthropocene 66
Levi R. Bryant

5 Cultural heritage and memory of the ecumene in the age of the Anthropocene 81
Laurent Olivier

6 Oil matters 92
Esther Breithoff

PART III
Affect **105**

7 Emergent images: matters of affect in heritage photography 107
Colin Sterling

8 Affective encounters in museums 129
Marzia Varutti

9 A gentle shock of mild surprise: surface ecologies and the archaeological encounter 145
Tim Flohr Sørensen

10 From-the-hip: rocks and critical heritage ecology in the Western Australian Pilbara 165
Janice Baker

PART IV
Memory **183**

11 Mending shattered time: 22 July in Norwegian collective memory 185
Helge Jordheim

12 The remembrance of things: the industrial heritage of mining and the ecology of memory 208
Timothy LeCain

13 Interstitial heritage: *Industrienatur* and ecologies of memory 222
Torgeir Rinke Bangstad

14 Memory and redemption: lessons from a peasant ecology 246
Alfredo González-Ruibal

15 (Sm)all things remembered 264
Ingar Figenschau

PART V
Entanglements **283**

16 A positive passivity: entropy and ecology in the ruins 285
Caitlin DeSilvey

17 Heritage ecologies as worlding practices 306
Christina Fredengren

18 Mold, weeds and plastic lanterns: ecological aftermath in a derelict garden 327
Stein Farstadvoll

19 Heritage and the visual ecology of the Plantationocene 350
Rui Gomes Coelho

20 I shed tears, left, and forgot: the common frog, mosquitoes, and Grandmother Pine stayed 369
Suvi Tuominen and Marko Marila

PART VI
Epilogues/reflections **381**

21 Inheritance 383
Timothy Morton

22 Ecotone 391
Þóra Pétursdóttir

Index *403*

FIGURES

1.1	Heritage in volatile landscapes	8
1.2	Plastiglomerate – the alleged petrological marker of the Anthropocene	11
1.3	Human heritage is always entangled with the unpredictable actions and capacities of non-human actors	18
1.4	Clearings. An old forest road runs through a pine plantation in Øvre Pasvik Landscape Protection Area, Norway	23
2.1	Inside the SFR at Forsmark, Sweden	36
2.2	The Svalbard Tube, inside the SGSV, Norway	36
2.3	Traces of the former Berlin Wall conserved as "heritage" near Potsdamer Platz, Berlin, Germany	38
2.4	Plastic and rubbish floating in the ocean	39
2.5	Interpretive heritage signage at the Ongarue Spiral Trail, Oregon, USA	41
2.6	ISO nuclear radiation warning sign	41
3.1	The Chino copper mine, New Mexico, USA	51
3.2	Mass grave, originating from the Spanish Civil War, Estépar, Spain	52
3.3	Lowestoft Station, Lowestoft, UK	55
3.4	Lebbeus Woods, 1993. "War and Architecture: SCAB Construction"	56
3.5	Coral reef, Andaman Islands, India	58
3.6	Verdun, France	59
6.1	Back wall of a restaurant covered in oil memorabilia in Cayucos, California, USA	96
6.2	Ruined Mennonite house with roof partly made out of flattened oil canisters, Paraguay	97
6.3	Butter churn on display in the Koloniehaus Museum, Fernheim, Paraguay	98
6.4	Baking tray from the Geschichtliches Archiv Kolonie Menno, Loma Plata, Paraguay	99
6.5	Village fence partly made of oil drums at Long Tieng, Central Laos	99
6.6	Inside the Norsk Oljemuseum in Stavanger, Norway	100

6.7 Geopark "junk playground" outside the Norsk Oljemuseum in Stavanger, Norway 100
6.8 Oil-covered Kemp's ridley sea turtle rescued by the Louisiana Department of Wildlife and Fisheries, USA 101
6.9 Artificial reefs 101
7.1 John Thomson, 1866. *A Tower at Nakhon Thom [Angkor Thom], Cambodia* 107
7.2 John Thomson, 1866. *A Tower at Nakhon Thom [Angkor Thom], Cambodia* 111
7.3 John Thomson, 1866. *Prea Sat Ling Poun, Cambodia* 116
7.4 Woodcut engraving based on John Thomson's photograph 119
7.5 Screenshot from online British Library print shop: John Thomson, *Tower of Prea Sat Ling Poun* 122
7.6 Digital scan of photocopied page showing *Prea Sat Ling Poun* from John Thomson's *The Antiquities of Cambodia*, 1867 123
7.7 Digital scan of *Prea Sat Ling Poun* from John Thomson's *The Antiquities of Cambodia*, 1867 124
8.1 The Divine Cat 130
9.1 Life in the dark 145
9.2 Cave-in 150
9.3 Overgrowth 151
9.4 Window 152
9.5 Main entrance 153
9.6 Anna Rasmussen and Truelshøj 154
9.7 Collapse 160
9.8 Storage 161
9.9 Gravity 162
9.10 Skylight 163
10.1 Murujuga rock petroglyphs, Pilbara, Australia 167
10.2 Murujuga landscape. Rock petroglyphs and Karratha Gas Plant 172
10.3 Rock petroglyphs with Karratha Gas Plant in the background 175
10.4 Rock petroglyphs with Karratha Ammonia Plant in the background 179
11.1 Timeline at the 22 July Information Center in the Government Quarter, Oslo, Norway 191
11.2 Timelines in the report by the Gjørv Commission 192
11.3 The timeline from the newspaper *Verdens Gang* 194
11.4 The timeline from the Norwegian Broadcasting Corporation 195
11.5 The timeline from the 22/7 report by the Oslo police department 196
11.6 Jonas Dahlberg, "Memory Wound" 200
13.1 View toward the rumbustious *Industrienatur* at the northern end of Kokerei Hansa, Dortmund, Germany 224
13.2 The "black" side of Kokerei Hansa with conveyor belt on the right and coke ovens on the left 233
13.3 The interpretation sign outside the main entrance of Kokerei Hansa describes the site's metamorphosis into "*Industrienatur*" 235
13.4 Vegetation at Kokerei Hansa presented in the entrance of the former locker rooms 236
13.5 Interstitial species in anthropogenic settings 237

13.6	Habitable spaces open up in a decomposing brick wall building (former Sieberei-building)	238
13.7	Mnemonic environment at Kokerei Hansa industrial heritage site	240
14.1	The peasant landscape is better preserved in the mountains of eastern Galicia, Spain – despite afforestation	248
14.2	Some of the types of walls used to divide fields in Galicia	249
14.3	A heavily modified riverbank with bridge and paved way	251
14.4	An old path flanked by agricultural terraces peters out in the new-growth forest	253
14.5	Cleansing the past: a village with all the vernacular buildings demolished or disfigured	256
14.6	A traditional bedroom displayed in the Centro Etnográfico de Terra de Montes	260
14.7	Women threshing corn in the 1920s	261
15.1	Kitzbühel, Austria, 2016. The PoW camps appear as circles in the landscape	265
15.2	Two German M35s smashed and camouflaged in a small scree in Kitzbühel	266
15.3	A mountain of barbed wire percolating the ground, Spittal, Austria, 2016	269
15.4	A German helmet lingering in the leftovers of a barrack, Spittal, 2016	270
15.5	Canned food, Kitzbühel, 2017	271
15.6	Kitzbühel, 2016: excavation work	276
15.7	Post-excavation work: new care and new reminiscence, Spittal, 2016	277
15.8	The red cabin, Spittal, 2017	279
16.1	*A section of buddleia trunk as thick as my wrist had fused into the grid of steel*	290
16.2	*Silver birch trees grew in incongruous places*	291
16.3	*The meadow was in its high summer glory, abundant and sovereign*	294
16.4	*When I returned the next day, the meadow had been mown down, and workers were spreading fine gravel on the shorn ground*	295
16.5	*A high fence of steel slats, each sharpened to a threatening point*	299
16.6	*We began with the area around the cracked granite altar*	300
16.7	*One cypress specimen had its roots tangled in a snarl of red and white warning tape*	301
16.8	*The ruin will be ruined*	303
17.1	Fiskartorpet and the oak, Norra Djurgården, Stockholm	314
17.2	Lyell's Oak, cross-section, Djurgården, Stockholm	315
17.3	Installation of the artwork *The Flesh of a Female Goddess* at Husarviken, Stockholm	316
18.1	Wildlife in the thicket	327
18.2	A muddy garden path	329
18.3	Jelly fungus on a painted stick	332
18.4	Looking up at a moldy roof in the gardener's residence	335
18.5	Crawling under the Japanese knotweed	337
18.6	Spanish slug (*Arion vulgaris*)	339
18.7	A plastic lantern and tree litter	341
18.8	An ephemeral pond in the curl of the hippocamp's tail	343

19.1 Fazenda Boa Esperança. Domingos Braga, Minas Gerais 352
19.2 Fazenda Boa Esperança. Domingos Braga, Minas Gerais 354
19.3 Fazenda Boa Esperança. Domingos Braga, Minas Gerais 355
19.4 Fazenda Boa Esperança. Domingos Braga, Minas Gerais 356
19.5 Fazenda Boa Esperança. Domingos Braga, Minas Gerais 357
19.6 Fazenda Boa Esperança. Domingos Braga, Minas Gerais 358
19.7 Fazenda Boa Esperança. Domingos Braga, Minas Gerais 361
19.8 Rui Gomes Coelho, 2019. *The Road to Massambará* 362
19.9 Rui Gomes Coelho, 2019. *The Road to Massambará* 363
19.10 Rui Gomes Coelho, 2019. *The Road to Massambará* 364
19.11 Rui Gomes Coelho, 2019. *The Road to Massambará* 364
19.12 Rui Gomes Coelho, 2019. *The Road to Massambará* 365
19.13 Rui Gomes Coelho, 2019. *The Road to Massambará* 365
20.1 Barbed wire is a ubiquitous material commonly found at Second World War heritage sites 370
20.2 Suvi practicing somatics with Grandmother Pine 371
20.3 Remains of a wooden cargo crate at Haukkapesäjoki, Inari, Finland 372
20.4 Remains of an unknown structure hide in a small pond next to the ruined Kankiniemi PoW camp 373
20.5 A common frog at the Haukkapesäjoki PoW campsite 373
20.6 Lapland is notorious for the huge amount of mosquitoes in the summer 374
20.7 Leaf beetles make love in the midst of the remains of an unruly past 375
20.8 Remains of a wooden cargo crate provide conditions for fungi to thrive 376
20.9 Suvi repeating and reconstructing postures known to have belonged to the choreographic repertoire of Elsa Puolanne 377
20.10 Suvi repeating and reconstructing postures known to have belonged to the choreographic repertoire of Elsa Puolanne 378
20.11 A common frog contemplating the ecology of a Second World War heritage site near Haukkapesäjoki, Inari 379
22.1 Writer Andri Snær Magnason talks to media by the foot of Ok, SW Iceland 394
22.2 Trampled pathway leading between the "science station" and the seagull colony 396
22.3 Drift matter, wood and plastic on Surtsey's shore 398
22.4 Lush grasses enclose the helicopter platform in Surtsey 399

CONTRIBUTORS

Janice Baker teaches visual culture and creative practice research at Curtin University in Western Australia in the School of Media, Creative Practice and Social Inquiry. She supports art and material encounters that engage in the production of subjectivities for sustainable futures. Taking an interdisciplinary approach to research, she addresses current concerns in creative practice, post-humanism and the geo-humanities within a framework of critical theory, art history and cultural studies. To engage in current social, cultural and political issues, Baker explores innovative encounters and speculates on better futures through their manifestation as fictive and creative encounters.

Torgeir Rinke Bangstad is a researcher at the Department of Archaeology, History, Religious Studies and Theology at UiT – The Arctic University of Norway. He is affiliated with the research project Unruly Heritage: An Archaeology of the Anthropocene, and his research interests include contemporary archaeology, museology, memory and heritage studies and the preservation of twentieth-century architecture and industrial heritage. During his postdoctoral research (2016–2019), he investigated the memory and heritage of post-war reconstruction housing in Finnmark, Norway. Bangstad holds a PhD in cultural heritage studies from the Norwegian University of Science and Technology in Trondheim.

Esther Breithoff is a lecturer in contemporary archaeology and heritage and a UKRI Future Leaders Fellow in the Department of History, Classics and Archaeology at Birkbeck, University of London. She previously held postdoctoral positions at UiT – The Arctic University of Norway and the UCL Institute of Archaeology, London. Her research spans the fields of contemporary archaeology and critical heritage studies and has ranged across a number of different topics – including war, natural and cultural heritage, nuclear and petroleum industries, dictatorships and biobanking – but traces a common set of interests in the relationships between conflicts, resources, recycling and rights across the human/non-human divide in the Anthropocene era.

Levi R. Bryant is a professor of philosophy at Collin College outside of Dallas, Texas. He is the author of *Difference and Givenness: Deleuze's Transcendental Empiricism and the Ontology* *of Immanence* (2008), *The Democracy of Objects* (2011), and *Onto-Cartography: An Ontology of Machines and Media* (2014). He has written widely on post-structuralist thought, object-oriented ontology and speculative realism. When he is fortunate, he occasionally accompanies archaeologists on their fieldwork.

Rui Gomes Coelho is Assistant Professor in the Department of Archaeology at Durham University, United Kingdom. He is also affiliated with the Centre for Archaeology, University of Lisbon, Portugal. Coelho is interested in historical archaeology and committed to the ongoing struggle to decolonize the discipline. He is currently working on two main projects: the archaeology of slavery in Cacheu, Guinea-Bissau, and the archaeology of anti-fascist resistance in Drežnica, Croatia.

Caitlin DeSilvey is Professor of Cultural Geography, based at the University of Exeter's Cornwall campus, where she is Associate Director for Transdisciplinary Research in the Environment and Sustainability Institute. Her research into the cultural significance of material change has involved extensive collaboration with heritage practitioners, archaeologists, ecologists, artists and others. From 2015–2019, she was co-investigator on the Heritage Futures project (funded by the Arts and Humanities Research Council), and in 2016–2017, she was a fellow at the Centre for Advanced Study, Oslo. Her publications include *Anticipatory History* (2011, with Simon Naylor and Colin Sackett) and *Visible Mending* (2013, with Steven Bond and James R. Ryan). *Curated Decay: Heritage Beyond Saving* (2017) received the University of Mary Washington's Center for Historic Preservation 2018 Book Prize.

Stein Farstadvoll is an archaeologist and postdoctoral researcher at the Department of Archaeology, History, Religious Studies and Theology at UiT – the Arctic University of Norway. Farstadvoll has extensive experience from archaeological excavations and surveys, including sites dating from the Mesolithic and up to the recent past, such as WWII POW camps in northern Norway. His research interests include archaeology of the contemporary past, critical heritage studies, thing theory and archaeological reconstructions. Recent publications include a paper on plants and contemporary archaeology with the title "Growing Concerns: Plants and Their Roots in the Past" (*Journal of Contemporary Archaeology* 2018) and an article on vestigial and rudimentary things called "Vestigial Matters: Contemporary Archaeology and Hyperart" (*Norwegian Archaeological Review* 2019).

Ingar Figenschau holds a PhD in archaeology from the Department of Archaeology, History, Religious Studies and Theology, UiT – The Arctic University of Norway. He has worked as a field archaeologist for Tromsø Museum, Sogn og Fjordane County, UiT and The Sami Parliament. His research interests include studies of medieval and early modern axes and craft production in northern Norway, with a focus on the interaction between craftsman, raw material and tools in both production and use. In his PhD project, Figenschau has investigated prisoner of war camps from the Second World War in Kitdalen, Troms County, with a particular emphasis on how an archaeological approach may shed light on untold stories of life in the camps.

Christina Fredengren is Associate Professor at the Department of Archaeology and Classical Studies, Stockholm University, and is a founding member of the Stockholm University Environmental Humanities network. Her research deals with archaeology, heritage and matters of intra-generational ethics and care. Christina's current research focuses on more-than-human relations and water and sacrifice, as reflected in her research projects, "Waters of the Time" and "Checking-in-with-Deep Time", financed respectively by the Swedish Science Council and Formas. Christina is Scientific Leader of the Deep Time research cluster at the Formas/Mistra-funded SeedBox at Linköping University and Affiliated Researcher at the Posthumanities hub.

Alfredo González-Ruibal is a researcher with the Institute of Heritage Sciences of the Spanish National Research Council. He specializes in the archaeology of the contemporary past and African archaeology. His research interests include the darker side of modernity (conflict, dictatorship, capitalism, mass migration) and resistance to modernity and the State. He has explored these topics in Spain, Brazil, Equatorial Guinea, Ethiopia and Somaliland. Recent books include *An Archaeology of the Contemporary Era* (2019) and *The Archaeology of the Spanish Civil War* (2020), both with Routledge.

Rodney Harrison is Professor of Heritage Studies at the Institute of Archaeology, University College London. He was AHRC Heritage Priority Area Leadership Fellow (2017–2020), Principal Investigator of the Heritage Futures Research Programme (2014–2019) and led the Work Package on "Theorizing heritage futures in Europe: heritage scenarios" as part of the EC-funded Marie Skłodowska-Curie action (MSCA) Doctoral Training Network CHEurope: Critical Heritage Studies and the Future of Europe (2016–2021). He is the founding editor and editor-in-chief of the *Journal of Contemporary Archaeology* and was a founding executive committee member of the Association of Critical Heritage Studies. He is (co)author or (co)editor of 17 books and guest edited journal volumes and over 80 peer-reviewed journal articles and book chapters. His research has been funded by the Arts & Humanities Research Council (AHRC), UKRI Global Challenges Research Fund, British Academy, Wenner-Gren Foundation, Australian Research Council, Australian Institute of Aboriginal and Torres Strait Islander Studies and the European Commission.

Helge Jordheim is Professor of Cultural History at the University of Oslo. He received his PhD in German literature from the University of Oslo in 2006 for a work on genre and politics in eighteenth-century Germany ("Der Staatsroman im Werk Wielands und Jean Pauls", 2007). He has published extensively on eighteenth-century intellectual culture in Europe. Among his latest books are global histories of the concepts of civility and civilization (*Civilizing Emotions*, 2015) and the world (*Conceptualizing the World*, 2018), as well as a history of universal history (*Universal History and the Making of the Global*, 2018, Routledge), written with international teams of scholars. Jordheim is heading several projects exploring times in the plural at the interface between cultural and natural history. He has written extensively on time and concepts in journals such as *History & Theory* and *History of the Present*.

Timothy LeCain is currently Professor of History at Montana State University, USA. He is the author of *The Matter of History: How Things Create the Past* (2017), which argues for a less anthropocentric theory and method that emphasizes the central role of things like cattle, copper and silkworms in making both humans and history. LeCain's first book, *Mass Destruction* (2009), won the 2010 best book prize from the American Society for Environmental History.

LeCain has been an invited research fellow at the Rachel Carson Center in Munich, Germany, and the Center for Advanced Study in Oslo, Norway.

Marko Marila is a postdoctoral researcher at the University of Helsinki. Marila has published and taught on topics related to archaeological theory, cultural heritage and semiotics. Within these fields, his primary research interests have included the history and philosophy of archaeology, the ontological and epistemological relevance of vagueness and speculation in archaeological theorizing, the nature and limits of multidisciplinarity in the context of archaeology's scientific revolutions and the intersection of archaeology and art. In his future research, Marila hopes to delve deeper into the development of speculative methods, especially through the amalgamation of archaeological philosophy with artistic practice.

Timothy Morton is Rita Shea Guffey Chair in English at Rice University, Texas, USA. They are part of the object-oriented philosophy movement, and their work is focused on the intersections of object-oriented thought, ecological studies and the arts. They have collaborated with, amongst others, Björk, Ólafur Elíasson, Laurie Anderson and Jeff Bridges. Their publications include *Being Ecological* (2018), *Dark Ecology: For Logic of Future Coexistence* (2016), *Hyperobjects: Philosophy and Ecology after the End of the World* (2013), *The Ecological Thought* (2010) and *Ecology Without Nature* (2007).

Laurent Olivier is Curator in Chief of the Department of Celtic and Gallic Archaeology at the National Museum of Archaeology in Saint-Germain-en-Laye, France. His work is focused on the social and anthropological approach of European Iron Age archaeology, especially through archaeology of salt and early Celtic art. Olivier's research interests include history and theory of the archaeological discipline, as well as the archaeology of the contemporary past. He has published *Le Sombre abîme du Temps. Mémoire et archéologie* (2008), which is available in English (*The Dark Abyss of Time. Archaeology and Memory*, 2011). One of his last books is devoted to the "germanization" of European archaeology during the Nazi period (*Nos ancêtres les Germains. Les archéologues français et allemands au service du nazisme*, 2012). *Le Pays des Celtes* (2018) is focused on the memory of Celtic and Gallic pasts.

Þóra Pétursdóttir is Associate Professor of Archaeology at the University of Oslo, Norway. Her research interests are mainly within contemporary archaeology, critical heritage studies, memory studies and archaeological theory and practice. She has published widely on these topics, including the volumes *Ruin Memories: Materialities, Aesthetics and the Archaeology of the Recent Past* (Routledge 2014, together with Bjørnar Olsen) and *After Discourse: Things, Affects, Ethics* (Routledge 2020, together with Bjørnar Olsen, Caitlin DeSilvey and Mats Burström). She is currently developing a research project on natural heritage in the Anthropocene.

Tim Flohr Sørensen is an associate professor of contemporary archaeology and archaeological theory at the Saxo Institute's department of archaeology, University of Copenhagen. He has been employed there since 2014, after working at Aarhus University (2012–2014) and the University of Cambridge (2010–2012). His research focuses on archaeological epistemology, ruins and empty places, in addition to objects with little or no culture-historical significance or context. Also, he is the director of *The Hub for Speculative Fabulations Upon Incidental Observations.*

Colin Sterling is Assistant Professor in Memory and Museums at the University of Amsterdam. His research focuses on critical-creative approaches to heritage, memory and museums. He is the author of *Heritage, Photography, and the Affective Past* (Routledge, 2020) and co-editor with Rodney Harrison of *Deterritorializing the Future: Heritage in, of and after the Anthropocene* (2020). Sterling was previously a Research Associate and AHRC Early Career Leadership Fellow at UCL Institute of Archaeology, London. His current work explores questions of experiential design, the posthumanities, the Anthropocene and the role of museums in climate action.

Anna Storm is Professor of Technology and Social Change at Linköping University in Sweden. Her research interests are centred on industrial and post-industrial landscapes and their transformation, in the physical as well as imaginary sense, and comprise both cultural and "natural" environments. Such landscapes not only trigger perspectives of power relations but challenge our understandings of ecology, aesthetics, memory and heritage. Recent work more specifically deals with nuclear nature imaginaries, deployed through human relations to animals in nuclear settings, such as fish, snakes, bison and wild boar. Anna Storm is currently leading the multidisciplinary project "Atomic Heritage Goes Critical: Waste, Community and Nuclear Imaginaries" and is Senior Researcher in "Cold War Coasts: The Transnational Co-Production of Militarized Landscapes" and "Nuclearwaters: Putting Water at the Centre of Nuclear Energy History". She is the author of *Post-Industrial Landscape Scars* (2014), which was shortlisted for the Turku Book Award 2015.

Suvi Tuominen is an artist, archaeologist and dancer based in Helsinki, Finland. She holds degrees in archaeology and contemporary dance and is currently pursuing one in live art and performance studies at the Theatre Academy of the University of the Arts, Helsinki. Her research circles around such concepts as orientation, intimacy, heritage, aliens, proximity and withdrawal. Tuominen is interested in the dramaturgies and dynamics of bodily archives and the corporeal dimensions of materiality. Her artistic works operate mostly in interdisciplinary and collaborative settings. Tuominen's artistic works take shape in various mediums and platforms, from live performances to installation works.

Marzia Varutti is Associate Professor in Cultural History and Museology at the Centre for Museum Studies, IKOS, University of Oslo. Her research has focused on relationships between museums and Indigenous Peoples (specifically in Taiwan) and various aspects of the politics of representation in museums. She is the author of *Museums in China: The Politics of Representation after Mao* (2014). Her recent articles include "Authentic Reproductions: Museum Collection Practices as Authentication" (*Museum Management and Curatorship* 2018), "Crafting Heritage: Artisans and the Making of Indigenous Heritage in Contemporary Taiwan" (*International Journal of Heritage Studies* 2015) and "Polysemic Objects and Partial Translations" (*Museum Anthropology* 2014).

ACKNOWLEDGMENTS

The idea of compiling a book on *Heritage Ecologies* was first born during a research stay at the Centre for Advanced Study (CAS) at the Norwegian Academy of Science and Letters in Oslo in 2016–2017. The research project, After Discourse: Things, Archaeology, and Heritage in the 21st Century, led by Bjørnar Olsen, and the many international scholars who visited the Centre during this year, engendered a dynamic and fruitful environment where diverse matters of heritage were tackled, debated and redrawn. We feel greatly indebted to all these friends and colleagues. Additionally, we express our gratitude to the administrative staff at CAS for facilitating our stay and for making it a rewarding experience for all.

Further, this book has gained greatly from two consecutive research projects hosted at UiT – The Arctic University of Norway. These are Object Matters: Archaeology and Heritage in the 21st Century (2014–2018) and Unruly Heritage: An Archaeology of the Anthropocene (2017–2023), both funded by the Norwegian Research Council and directed by Bjørnar Olsen. Participating in these research projects has enabled us to meet with many brilliant researchers over the years, and we extend our gratitude to the project teams and other contributors, many of whom are represented among the chapters in this book. Special thanks go to Bjørnar Olsen – without his commitment to establishing and maintaining these unique research milieus, this book would not have seen the light of day.

We are grateful to Sofie Scheen Jahnsen, research assistant at the Department of Archaeology, Conservation and History and research and outreach coordinator for HEI: Heritage Experience Initiative at the University of Oslo for her assistance in the final stages of editing and formatting individual chapters. Margrethe K. H. Havgar, Research and Outreach Coordinator at HEI: Heritage Experience Initiative, has compiled the book's index, and we are very grateful for her meticulous work. Thanks also go to our good friend, Stein Farstadvoll, postdoctoral researcher at UiT – The Arctic University of Norway, who has kindly permitted us to use his wonderful photographs in our introductory chapter, on the title pages of subchapters and on the front cover of this book. We would also want to thank the series editors, Gavin Lucas and Christopher Witmore, for believing in our initial proposal and for their assistance during the process.

Finally, but most significantly, we would like to extend our thanks to all the researchers who have contributed with chapters to this book. Thank you all for sharing your work and for showing us patience along the way.

Þóra Pétursdóttir
University of Oslo
Torgeir Rinke Bangstad
UiT – The Arctic University of Norway

PART I
Introduction

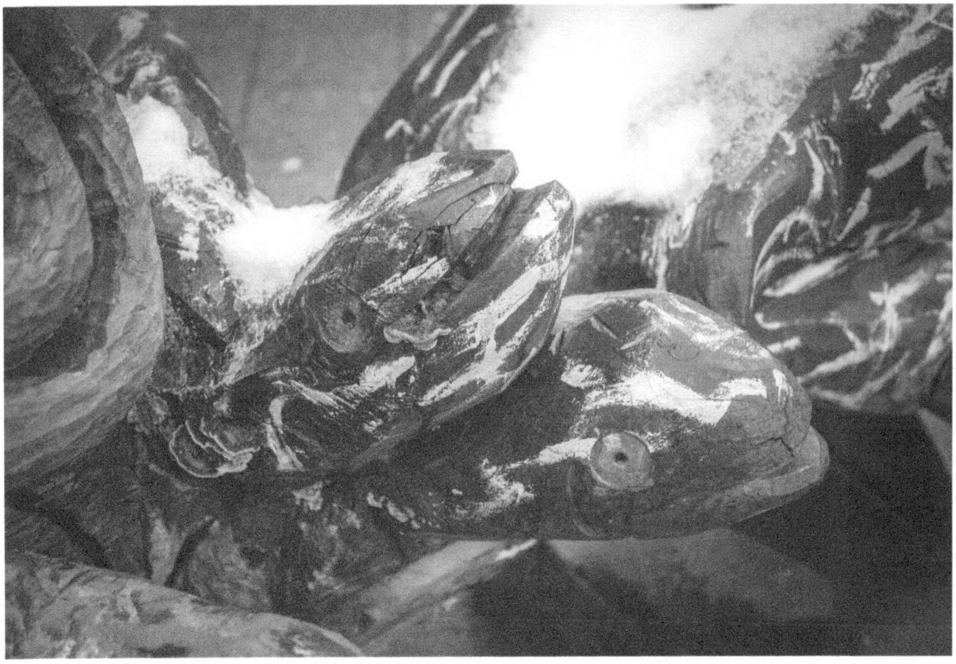

1

AN ECOLOGICAL APPROACH TO HERITAGE

Torgeir Rinke Bangstad and Þóra Pétursdóttir

The naming of the Anthropocene as a new geological epoch suggests some fundamental changes in the relationship between humans and their environment. The term itself has resonated widely in scientific and popular debate in recent years and has encouraged a deeper reflection on phenomena that cross the nature-culture divide and challenge categories we have long taken for granted. Classic issues in the humanities, like memory, aesthetics, experience and meaning, have been reconfigured and are often treated as embedded in a wider ecology of worldly material and organic environmental relations, rather than transcendent properties that set our species apart from the world. One manifestation of this is how, in recent decades, ecocriticism and an ever broader use of various ecological frameworks have rendered *ecology* a mainstay across the humanities and social sciences, represented for example by media ecology (Fuller 2007), ecology of literature (Zapf 2008) theatre ecology (Kershaw 2007), memory ecology (Hoskins 2016) and here: heritage ecology. Anchored in different disciplinary traditions, these ecological approaches all share an interest in the entangled compositions and workings of mediatic environments, configured through an "indissoluble interconnectedness and dynamic feedback between culture and nature, mind and matter, text and life" (Zapf 2008, 86).

The term "ecology", however, has a much longer pedigree and was first used in 1866 by the German zoologist and evolutionary biologist Ernst Haeckel (1834–1919), who understood it as "the whole science of the relations of the organism to the environment including, in the broad sense, all the 'conditions of existence'" (Haeckel 1866, 286, translated by Stauffer 1957, 140). Still to this day, this seems to be the most commonly accepted and widespread definition of ecology. For Haeckel, ecology belonged under the general physiology of organisms alongside chorology, or zoogeography, which was concerned with the spatial distribution of species. Ecology would for Haeckel mean that the *relations* of organisms to their organic and inorganic environment were placed front and centre. *Oikos*, which literally means "house" or "household", is the etymological root of "eco" in both ecology and economy. Similar to other naturalists such as Carl Linnaeus, Charles Lyell, Alexander von Humboldt and Charles Darwin, Haeckel referred repeatedly to the term "economy of nature" (Pearce 2010), implying a greater order maintained through a finely tuned balance of forces in nature known today as "dynamic equilibrium" (Stauffer 1957).

This idea of systemic *balance* has largely persisted throughout the later development of ecosystems theory. While modern ecosystem thinking is often described as the study of energy flows within a system, disturbances observed are, nevertheless, over time expected to develop towards the reinstating of an equilibrium, of nature's fundamental balance. This is one of the enduring legacies of ecological thinking in the nineteenth and twentieth centuries and one that has proven highly influential in the humanities and social sciences as well. Social scientists of the 1960s and 1970s approached human cultural systems in a similar way, drawing attention to analogous properties with "habitats" (Geertz 1963) and "niches" (Barth 1956) in the natural world. "Human" or "cultural ecology" were the names given to approaches in anthropology that treated cultural characteristics, such as different forms of subsistence technology and land ownership, as key to understanding how human societies adapted to a specific environment or habitat (e.g. Steward 1972).

Characteristic of these early approaches, however, is a clear distinction between culture and nature, where anthropogenic environmental change is not yet a significant factor and where the environment is treated as a given, a more or less unchanging set of natural conditions that human cultures adapt to through technological development. Hence, cultural ecology was eventually criticized for not being able to account for the changes that modern societies caused to environments. Adaptation, as a term, did not reflect the large-scale intentional manipulation of landscapes through agriculture, deforestation, urbanization and the extractive industries, nor how "nature" was rendered a "sediment of human practice, a construction in the material and not merely semantic sense of that word" (Biersack 1999, 9): "nature" *qua* artefact. It is, therefore, perhaps only in retrospect, looking back from a time when the aggregated actions of successive generations of human beings are finally being recognized as a geo-physical force of planetary reach, that we fully appreciate Haeckel's contribution to an *other* ecology, where the history of human beings was already legible in Earth's sedimented heritage. In 1868, he described the most recent phase of geological time as the "era of man", "the Anthropolithic" or "the Anthropozoic age", characterized by the wide dispersal of the human species and of human culture:

> Man has acted with a greater transforming, destructive, and modifying influence upon the animal and vegetable population of the earth than any other organism. For this reason, and not because we assign to man a privileged exceptional position in nature in other matters, we may with full justice designate the development of man and his civilization as the beginning of a special and last main division of the organic history of the earth.
>
> *(Haeckel in: Kutschera and Farmer 2020, 2)*

Interestingly, this is not far from contemporary approaches where ecology encompasses not only how human culture is conditioned by natural forces, but also how humans shape natural history on a local, regional and even planetary scale. Through technological means and conscious, intentional appropriation of nature, the human animal has long challenged any natural limits of a given habitat, and in comparison to other animal species, human beings seem far less conditioned by their natural surroundings. Humans have demonstrated "an adaptive capacity unrivalled among vertebrates" (Sullivan and Malkmus 2016, 5), which also implies that humans in comparison with other animals are "*under*-determined by nature and that the consequences of their actions are more significant and far-reaching" (Soper 2012, 366). Is it possible, with this crucial caveat in mind, to regard humans as simply a species among species, or is ecological thinking required to appreciate the disproportional and de facto exceptional status of humans as an unusually destructive force? Political ecology, ecocriticism and ecophilosophies increasingly venture beyond a notion of ecology as metaphor or analogy to raise questions regarding how to "best address – and improve upon – our current ecological situation" (Soper 2012, 366). The term "ecology" was popularized in the 1960s due to the growing public awareness of the environmental costs of human growth. Rachel Carson's *Silent Spring* (1962), not the least, contributed strongly to this by drawing attention to the long-term effects of pesticides on soil ecology, wildlife and human health. The further development of "deep ecology", devised by the Norwegian philosopher Arne Næss (1973), became central in the critique of the anthropocentrism of Western Enlightenment philosophy which posited the self-contained, rational human subject as detached from the natural world.

And further yet, in contrast to earlier forms of ecocriticism and certainly in contrast to the deep ecology movement, several contemporary ecological thinkers have attempted to broaden the domain of the ecological beyond a rigid environmentalist discourse rooted in the idea of nature as an autonomous domain outside of cultural influence. The appreciation of large-scale processes that cut across these ideal domains reflects a need to reckon with what ecology might mean "after nature" (Escobar 1999; Morton 2007). Advances in biotechnology, molecular biology, nanotechnology and synthetic biology present several challenges to the conventional notion of natural organic life, and it has become more critical to bring ecological thinking past the notion of nature as a redemptive outside realm (Morton 2007, 2015, 2016).

This was also recognized, among others, by the French philosopher Felix Guattari, who in contrast to the naturalism of the deep ecologists, advanced a transversal thinking to enable an understanding of the interactions between natural ecosystems, the mecanosphere and the human, social universe. Guattari claimed that, "[n]ow more than ever, nature cannot be separated from culture ... " (2000, 43). These kinds of observations have become more common with the recognition of the profound human impact on earth's ecosystems, which makes the insistence on the fundamentally distinct realms of culture and nature difficult to sustain. This, we argue, is also why an ecology of heritage should attempt to exfoliate the binaries of culture and nature, human and non-human and make room for the appreciation that heritage phenomena are entangled in more-than-human material and environmental processes.

The contributions collected in this volume do not represent a uniform methodology or an ecology where tensions and provocations resolve and ensure a higher-level stability. Rather, the key insight of the diverse perspectives offered here, is the need to reflect critically on the anthropocentric assumption that heritage is an insular human property detached from its messy worldly relations.

To realize this – and while also embracing that the volume itself has grown into a rich and diverse ecology of cases and ideas – we particularly encourage a view of heritage from the perspective of ecological frameworks abounding in recent years under headlines such as object-oriented ontology, new materialism and post-humanism. As explained by Levi R. Bryant and Eileen A. Joy "the post- and nonhuman thingly turn is essentially an ecological vision of the world" (2014, x). Ecology, as they propose, implies a concern for "relations between entities of all kinds" (ibid, xi) – be they mineral, animal, human, plant or other – and how these act and interact. As such, thinking ecologically is to avoid anthropocentric understandings of agency and explore instead the distributed quality of emergent heritage phenomena. Importantly, this does not mean that entities are reduced to their relations or that their differences are replaced with symmetrical or flat ontologies. Rather, an ecological understanding of heritage is here concerned with the entanglements that take part in the formation of heritage and engender certain forms of significance, but where no one species, kind or relation is favoured by default. This is also essential in order to address issues of power vested in the humanist idea of the conscious and knowing subject that has foreclosed many non-discursive, affective, material, mnemonic and environmental accounts from the history of heritage preservation. For example, accounts of how non-human organisms make meanings by sensing patterns and recognizing novelty in their surroundings (Parikka 2010), how many biological systems are capable of "remembering", without the aid of a conscious mind (Favareau et al. 2017), or how memory becomes enfolded in the world's ongoing material reconfiguration and "sedimenting historicity" (Barad 2018, 84).

Understandings of humans as the prime makers of worlds and meanings have become inadequate, and there is, we believe, a need to replace linguistic, social constructivist and anthropocentric theories with a more materially oriented, environmental consideration of the extra-discursive conditions that enable, delimit and transform heritage. Thus, while the authors contributing to *Heritage Ecologies* are in no way a univocal group, they do signal a conscious effort to populate heritage with a broader range of affects and non-human actors. To us, ecological thinking represents a way of questioning the categories, scales and temporalities we have grown accustomed to as part of a long tradition of social-constructivist, representationalist, anthropocentric approaches. Here, the aim of ecological heritage research is to stress the interrelations between myriad actors and extras in our "theatres of memory" (Samuel 1994) in order to explore processes of heritage making, engagement and preservation.

Heritage and humanism

The study of heritage concerns fundamental questions about the human condition in the contemporary world. The right to control, define and actively engage with one's own heritage is increasingly framed as a fundamental human rights issue, as visible in twentieth-century articulations of an ever wider, political commitment to preserve the world's cultural and natural diversity to make it accessible for present and future generations (Meskell 2010, 2015).

Most formal definitions of heritage foreground the role it plays as a *resource* that may help sustain the life and wellbeing of human communities in the present and for the future. As such, heritage in its variety constitutes a significant building block of the human *oikos*, or human "household". This not only pertains to natural landscapes, ecosystems or wildlife habitats but just as much to value priorities and matters of culture. Hence, the understanding of heritage as resource can be seen as ecological insofar as it recognizes that human lifeworlds rely on a precarious balance between the organic and inorganic environment of a system that requires precautionary measures and care to remain healthy.

Here, the idea of sustainability has also had significant leverage on the understanding of heritage as a limited and non-renewable resource (Holtorf 2008). A responsible and considerate use of heritage resources is seen as integral to sustainable development at a more general level. This is, for example, evident in the preamble to the Faro Convention (Council of Europe 2005), where the role of heritage in engendering sustainable development and quality of life in a constantly evolving society is emphasized. At the same time, present-day communities should manage their use and engagement wisely to ensure that it will not deny future communities the same potential for self-expression and -realization. Heritage is here and elsewhere seen as a fundamental part of human values, beliefs, identities, knowledge and traditions, as well as the manifestation of interaction between people, land and environment over time.

It would therefore not be a gross exaggeration to say that heritage has become an indissoluble part of the Enlightenment humanist question of what it means to be human, "a concept of that which is proper to man" (Derrida 2001, 25). Moreover, the distinction between nature and culture reflected in most general taxonomies of heritage also mirrors the modern Western idea of humanity as ontologically distinct from nature. In modern Enlightenment thought, nature constituted an ontological negativity that humans would transcend and control through technological advance and cognitive signification, thus engendering the notion of the human being and, more broadly, humanity as *a priori* opposed to animality (Bignall and Rigney 2019, 166). Contemporary heritage practices maintain this fundamental distinction, and the efforts out into securing the physical integrity of objects handed down through time is evidence of the strenuous work of keeping natural elements at bay to stabilize a dynamic heritage prone to change and transformation.

The world of heritage abounds with cultural creations that seem to defy time and gravity, and it is not uncommon to think of heritage as an expression of the existential reassurance human beings seek in the relative stability of the material world. "We can control our collections as we can few other matters in this world", Susan Pearce (1995, 178) has claimed, and she relates the acquisitive nature of the collector to a desire to transcend corporal decay, fragility and vulnerability. The imagined equilibrium in the world of resource management is, however, primarily understood from the point of view of the human being and the value things have for us. And while it acknowledges the external, natural limits to growth, it is built on the assumption that such resources exist as discrete entities, given in time and space, and as components of an ideally balanced and manageable system.

Anthropocene

Assumptions of systemic order are increasingly questioned in non-equilibrium models of ecology, a theoretical shift where the notions of stability and balance are traded for "forms and processes with multiple and often divergent trajectories, wherein single events can have significant and unforeseeable consequences" (Lorimer 2015, 29). This involves acknowledging non-linear developments and more-than-human agency, which make it difficult to predict and foresee what will occur in an ecosystem, for instance, after human interference. Current ecological thinking thus often foregrounds disruption, randomness and stress resilience instead of homeostatic stability. Needless to say, this may present a tremendous challenge to conservation regimes where human intervention is based on prediction and an aim for balance and stability. Nevertheless, a deeper reckoning with the unpredicted, with instability, discontinuity, loss and change, has arguably become inevitable to management schemes in the geological era of the Anthropocene. While an idea of lingering threat may be seen as essential to heritage designations generally, heritage in the contemporary world is exposed to unprecedented challenges that will require a fundamental rethinking. We may even be forced, some have suggested, to give up the "dream of preservation" as something forever out of reach (Harvey and Perry 2015, 12) and instead acknowledge that even our collections are in fact subject to the same forces that continuously shape and reorder "other matters" in the world.

The heightened sense of vulnerability apparent in the contemporary world has engendered different ways of dealing with rapid and anticipated heritage loss (DeSilvey and Harrison 2020). As Bruno Latour has remarked, Earth itself has become the active subject, a "quaking, and easily tickled envelope" (Latour 2014, 4), and it seems increasingly difficult from the point of view of heritage preservation to keep pace with this rampant force. In October 2019, a lighthouse at Rubjerg Knude in northwestern Jutland, Denmark, first lit in the year 1900, was moved about 80 metres inland to a location safe from the eroding cliff edges (Associated Press 2019). The complex and expensive operation is said to keep the lighthouse, a treasured national heritage object, safe for at least another 40 years (Figure 1.1). Effectively passing the obligation on to the next generation, the relocation of the lighthouse reflects the temporary arrangements that are set up to mitigate the effects of large-scale physical change on monuments, cultural landscapes and the heritage of the built environment. Dunes are highly dynamic landscapes that change at a rapid pace, and the lighthouse at Rubjerg Knude illustrates the lengths that contemporary societies go in order to save heritage for future generations.

FIGURE 1.1 Heritage in volatile landscapes. The lighthouse Rubjerg Knude Fyr in northern Jutland, Denmark was built in 1900 and moved further away from the eroding cliff edge in 2019

However, this example begs the question: is a stable and predictable material environment a necessary precondition for public affirmation and valuation of heritage, or are there alternative ways to perceive of and respond to *heritage in change*?

The anticipation of loss and a generalized sense of endangerment is often reflected in heritage policies and conservation practices that address both long- and short-term futures (Harvey and Perry 2015; DeSilvey and Harrison 2020). However, in times of more frequent extreme weather events, thawing ice and sea level rise, traditional virtues of heritage management, such as material stability and preservation, may require more radical rethinking (Harvey and Perry 2015) and, moreover, a decoupling from "endangerment narratives" (DeSilvey and Harrison 2020) where (climate) *change* inevitably equals *loss* (Harvey and Perry 2015). Against this conservative ideal of permanence and preservation, David C. Harvey and Jim Perry (2015, 13) have claimed that adaptation requires a fundamental rethinking of heritage where "natural processes" are no longer seen as separate from "cultural context". There is, in other words, good reason to reflect on how different "regimes of care" (Harrison 2015) are constituted through and affected by natural change, contingency and formation processes. Without claiming that our current times are essentially exceptional, we believe that an ecological approach to heritage in the Anthropocene is apt because the rigid distinction between human history and natural history is becoming ever more blurred and impossible to maintain. The anthropogenic material assemblages that this age has spawned are already part of the planet's deep geological futures and literally contest against attempts to distinguish beings as either biological, archaeological or geological.

This is also why a notion of the human *oikos* as composed of natural and cultural *resources* dependent on our management is not ecological quite in the sense we want to espouse in *Heritage Ecologies*. The limit of this particular idea is that it often fails to account for the more sombre qualities of environmental entanglement that define the human condition and how it is intricately bound up with other existences. This was, for example, made manifest by the sudden and violent interruption of the COVID-19 pandemic in 2020, which has forced us to recognize that human health and wellbeing are inextricably connected to animal health and not least the loss of wildlife habitat. It is, in other words, becoming increasingly difficult to think of *oikos* in spatial terms, as an *external* environment that encircles and supports human lifeworlds. "[W]ildlife lives among us" (Lorimer 2015, 7) and it comprises everything from the bacteria in our guts to feral plants and roaming animals in novel, urban ecosystems (see Farstadvoll, chapter 18, and also Fredengren, chapter 17). It also includes the lively reefs around abandoned oil rigs (see Breithoff, chapter 6), deadly pathogens that transform human lifeways or the more benign brown carpet beetle that feeds on the woollen fabrics of old floor carpets. "[S]pecies other than humans shape the places, landscapes, and practices we encounter on a daily basis" (Rutherford 2018, 208), and this certainly applies to historic sites and heritage landscapes as well. Historically, preservation of wooden buildings has relied on a wide variety of industrially manufactured chemical agents, like pesticides, fungicides and insecticides, to facilitate forms of heritage perpetuity by stalling ecological decomposition. The boundaries enforced through such toxic practices reflect that heritage is host to a wider non-human constituency, a "heritage biota" – "communities of organisms living on and in built heritage" (Viles and Cutler 2012, 2407; see also Farstadvoll, chapter 18) that also partake in its transformation.

The assumption that archives, libraries, museum collections and banks of biological material may act as eternal refuges is challenged in Rodney Harrison's chapter in this volume entitled "Legacies: Rethinking the Futures of Heritage and Waste in the Anthropocene" (chapter 2). Even archives that were designed as secure, permanent storages are affected by sudden, unpredictable events, political unrest and climate change. The Svalbard Global Seed Vault that Harrison deals with in his chapter currently stores more than one million different seed samples at −18 degrees in earthquake-proof underground facilities and is the world's largest collection of crop biodiversity. Already in 2015, the first requests to withdraw seed samples from the vault were made by scientists of the International Centre for Agricultural Research in Dry Areas (ICARDA) who had lost access to the genebank in Aleppo. Withdrawal of seeds that was originally thought to happen in a distant future occurred within the first decade of the vault's operation. The seed vault at Svalbard plays a significant role, Harrison argues, in reversing processes of biodiversity decay and is effectively a way of harvesting forms of biocapital in an emerging late-capitalist bioeconomy. These practices, however, relate closely to questions of what a specific Anthropocene heritage might look like and what kind of futures these different management technologies prescribe and engender. The material legacy of the Anthropocene clearly does not stop with old objects, sites and landscapes that we value and recognize but comprises a mass of waste products that will "inevitably overshadow and outlast any intentionally preserved form of cultural heritage" (Harrison, chapter 2).

Moreover, the study of heritage in the Anthropocene seems to call for both ethnographic and ecological attention to ongoing landscape transformation (Gan, Tsing and Sullivan 2018) that engenders new relations between human and more-than-human agents (DeSilvey, Bartolini and Lyons 2020, 360). The wide range of lifeforms inhabiting places shaped by human presence, calls for a new sensibility of the "wild" that is attuned to novel ecosystems that develop in proximity to urban clusters, and that appreciates feral objects like the plastiglomerate – the alleged petrological marker of the Anthropocene (Figure 1.2). The plethora of enmeshed species relations intersecting and overlapping with human lives makes it difficult to approach humans in isolation as detached from their messy relations with the wider world (Haraway 2016). The Anthropocene has further actualized such connections, particularly through the emergence of ecologies of "global weirding" (Canavan and Hageman 2016), seen, for example, in the proliferation of plant or animal hybrids that are caused by human-assisted species migration and climate change.

In Anna Storm's chapter entitled "Scars: Living with Ambiguous Pasts", the metaphors of wounds, scabs and scars that often appear in reference to anthropogenic landscape change in the Anthropocene, are significantly broadened through an exploration of what these different bodily processes literally imply and what they might hold for our treatment of damaged landscapes. A wound on the human skin, Storm points out, is healed through a form of repair that creates something new and different rather than the regeneration of a prior and identical skin tissue. This fact is brought to bear on the ecology of scarred industrial landscapes where the connections between past human activity, climate, soil conditions and emerging ruderal, plant communities partake in the formation of a new ecosystem that incorporates prior events of disruption and damage. Instead of ecological regeneration of pristine landscapes and the restoration of an imagined, prior equilibrium, we live with and acknowledge the ambiguous legacies of past exploits and abuses, only to realize that processes of healing, whether metaphorical or material, are never "linear, predetermined and unproblematic" (Storm, chapter 3).

FIGURE 1.2 Plastiglomerate – the alleged petrological marker of the Anthropocene

Photo: Þ. Pétursdóttir.

It may be, as Levi Bryant argues (in chapter 4), that the physical reality of the earth has been transfigured by human thought and practice to the extent that humanity no longer recognizes itself in its own image. However, to admit the extent to which humans have impacted the planetary ecosystem, must not equal claiming that the human species belongs at the centre of the universe. As a new geo-chronological era and conceptual framework, the Anthropocene involves both the centring and decentring of human beings (Pétursdóttir 2020). It acknowledges the deep ecological and atmospheric impact of a terraforming human species, but it also involves the realization that humans are no longer (or never were) able to fully control the trajectories of the things of their own making. The challenge to classic humanist thought is obvious; it requires that we think through the planetary reach of a material heritage that carries our faded signature but which nevertheless has embarked on journeys that were neither planned nor anticipated in advance. Thus, if this "unruly heritage" (Olsen and Pétursdóttir 2016) in and of the Anthropocene is to be conceptualized, valued and "preserved", this will hardly be realized through stabilization, purification or identification as either "natural" or "cultural". What is needed, we argue, is a truly ecological approach, which is willing and able to account for the more-than-human processes involved in different heritage formations.

The idea of a slow and timeless nature serving as backdrop to human actions, is challenged by the frenzied pace of current biological and climatological changes known as the Great Acceleration. In his chapter, Laurent Olivier refers to this as a tectonic upheaval that shakes us on an ontological level as we realize the force and agency of a natural environment that is outpacing the human intention or ability to act. How are we to understand heritage in this age of global warming and acceleration? Can heritage act as a means of appreciating the heterogeneous temporalities present in the Anthropocene, as the "coexistence of all of the pasts that survive materially in the present" or as "things being reactivated and brought back into play" (Olivier, chapter 5)?

The clear distinction between the artificial and the natural fades if we consider forms of habitation from the perspective of nonhuman animals. In her photo essay "Oil Matters", Esther Breithoff traces the widely and wildly distributed material legacies of oil production across the globe. The many forms that this sticky oil heritage assume in the present range from vernacular material culture in Paraguay where roofs and fences are made from residual oil drums, and all the way to rarely seen underwater worlds where in-situ steel foundations from disused oil rigs resemble natural reefs and attract a variety species of fish, crab, mussels and clamps. These examples could be seen as instances of the specific ecological convergence we encounter in Anthropocenic settings. Humans and non-humans are tangled up in the materiality of oil production, which perseveres as part of the everyday environment of fish and humans alike.

Affect

During the past two decades, an understanding of heritage as process has gained influence in heritage studies, and it is now widely asserted that heritage can be thought of as a verb rather than a noun (e.g. Logan and Wijesuriya 2015; Harvey 2001; Smith 2006). Heritage is treated performatively as a specific way of *doing* and as a wide set of practices where the relation between past and present is entertained and negotiated through active social engagement. Following this, the significance of intangible aspects of folklore, music, public rituals, festivals, performative arts and traditional craftsmanship has been highlighted through more emic understandings of heritage. These practices are embedded in dynamic, popular practices and ongoing acts of meaning-making, rather than in scientific values attributed to ostensibly static material things. Whereas early to mid-twentieth-century international charters of conservation were embedded in a scientific framework that focused on the value of historical fabric established through seemingly transcendent and universal criteria, later charters have articulated a greater sensitivity to culturally specific values and meanings (Hølleland and Skrede 2019). The stress on broader public participation in heritage management has also been articulated within the UNESCO system, not least the recognition of the role of communities in creating, maintaining and transmitting intangible heritage (UNESCO 2003, article 15). Importantly, according to the Burra Charter, revised in 2013, the process of interpretation, conservation and management should involve the people whose cultural identity, sense of belonging and lived experience are tied to the particular place in question (Australia ICOMOS 2013, article 12).

While this important shift in perspective has encouraged research into cultural expressions that differ from the official and canonized monumentalism of established heritage, this emic account of heritage as a pre-eminently cultural process has also unwittingly reiterated the modernist duality between the social and the material in heritage discourse. In cultural contexts where sacred objects *are* subjects with divine agency, we encounter the shortcomings of a modernist schema where the ideal and the material have been distinguished and "purified" (Latour 1993). The recent focus on intangible as opposed to tangible heritage systems has also in many ways reinforced social constructivist thinking in heritage studies (Pétursdóttir 2013).

The conceptual divide in heritage taxonomy between the tangible (material) and intangible (immaterial) aspects of people's lifeworlds is crucially at odds with how many cultures experience and interact with things that matter to them. The so-called ontological turn that has inspired much debate in anthropology over the past two decades attempts to develop an ethnographic sensibility that does justice to people's realities even where they differ radically from our assumptions and conceptual repertoire. For anthropologists who have studied divinatory practices of the Cuban Ifá cult and Mongolian shamanism, for instance, it implies that things may in fact make "their own kinds of difference" to how they are conceptualized by the researcher (Holbraad and Pedersen 2017, 216). That is, when things change, they differentiate and produce a difference – they invent themselves (Holbraad and Pedersen 2017, 219). The key is how that difference may inform and act upon the repertoire of concepts employed to understand the variety of realities where also things affect and have real power and agency.

Such post-humanist understandings of things render them qualitatively different from the mute derivatives of meaning-making that begin and end with human subjects. This is also an analytically fruitful position, we argue, because it considers *process* more broadly in terms of reciprocity between human invention and creativity and the agential and affective capacities of other-than-human beings. It opens for, in other words, an ecological view where heritage *qua performance* encompasses the difference that non-human beings make in the world as parts of the ongoing reconfiguration of the past in the present. Usually however, when the processual dimensions of heritage are considered, this does not seem to extend to things as agential, as having the capacity to transform the course of events. "Within this binary structure, tangible matter is made meaningful by being brought into a world of intangible meanings which are the property of human culture and experience" (Harrison and Rose 2010, 250). The binary structure here referred to, is the divide between the tangible, material world of monuments and artefacts and the ostensibly intangible domain of traditions, rituals, performing arts and music, which has become a standard trope in heritage discourse.

A clear distinction between the material and social world is, however, difficult to sustain. Most forms of intangible heritage are veritably steeped in materiality, and any experience of tangible heritage is equally based on affective exchange between humans and their physical environment. As Denis Byrne has claimed with regard to popular religious practices in Southeast Asia, they are in reality intensely materialist, but the secular and rationalist humanist framework of heritage governance would likely label such practices intangible for the simple reason that the divine agency and supernatural power of objects "cannot be seen" (Byrne 2019, 4). The notion of affect is central in attempts to explore what Byrne calls "human-nonhuman resonance and interpenetration" and how external bodies become present in us as an affective force (Byrne 2020, 863).

The affective power of particular heritage objects may thus be understood relationally as a force that works across human and non-human assemblages without being restricted to the range of internal human, subjective emotions. In his chapter called "Emergent Images: Matters of Affect in Heritage Photography", Colin Sterling analyses a wide range of agencies active in the production, circulation and use of photographs of the temple complex Angkor in Cambodia that began to appear in Europe in the 1860s. Instead of analysing heritage photography primarily as a mode of representation, Sterling brings to our attention how a specific photographic event unfolds in a material-discursive environment where semiotic, atmospheric, chemical, algorithmic and computational ecologies co-determine the heritage trajectories of a specific photo. Whether it collapses due to material decay or re-awakens through digital forms of circulation, the non-human agencies condition the possibilities for heritage photos to act and affect. Sterling notes that the power dynamics and human dimensions remain important, though "the affective capacities of the image are not coterminous with social practices and personal mediations". The emergent image holds an "always untapped potential for different ways of being and acting in the world" (Sterling, chapter 7).

The sense that human bodies are permeable to affects in encounters with heritage and museum exhibitions is central to Marzia Varutti's chapter "Affective Encounters in Museums". Here, she argues that the production of affect in the museum environment is engendered through a pre-subjective and transpersonal exchange of intensities between mutually inflicting bodies. Affect is fundamentally relational, and a museum environment may mobilize an entire ecology of things, emotions, memories and atmospheres to move us and allow for museum objects to get under our skin. Varutti's auto-ethnographic account of her encounter with the ancient Egyptian Divine Cat exhibited in the British Museum, raises the fundamental question of how and why humans are moved in the presence of inanimate, nonhuman others.

A rich material life often unfolds in places that are not actively maintained by human beings, and some of the contributions in *Heritage Ecologies* deal specifically with such ruinous places and things that normally "fall way off the heritage grid" (Sørensen, chapter 9). Tim Flohr Sørensen is interested in the enigmatic and affective power of such places that seize us and stick with us despite their lack of an obvious culture-historical value. Part of the explanation, he claims, lies in the formation processes themselves: the spontaneous reordering of materials that is not orchestrated by human beings, but which consciously or unconsciously affects our experience of structures and sites. For Sørensen, a heritage ecology proposes a departure from the prescribed and analytical focus on meaning, purpose and linear narratives by admitting the curiosity and confusion that encounters with ruins often elicit. The ecological approach in Sørensen's chapter is understood as a mode of caring for and attending to objects "that otherwise go unnoticed as 'irrelevant' to humans" (Sørensen, chapter 9).

A similar take on affective materials that move bodies, albeit on a larger scale, is evident in Janice Baker's chapter, where she points to how the agential capacity of earth itself and the earth-human assemblages of rock art in the Australian Pilbara region "rejects the notion of earth as dead matter" – a notion so fundamental to the mindset of extractive industries in the area (Baker, chapter 10). Rather than consigning rocks and stone to deep and unassailable, timeless pasts, their very present-ness and materiality shapes us as we manoeuvre across stony lands. And through such engagements, Baker argues, the earth enters the bones of people, forming an affective, corporeal geology. By bringing rocks into the present as felt and embodied beings, as proper *con*temporaries, Baker's earth-rock-bone ontology demonstrates that a material, earthly world can be framed differently, not as an "out there" – elsewhere or elsewhen – or as the backdrop of our heritage engagements, but as a shared formation of affective earth-human assemblages and processes.

Memory

Things may persist and outlive the political and cultural ruptures that constitute human-centred history, and this material continuity renders the memory encapsulated in heritage objects and sites something more than *ours* – something shaped also by other-than-human histories and object biographies. Working with the difficult memory and the contested memorials of victims of the 22 July terror attacks in Oslo and at Utøya, Helge Jordheim uses the concept of "lifetimes" to discuss current understandings of heritage and conceive ways of bridging natural and human histories:

> To understand what heritage is in the 21st century, we need to radically expand our notions of time. "History", "historical time", and indeed, "heritage" need to be opened up to include other "lifetimes", in the sense of temporal durations, rhythms, and speeds embodied in different forms of life. The concept of "lifetimes" bridges the nature-culture distinction and replaces the dichotomy of natural time, on the one hand, and social and historical time, on the other, by a continuity of scales. Life – both human and non-human – unfolds at various time scales, from seconds to millions of years, which converge in events and their aftermaths.
>
> *(Jordheim, chapter 11)*

Lifetimes is a profoundly ecological concept insofar as it covers a wide range of intersecting, temporal trajectories that become apparent when the exclusive time of human history is replaced with a more general, and indeed generous, life perspective. The 22 July terror attack shattered the broader temporal fabric that connects past, present and future as history, and the issue of how this traumatic event could materialize as public memorials and artworks has, interestingly, involved ideas of permanent landscape wounds, deep cuts in geophysical mass that would render eternal the memory of the victims. The injunction of deep geological time and the time of "nature itself" to facilitate the memory of rupture and absence also points to the work of synchronization engendered through heritage: "'[H]eritage' is at present probably the concept of historicization most able to assemble and align various lifetimes" (Jordheim, chapter 11).

The appeal to permanence and geological stability at modern memorials to victims of terrorism does not always reflect the difference between geological formations young and old, where, for instance, the sandstone of the Sørbråten peninsula would crack and split when exposed to external force. More broadly, this invites a critical re-assessment of the alliance between cultural memory and material stability. What kinds of memory would be engendered if these connections were severed? Could the geophysical formation processes we are currently witnessing be appreciated not only as the collapse of natural and human history as separate concerns (Chakrabarty 2009), but also as a way to reconfigure the study of memory as a concern traces of past encounters where different lifetimes converged? After all, this was how Haeckel first conceived of the Anthropozoic age using an approach that would integrate vastly different temporal scales to make the brief lifetimes of the human species palpable in the organic historical evolution of earth itself.

Indeed, it has been acknowledged that the growing awareness of the Anthropocene may also instigate a spatiotemporal, scalar expansion of memory research. Claire Colebrook has suggested that this shift uncouples memory from the idea of a past being held over time and brought into the future to instead consider series of many possible pasts with different amplitudes:

> It would seem then that the Anthropocene is an expansion or dilation of memory taking us beyond those forms of inscriptions authored by humans, and events intended by humans, and towards a "memory" where we will have been an agent in an epochal shift that we only recognized after the event.
>
> *(Colebrook in: Craps et al. 2018, 507)*

Faculties of transmission, recording and connecting, while often associated with human memory media technologies, also find their corollaries in the natural world, whether expressed through geological memory of earthly transformation or by insects and plants coupled with the environment through "interactional sensing, movement, and memory of their surroundings" (Parikka 2010, xxvii). These processes occur in settings other than the specifically human media environments and thus challenge the archival conception of memory as static imprints on a substrate (Burton 2008) by demonstrating the permanent transfer between the environment and various sensing bodies. In this understanding, memorizing forms part of a more general autopoeisis of adapting to changes, where long- or short-term retentions enable certain futures (Craps et al. 2018).

As remarked by Knittel and Driscoll (2017) in their post-humanist approach, memory is stored more effectively in the fleshy, animal body than in any one of our sophisticated technological and cultural artefacts or prosthetic media. For example, even if human beings are themselves effective terraforming species, minerals of the Earth's crust similarly inhabit us, shape us and move us in a geo-physiognomy that is also a memory of co-becoming. The calcium phosphate in our bones and the salt in our sweat recall the etymological (and ontological) common ground of humus and human (Farstadvoll 2018; Haraway 2016). In proposing that human memory extends to the material environment (LeCain, chapter 12) and that the things we have created "override the trajectories and destinations defined and anticipated for them" (Pétursdóttir 2020, 91), it is also worth bearing in mind the autopoesis of algorithmic memory processes that need no human interlocutors to operate. Memory technologies are radically reconfigured by digitization through pattern recognition algorithms, which mine big data looking to establish similarities between, for instance, melodies or images. Once digitized, ethnographic collections of traditional folk songs may, through algorithms, visualize patterns, rhythms and regularities and "reveal a past that has never been accessible to historiography" (Ernst 2017, 53). The digitized archive operates according to its own laws, and instead of thinking contemporary archives as order engendered through classification, Ernst suggests that we think of them entropically as systems where disorder also contains the highest (potential) degree of information (Ernst 2010, 60, 69).

This closely resembles the sense that entropic processes and loss of material stability in the heritage domain are in no way antithetical to memory, as Caitlin DeSilvey has argued (2014, 2017). The fact that heritage is conditioned by entropy and forces that compromise the physical and symbolic coherence of objects is not necessarily tantamount to loss of form, meaning or memory but may also signal the generative potential of materials that decay (DeSilvey 2006, 2017). Dissolution and decomposition are integral parts of energy flows in ecosystems, just as decay and ageing are part of the life-cycle of artefacts and the materiality of people and things (Shanks 1998). Material loss also engenders forms of memory that go unnoticed because they do not resemble the stable, legible material imprint that we have come to associate and trust with modern notions of cultural memory (DeSilvey 2017). Similarly, environmental historian Timothy LeCain (chapter 12) claims that the removal of human-made buildings and artefacts at former copper mining sites and smelters might even *contribute* to the mnemonic process, as it would permit other material memories to emerge more clearly. Though continuously reshaped by ecological and geophysical processes, the toxic legacies of long-bust industrial operations live on in vicious and viscous forms and demand attention for generations to come. As a consequence, one can see how, for instance, industrial heritage is equally expressed in the novel ecologies of ruderal plants that settle in disturbed, anthropogenic soils. As vegetal media that articulate past events and past production lines with great precision, we could think of them as "mnemophytes" – or memory plants of the industrial landscape (Bangstad, chapter 13; see also Storm, chapter 3, and DeSilvey, chapter 16). May we approach the gradual succession from brownfields to urban-industrial woodlands as a complement to cultural memory work rather than a "greenwashed" way of glossing over past industrial exploitation and fostering of collective amnesia?

In Alfredo González-Ruibal's chapter, "Memory and Redemption: Lessons from a Peasant Ecology", one important caveat is that while ecological approaches tend to the indissoluble mixture of humans and non-humans in the world, they often obscure social struggle and conflicts between, in his case, peasants and their dominators. The peasant world of Galicia was, he reminds us, a landscape of tension forged through the opposition between peasants and landlords or aristocrats, and that peasant world was sustained for centuries through "thick meshworks of human solidarity" (González-Ruibal, chapter 14). The collapse of the peasant world and its dense web of terraces, pathways and enclosures, manifests in the afforestation of formerly cultivated lands and the abandonment of villages, but also in the disappearance of a sensorial landscape of sounds and smells. The redemptive memory which González-Ruibal contrasts with the limitations of traditional heritage preservation, is one that accepts loss while retaining the potential of rekindling unfulfilled hopes. The encounter with fragments and traces – the plough, the winder for flax-weaving, the mallets – may reveal "indexes of redemption in the physical presence of vanished worlds" (González-Ruibal, chapter 14).

The duration of material memory is also explored in Ingar Figenschau's chapter, as expressed through the archaeology of two prisoner of war camps, Spittal and Kitzbühel, from World War II which are located in Norddalen in northern Norway. The material remains of war have here acquired a unique capacity to enable the memory of stories that are largely forgotten and marginalized in national war history. Conveyed through things that have outlived their human owners to become part of this mountainous landscape, the presence of Austrian and German soldiers is here continued in the form of mundane objects: shoes, shards of porcelain, chess pieces and barbed wire, remains that seem to have merged with their surroundings in a "natural state of conservation" beyond human interference and intention.

Entanglements

Humans, as Jane Bennett argues, "are always in composition with nonhumanity, never outside of a sticky web of connection or an ecology" (2004, 365). Moreover, material practices do not simply give meaningful form to things but receive something from them as well. This amounts to a form of "co-respondence" that Tim Ingold (2012) proposes as an alternative to the anthropocentric notion of hylomorphism, where humans inevitably impose form on materials. Matter always presents with an unforetold capacity, he argues, an opening to a different becoming, a potential and a vibrancy that are not exhausted by any meaning we may attribute to it. This irreducible element of vitality and unpredictability, and the constant tension between change and durability, constitutes an ecology of forces that contribute to the complex material processes through which cultural objects come into being and through which they extend or fail to extend meaning across time and space.

Heritage management and preservation involve a brigade of technologies and agents to keep things at bay, as an ecology of forces is at all times involved in the subtle transformation

FIGURE 1.3 Human heritage is always entangled with the unpredictable actions and capacities of non-human actors

Photo: S. Farstadvoll.

of sites, things and places. The sense in which things are constantly falling out of place or asunder is important to an ecological view on heritage practices, because it emphasizes the precarious, indeterminate and contingent as properties of rather than (merely) threats to heritage objects (Domínguez Rubio 2016). Nevertheless, in a system that understands heritage as an ultimately social rather than material process, the physical transformation of matter over time is seen as a threat to the objecthood that museums, archives and heritage organizations work so hard to uphold.

However, by attending to the work of nonhuman agents like moss, weed, limestone, concrete, acid rain, mushrooms, rust, geese, black mould and mud, we may begin to recognize the multifarious nature of sites and the contribution of the other-than-human world in the materialization of memory and heritage. As Colin Sterling recently noted in his review of post-human approaches in critical heritage studies;

> The point here is not to eradicate or marginalise human agency completely, but instead to acknowledge that any human action or intentionality is always entangled with other, non-human materialities and capacities; forces which may support, challenge, undermine or fundamentally reverse the work of "human" actors.
>
> *(2020, 1039)*

This is why the relation between a particular matter of concern in heritage and the wider technological, discursive, material and symbolic field is not exhausted by human reason and intention.

In light of the material turn where objects are seen as ontologically generative mediators with the capacity to "translate what they transport, to redefine it, redeploy it, and also to betray it" (Latour 1993, 81), the practice of heritage-making can be approached as onto-epistemological in the sense that things partake in forms of world-making. Karen Barad has stated that "practices of knowing and being are not isolatable, but rather they are mutually implicated. We do not obtain knowledge by standing outside of the world; we know because 'we' are of the world" (2003, 829). Thus, when getting to know an object, or when articulating such knowledge, it is hard to discriminate between purely epistemic practices and notions and those that are of ontological nature. As follows, in foregrounding materials and things as agential constituents in heritage processes, one of the aims of *Heritage Ecologies* is to explore the frictions between the specific identity assigned to heritage objects and the characteristics acquired through other-than-human forces. An ecological understanding of heritage, thus, is an attempt to appreciate that historic environments, cultural heirlooms, built landscapes and coral reefs are part of larger ensembles and processes that are not entirely sanctioned by resource management or cultural memory practices. The "ecologies" in the title of this book is a tribute to this and a motion toward approaching heritage as profoundly entangled phenomena where processes of care, conservation and commemoration are not determined by one kind of actor alone.

Rodney Harrison (chapter 2) refers to heritage making not as one uniform and prefigured line of action but as different domains of contingent practices that assemble specific bodies, techniques, materials, values, temporalities and spaces in certain ways in order to conserve and care for equally distinct heritage phenomena (e.g. linguistic, biological or historical diversity). Heritage is not one thing but many, and this is why Isabelle Stengers' notion of an ecology of practices (2005) may serve as a fundamental stepping stone in approaching heritage practices as heterogeneous and divergent (see Harrison, chapter 2). Following Stengers, one may see heritage as circumscribed by a particular "ethos of practitioners", where practitioners inhabit a particular domain that is delineated through the identification of specific problems and obligations (Stengers 2005). This could include the maintenance and preservation of biodiversity, natural landscapes, architecture or artworks with vastly different temporal horizons. Different heritage domains, in other words, should be understood in terms of their peculiarity and the specific cause which connects practitioners and forces them "to think, to act, to invent, to object" (Stengers 2015). Practitioners are engaged, Stengers claims, by a type of achievement proper to each field which obliges them to work together. First, an ecology of practices aims at the construction of new "practical identities", that is, "new possibilities for them to be present, or in other words to connect" (Stengers 2005, 186). The second implication of an ecology of practices is that it works with specific surroundings or habitats; the aim is not to arrive at a general theory or universally valid principle which grants the "power to disentangle something from its particular surroundings" (Stengers 2005, 187). And, finally, it aims not at describing practices "as they are" "but as they may become" (ibid).

The idea of new, practical identities which connect different and highly specific competencies within heritage work is clearly illustrated in management plans that draw on a range of distinct, but not necessarily incommensurable, scientific and non-scientific practices. It may seem that the specific nature of some heritage sites, when considered as particular habitats for practices, requires the sort of pragmatism that Stengers (2005, 186) attributes to work carried out in the "minor key". The practical identity that is formed through creative engagement arises from an encounter with a problem that is specific to a given situation or a particular habitat. It is not a problem that matters to everyone everywhere, but a local concern that draws together practitioners to invent. In her chapter, "A Positive Passivity: Entropy and Ecology in the Ruins", Caitlin DeSilvey addresses the issue of experimental practice in heritage management contexts, where preservation practices have expressed an unusual tolerance for the incomplete, indeterminate and improvised. Here, she presents different heritage management policies of a former Huyssen steel plant, now Landschaftspark Duisburg Nord in Germany, and the long-abandoned St. Peter's Seminary near Cardross in Scotland. Between complete non-intervention and active management, experiments *do* take place that may point toward a messier form of coexistence and a "willingness to collaborate with plants and other agents to interpret shared histories and craft shared futures" (DeSilvey, chapter 16). DeSilvey explores the notion of "positive passivity" as a potential attitude where the lack of intensive human management yields a powerful affective, aesthetic and ecological openness as materials are unmade and other forces become more apparent. The "productive indeterminacy" that a site like Landschaftspark Duisburg Nord articulates between intervention and abandonment, process and preservation, intention and accident is an important part of what attracts people to visit sites of "intentional ruination" (DeSilvey, chapter 16). Practical experiments like these, where the succession of plants on industrial soils intersects with and inhabits industrial architectures, are important because they admit that change, transformation and even loss are integral to material history (DeSilvey 2017), to ecological life in general and to our experience of heritage. "On the other side of loss", DeSilvey has written elsewhere, "there is always the possibility that new relations, and new subjects, will emerge" (2017, 171).

Moreover, as she notes in her chapter, a deliberate accommodation of transformative processes does occur in some heritage contexts in cases where memory work relies less firmly on the material permanence of sites and objects. If we were to admit the ambiguous, post-industrial entanglements of the biotic, the social and the chemical a more pronounced place in future heritage practices, it would seem to require a willingness to surrender some control for the sake of indeterminacy, ambivalence and, not least, experiment. In such an ecology of practices, the specific cause which connects practitioners in a field is not established in advance, but open for negotiation and improvisation to solve a problem locally by acting on the vital cues that emerge from the ongoing transformation of landscapes beyond our control. This "productive indeterminacy" of weird, ambiguous and contradictory landscapes that are neither culture nor nature makes it all the more worthwhile to re-envision heritage from an ecological and non-anthropocentric point of view. While the generative facets of material decay (DeSilvey 2017), and the tolerance for novel, urban-industrial ecosystems, may remain at the conceptual fringe of established historic preservation and nature conservation, these are vital building blocks of an ecology of practices that engenders situated, tentative and improvised approaches to practical problems without claiming universal validity.

With the practices that operate across the fields of nature conservation and cultural heritage preservation, we may regard heritage affirmatively as a "connector" that opens practice unto a wider ecology. As Christina Fredengren notes in the chapter "Heritage Ecologies as Worlding Practices", heritage phenomena are complex entanglements of natureculture where it may be difficult to isolate, unambiguously, once and for all, the source of agency in different "worldings". To consider heritage ecologically entails, for Fredengren, unfolding the spatiotemporal entanglements that include non-human othered others that are equally affected by our dubious material inheritance. Fredengren also touches on a central point when she states that heritage practice may engage in a more affirmative and creative mode of crafting new places and habitats, refuges that enhance life in challenged worlds where "hospitality needs to be practiced" (chapter 17). The reparative work that the heritage sector may take on in broken places entertains the notion that particular geographies are always works-in-process, always preliminary outcomes of processes of worldings where humans are not the sole crafters.

What happens with the architectures, objects and landscapes of human culture when human presence and anthropogenic influence and creativity is only one presence among many others? With a clear commitment to nonhuman agencies active in the ceaseless transformation of landscapes, Stein Farstadvoll (chapter 18) explores the aftermath of the derelict landscape garden Retiro in the city of Molde in Norway. This case study is ecological in the sense that it allows the relations forged by organisms, feral artefacts and the radical openness of ruinous places to articulate a distinct heritage presence, which persist despite *and* because of the lack of human heritage maintenance. Based on how these non-human beings bring attention to themselves through repeated site visits, the complex ecology of the Retiro landscape garden in Farstadvoll's view comes to encompass modern trash, spores, disposable plastic spoons, fungi, invasive organisms and "feral" artefacts. In this interface between the anthropogenic and the feral, heritage is seen not as a fixed thing but an "emergent presence" (Farstadvoll, chapter 18). The eclectic nature of this particular multispecies ecology, is crucial to understanding landscape transformations that undercut the neat categories of nature and culture. The black mould on Retiro's wooden buildings exhibits the entanglements of human-made structures and a larger "mycological topography", and while this is possibly not a harmonious form of cohabitation, fungi serves as "emphatic reminders that heritage can't be excised and placed alone on its own pedestal, elevated above and beyond the very stuff it consists of" (Farstadvoll, chapter 18).

If it cannot be excised and evicted, maybe there is an unrealized potential for heritage practice in learning from entropic processes and accommodating ongoing change rather than insisting on material stability (DeSilvey 2017)? Labels, categories and boundaries engendered through human curation and selection practices, are never able to fully capture the leaky or excessive properties of heritage places – feral connections that run through a specific site, or the slow material processes that set a place in motion over a timespan that dwarfs human lifetimes (see Fredengren, chapter 17). The advanced and sophisticated monitoring systems in cultural resource management, biodiversity and landscape conservation enable a meticulous patrolling of the borders between the wanted and the unwanted, the cultured and the unmanaged. These toolkits determine tolerated thresholds of change or exposure to adverse effects of biodegradation for cultural monuments or unchecked growth of invasive species for natural environments. The call for an ecology of heritage is not an uncritical celebration of "everything goes", a laissez-faire approach which lets nature run its course, serving as an excuse for defunding of the heritage sector. It is first and foremost a desire to explore "heritage making" as a broader process which involves the lifetimes, memories and unique trajectories of things, sites and deeply ambiguous Anthropocene landscapes that seem to escape the object identities we attribute to them. The appreciation of entanglement and co-becoming as a fact of life makes it difficult and, at times, undesirable to establish an unambiguous and discrete research object with clear outlines that removes it from the mess of the world and from our conceptual interferences. In the age of the Anthropocene, we no longer have the privilege of a practice of simply selecting the heritage we want or of naming, systematizing and isolating the components that we endorse or take responsibility for while ignoring how human creations are always enfolded in relations which yield results we can hardly comprehend, let alone predict.

This, and the way in which such heritage entanglements also necessarily include the legacies of power relations and violence, is discussed by Rui Gomes Coelho in his chapter, "Heritage and the Visual Ecology of the Plantationocene". Here he shows how the visual ecology of the plantation as heritage is reproduced and maintained through landscape archaeology and documentation practices like cartography and photography. Crucially, Coelho also explores a countervisuality that disrupts and counteracts the "visual-affective traps of the Plantationocene" (Coelho, chapter 19; see also Haraway 2015). A stack of black-and-white photographs recovered at Boa Esperança, a plantation in the state of Minas Gerais, Brazil, serves as the focal point in Coelho's exploration of historical acts of resistance in the plantation system and of the heritage of a distinct visual ecology of social and environmental exploitation.

In their photo essay in this book, Suvi Tuominen and Marko Marila explore the entangled heritage ecologies of World War II remains in Finnish Lapland and alternative ways in which researchers may engage with the material process of heritage formation at these sites. As they show, humans are not the sole inheritors of these landscapes of war. Other beings, be it mushrooms, grass, frogs or trees, also interact with and respond to the traces of past events in their own manner. By seeing the human body as a medium for the exploration of a surfaces where "different materials and actors intersect and bring new forms of knowledge into being" (chapter 20), Tuominen and Marila's ecological perspective acts out a central aim of

FIGURE 1.4 Clearings. An old forest road runs through a pine plantation in Øvre Pasvik Landscape Protection Area, Norway

Photo: S. Farstadvoll.

Heritage Ecologies as a whole: to move the reader from a knowledge *of* things to a making sense *with* things.

The lively relations of a heritage ecology

"Ecology" may mean a great number of things to different people, but the sense we have teased out in this introduction suggests that heritage phenomena extend more widely and wildly than the spatial and temporal coordinates of distinctive heritage categories seem to suggest. The process of making is not understood unidirectionally in terms of how humans attribute meaning to and enliven things but equally in terms of what kind of relations heritage objects enable and materialize by virtue of their specific properties, affects and memories. This change of perspective requires that new methods be developed for attending to heritage and to how specific objects unfold between and beyond concepts that frame heritage as a limited resource awaiting salvation and signification. Central to *Heritage Ecologies* is a broader ecological "life perspective" (Harrison and Rose 2010, 248) on the entangled processes and relationships that pass under the radar of such human-centred valuing and inferring.

An ecological understanding of heritage builds on the recognition that a heritage object often exceeds its historically specific figuration and is subject both to "creaturely" material change and dynamic practices of valuation. This implies that the making and unmaking of heritage involves a wide range of other-than-human actors and that a full appreciation of the processual dimension of heritage in the Anthropocene should also consider the affects, memories and landscapes that are engendered through such human-non-human entanglements. Heritage sites are meant to be open and shared spaces, and a recurring feature of the case studies presented in *Heritage Ecologies* is that this openness also includes feral presences, fungi, involuntary memories, weeds and wild affects that do not recognize the spatial perimeters and temporal order of sanctioned space. Openness always entails uncertainty. And, in the becoming of disorder and uncertainty, where things take unexpected turns and leaps, there is not merely peril and loss, but also a clearing for perceptive, experiential and informational plenitude to emerge. Such clearings are the vital signs of a heritage ecology.

References

Associated Press. 2019. "Danish Lighthouse Wheeled Away from Eroding Coastline." *The Guardian*. Accessed 4 December 2020. www.theguardian.com/world/2019/oct/22/danish-lighthouse-put-on-wheels-to-move-it-away-from-eroding-sea.

Australia ICOMOS. 2013. *The Burra Charter: The Australia ICOMOS Charter for Places of Cultural Significance*. Burwood: Australia ICOMOS.

Barad, K. 2003. "Posthumanist Performativity: Toward an Understanding of How Matter Comes to Matter." *Signs* 28(3): 801–831.

Barad, K. 2018. "Troubling Time/s and Ecologies of Nothingness: Re-turning, Re-membering, and Facing the Incalculable." *New Formations* 92: 56–86.

Barth, F. 1956. "Ecologic Relationships of Ethnic Groups in Swat, North Pakistan." *American Anthropologist* 58(6): 1079–1089.

Bennett, J. 2004. "The Force of Things: Steps Toward an Ecology of Matter." *Political Theory* 32(3): 347–372.

Biersack, A. 1999. "Introduction: From the 'New Ecology' to the New Ecologies." *American Anthropologist* 101(1): 5–18.

Bignall, S. and Rigney, D. 2019. "Indigeneity, Posthumanism and Nomad Thought: Transforming Colonial Ecologies." In *Posthuman Ecologies: Complexity and Process after Deleuze*, edited by R. Braidotti and S. Bignall, 159–181. London: Rowman & Littlefield.

Burton, J. 2008. "Bergson's Non-Archival Theory of Memory." *Memory Studies* 1(3): 321–339.

Byrne, D. 2019. "Prospects for a Postsecular Heritage Practice: Convergences Between Posthumanism and Popular Religious Practice in Asia." *Religions* 10(7): 1–13.

Byrne, D. 2020. "Divinely Significant: Towards a Postsecular Approach to the Materiality of Popular Religion in Asia." *International Journal of Heritage Studies* 26(9): 857–873.

Canavan, G. and Hageman, A. 2016. "Introduction: 'Global Weirding'." *Paradoxa* 28: 7–14.

Carson, R. 1962. *Silent Spring*. Boston, MA: Houghton Mifflin.

Chakrabarty, D. 2009. "The Climate of History: Four Theses." *Critical Inquiry* 35(2): 197–222.

Council of Europe. 2005. "Council of Europe Framework Convention on the Value of Cultural Heritage for Society." *Council of Europe*. Accessed 4 December 2020. www.coe.int/en/web/conventions/full-list/-/conventions/rms/0900001680083746.

Craps, S., Crownshaw, R., Wenzel, J., Kennedy, R., Colebrook, C. and Nardizzi, V. 2018. "Memory Studies and the Anthropocene: A Roundtable." *Memory Studies* 11(4): 498–515.

Derrida, J. 2001. "The Future of the Profession or the University Without Condition (Thanks to the 'Humanities', What Could Take Place Tomorrow)." In *Jacques Derrida and the Humanities: A Critical Reader*, edited by T. Cohen, 24–57. Cambridge: Cambridge University Press.

DeSilvey, C. 2006. "Observed Decay: Telling Stories with Mutable Things." *Journal of Material Culture* 11(3): 318–338.

DeSilvey, C. 2014. "Palliative Curation: Art and Entropy on Orford Ness." In *Ruin Memories: Materialities, Aesthetics and the Archaeology of the Recent Past*, edited by B. Olsen and Þ. Pétursdóttir, 79–91. London and New York: Routledge.

DeSilvey, C. 2017. *Curated Decay: Heritage Beyond Saving*. Minneapolis, MN: University of Minnesota Press.

DeSilvey, C., Bartolini, N. and Lyons, A. 2020. "Living with Transformation." In *Heritage Futures: Comparative Approaches to Natural and Cultural Heritage Practices*, edited by R. Harrison, C. DeSilvey, C. Holtorf, S. Macdonald, N. Bartolini, E. Breithoff, H. Fredheim, A. Lyons, S. May, J. Morgan and S. Penrose, 359–374. London: UCL Press.

DeSilvey, C. and Harrison, R. 2020. "Anticipating Loss: Rethinking Endangerment in Heritage Futures." *International Journal of Heritage Studies* 26(1): 1–7.

Domínguez Rubio, F. 2016. "On the Discrepancy Between Objects and Things: An Ecological Approach." *Journal of Material Culture* 21(1): 59–86.

Ernst, W. 2010. "Cultural Archive Versus Technomathematical Storage." In *The Archive in Motion: New Conceptions of the Archive in Contemporary Thought and New Media Practices*, edited by E. Røssaak, 53–66. Oslo: Novus Press.

Ernst, W. 2017. "'Electrified Voices': Non-Human Agencies of Socio-Cultural Memory." In *Memory in Motion: Archives, Technology, and the Social*, edited by I. Blom, T. Lundemo and E. Røssaak, 41–59. Amsterdam: Amsterdam University Press.

Escobar, A. 1999. "After Nature: Steps to an Antiessentialist Political Ecology." *Current Anthropology* 40(1): 1–30.

Farstadvoll, S. 2018. "Growing Concerns: Plants and Their Roots in the Past." *Journal of Contemporary Archaeology* 5(2): 174–193.

Favareau, D., Kull, K., Ostdiek, G., Maran, T., Westling, L., Cobley, P., Stjernfelt, F., Anderson, M., Tønnessen, M. and Wheeler, W. 2017. "How Can the Study of the Humanities Inform the Study of Biosemiotics?" *Biosemiotics* 10(2): 9–31.

Fuller, M., ed. 2007. *Media Ecologies: Materialist Energies in Art and Technoculture*. Cambridge, MA: MIT Press.

Gan, E., Tsing, A. and Sullivan, D. 2018. "Using Natural History in the Study of Industrial Ruins." *Journal of Ethnobiology* 38(1): 39–54.

Geertz, C. 1963. *Agricultural Involution: The Processes of Ecological Change in Indonesia.* Berkeley, CA: University of California Press.

Guattari, F. 2000. *The Three Ecologies.* Translated by I. Pindar and P. Sutton. London and New Brunswick, NJ: The Athlone Press.

Haeckel, E. 1866. *Generelle Morphologie der Organismen: Allgemeine Grundzügen der organischen Formen-Wissenschaft, Mechanisch Begrundet durch die von Charles Darwin reformierte Descendenz-Theorie, Volume II.* Berlin: Verlag von Georg Reimer.

Haraway, D. J. 2015. "Anthropocene, Capitalocene, Plantationocene, Chthulucene: Making Kin." *Environmental Humanities* 6(1): 159–165.

Haraway, D. J. 2016. *Staying with the Trouble: Making Kin in the Chthulucene.* Durham, NC: Duke University Press.

Harrison, R. 2015. "Beyond 'Natural' and 'Cultural' Heritage: Toward an Ontological Politics of Heritage in the Age of Anthropocene." *Heritage & Society* 8(1): 24–42.

Harrison, R. and Rose, D. B. 2010. "Intangible Heritage." In *Understanding Heritage and Memory*, edited by T. Benton, 238–276. Manchester: Manchester University Press.

Harvey, D. C. 2001. "Heritage Pasts and Heritage Presents: Temporality, Meaning and the Scope of Heritage Studies." *International Journal of Heritage Studies* 7(4): 319–338.

Harvey, D. C. and Perry, J. 2015. "Heritage and Climate Change: The Future Is Not the Past." In *The Future of Heritage as Climates Change: Loss, Adaptation and Creativity*, edited by D. Harvey and J. Perry, 3–21. London: Routledge.

Holbraad, M. and Pedersen, M. A. 2017. *The Ontological Turn: An Anthropological Exposition.* Cambridge: Cambridge University Press.

Hølleland, H. and Skrede, J. 2019. "What's Wrong with Heritage Experts? An Interdisciplinary Discussion of Experts and Expertise in Heritage Studies." *International Journal of Heritage Studies* 25(8): 825–836.

Holtorf, C. J. 2008. "Is the Past a Non-Renewable Resource?" In *The Heritage Reader*, edited by G. Fairclough, R. Harrison, J. H. Jameson Jnr. and J. Schofield, 125–133. London and New York: Routledge.

Hoskins, A. 2016. "Memory Ecologies." *Memory Studies* 9(3): 348–357.

Ingold, T. 2012. "Toward an Ecology of Materials." *Annual Review of Anthropology* 41(1): 427–442.

Kershaw, B. 2007. *Theatre Ecology: Environments and Performance Events.* Cambridge: Cambridge University Press.

Knittel, S. C. and Driscoll, K. 2017. "Introduction: Memory after Humanism." *Parallax* 23(4): 379–383.

Kutschera, U. and Farmer, S. 2020. "Ernst Haeckel, Ancient Forests, and the Anthropocene." *Plant Signaling & Behavior* 15(2): 1–5.

Latour, B. 1993. *We Have Never Been Modern.* New York: Harvester Wheatsheaf.

Latour, B. 2014. "Agency at the Time of the Anthropocene." *New Literary History* 45(1).

Logan, W. and Wijesuriya, G. 2015. "The New Heritage Studies and Education, Training, and Capacity-Building." In *A Companion to Heritage Studies*, edited by W. Logan, M. N. Craith and U. Kockel, 557–573. Hoboken, NJ: John Wiley & Sons.

Lorimer, J. 2015. *Wildlife in the Anthropocene: Conservation after Nature.* Minneapolis, MN: University of Minnesota Press.

Meskell, L. 2010. "Human Rights and Heritage Ethics." *Anthropological Quarterly* 83(4): 839–859.

Meskell, L. 2015. "Heritage and Cosmopolitanism." In *A Companion to Heritage Studies*, edited by W. Logan, M. N. Craith and U. Kockel, 479–490. Hoboken, NJ: John Wiley & Sons.

Morton, T. 2007. *Ecology Without Nature: Rethinking Environmental Aesthetics.* Cambridge, MA: Harvard University Press.

Morton, T. 2015. *Hyperobjects: Philosophy and Ecology after the End of the World.* Minneapolis and London: University of Minnesota Press.

Morton, T. 2016. *Dark Ecology: For a Logic of Future Coexistence.* New York: Columbia University Press.

Næss, A. 1973. "The Shallow and the Deep, Long-Range Ecology Movement." *Inquiry* 16(1–4): 95–100.

Olsen, B. and Pétursdóttir, Þ. 2016. "Unruly Heritage: Tracing Legacies in the Anthropocene." *Arkæologisk Forum* 35: 38–45.

Parikka, J. 2010. *Insect Media: An Archaeology of Animals and Technology*. Minneapolis, MN: University of Minnesota Press.
Pearce, S. 1995. *On Collecting: An Investigation into Collecting in the European Tradition*. London: Routledge.
Pearce, T. 2010. "'A Great Complication of Circumstances' – Darwin and the Economy of Nature." *Journal of the History of Biology* 43(3): 493–528.
Pétursdóttir, Þ. 2013. "Concrete Matters: Ruins of Modernity and the Things Called Heritage." *Journal of Social Archaeology* 13(1): 31–52.
Pétursdóttir, Þ. 2020. "Anticipated Futures? Knowing the Heritage of Drift Matter." *International Journal of Heritage Studies* 26(1): 87–103.
Rutherford, S. 2018. "The Anthropocene's Animal? Coywolves as Feral Cotravelers." *Environment and Planning E: Nature and Space* 1(1–2): 206–223.
Samuel, R. 1994. *Theatres of Memory*. London: Verso.
Shanks, M. 1998. "The Life of an Artifact in an Interpretive Archaeology." *Fennoscandia Archaeologica* 15: 15–30.
Smith, L. 2006. *Uses of Heritage*. London: Routledge.
Soper, K. 2012. "The Humanism in Posthumanism." *Comparative Critical Studies* 9(3): 365–378.
Stauffer, R. C. 1957. "Haeckel, Darwin, and Ecology." *The Quarterly Review of Biology* 32(2): 138–144.
Stengers, I. 2005. "Introductory Notes on an Ecology of Practices." *Cultural Studies Review* 11(1): 183–196.
Stengers, I. 2015. *In Catastrophic Times: Resisting the Coming Barbarism*. London: Open Humanities Press/Meson Press.
Sterling, C. 2020. "Critical Heritage and the Posthumanities: Problems and Prospects." *International Journal of Heritage Studies* 26(11): 1029–1046.
Steward, J. H. 1972. *Theory of Culture Change: The Methodology of Multilinear Evolution*. Urbana and Chicago, IL: University of Illinois Press.
Sullivan, H. I. and Malkmus, B. F. 2016. "The Challenge of Ecology to the Humanities: An Introduction." *New German Critique* 43(128): 1–20.
UNESCO. 2003. *Convention for the Safeguarding of the Intangible Cultural Heritage*. Paris: United Nations Educational, Scientific and Cultural Organisation.
Viles, H. A. and Cutler, N. A. 2012. "Global Environmental Change and the Biology of Heritage Structures." *Global Change Biology* 18(8): 2406–2418.
Zapf, H. 2008. "Literary Ecology and the Ethics of Texts." *New Literary History* 39(4): 847–868.

PART II

Anthropocene

2

LEGACIES

Rethinking the futures of heritage and waste in the Anthropocene

Rodney Harrison

Introduction

It could be argued that one of the requisite reorientations which the recognition of the Anthropocene epoch instigates for critical heritage studies is a temporal one – it necessarily forces a shift in attention away from the relations between past and present towards that of the present and future. In rethinking heritage studies *in* and *for* the Anthropocene (c.f. Harrison 2015; Harrison, Appelgren and Bohlin 2018; Harrison and Sterling 2020), one is immediately confronted with the material legacies of what Amitav Ghosh (2016) refers to as "The Great Derangement" and forced to consider the *unthinkable*: how the waste products of this epoch – its plastics, radionuclides, carbon monoxides, even climate itself – will form a persistent future archive of Anthropocene hyperobjects (c.f. Morton 2013) which will inevitably overshadow and outlast any intentionally preserved form of cultural heritage. The aim of this chapter is to begin to work through some of the implications of a critical rethinking of heritage ecologies through the lens of waste studies, framed within a broader emphasis on the value of such comparative approaches for critical heritage studies. In doing so, I draw initial inspiration from geographer Kevin Hetherington's (2004) framing of heritage as a category of both spatial and discursive placing which relates to broader, often cyclical practices of consumption and management of redundancy. In particular, I want to consider how bringing heritage and waste together helps us to retheorise both heritage and waste as forms of material and discursive *legacy* and to reconsider the ontological implications of living with, caring for and assembling futures out of both more and less persistent traces, residues and materials in the Anthropocene. In doing so, I map out a series of key conceptual terms which might provide the beginnings of a critical "toolkit of concepts" with which to navigate this newly expanded intellectual field. I draw on work undertaken as part of the collaborative, international *Heritage Futures* research programme (see Harrison et al. 2020). I begin by introducing this research programme and its framework before turning to look at some key concepts which I hope might prove productive in rethinking both heritage and waste in the Anthropocene.

Heritage futures

Heritage Futures was a large, collaborative, international research programme which involved ambitious interdisciplinary research undertaken by a team of 16 researchers to explore the potential for innovation and creative exchange across a broad range of heritage and related fields, in partnership with a number of academic and non-academic institutions and interest groups, undertaken over four and a half years from 2015 to 2019. The research was distinctive in its comparative approach which aimed to bring natural and cultural heritage conservation practices of various forms into closer dialogue with the management of other material and virtual legacies such as nuclear waste management. It was also distinctive in its exploration of different forms of heritage as distinctive future-making practices.

The research programme explored a large number of different practices within a range of divergent domains which are each dedicated to conserving and perpetuating ideas, words, objects, places, species, (both human and nonhuman) persons and things into more or less distant futures. In doing so, it sought to make a contribution to current discussions relating to the value of comparison in the humanities and social and historical sciences (e.g. van der Veer 2016), expanding them to the study of human and nonhuman social collectives. In developing the research, it was reasoned that a comparative analysis of different kinds of conservation and preservation practices might develop and open up the notion of heritage in creative and productive ways. Through a focus on conservation or preservation as a series of distinct creative, dialogical engagements between human and nonhuman persons, objects, places and practices, it aimed to explore heritage as distinctive processes rather than a series of "end products" of such practices (or, indeed, the values associated with those end products). And through an emphasis on heritage as forms of inheritance which may be positively, negatively or even ambiguously valued, it aimed to bring the field of critical heritage studies into conversation with the study of other hyperobjects (cf. Morton 2013; see also Breithoff 2020) and Anthropocene legacies, such as waste and climate. It contended that a reframed notion of heritage as material and discursive legacy would allow us to move beyond the uncritical valorising of certain objects, places and practices from the past in the present to reorient heritage studies more explicitly as a study of future making or worlding practices by showing how these legacies of the past in the present form templates for the organisation of new contingent realities and the construction of divergent future worlds. Working across natural and cultural heritage, the work was informed by Chakrabarty's (2009) observations of the ways in which climate change dissolves the distinction between natural and cultural history. Here the research programme joined a new critical engagement with nature conservation (e.g. Benson 2010) and extinction studies (e.g. Heise 2016) in exploring the distinct cultural frameworks which produce "natural heritage" (e.g. see Breithoff and Harrison 2018, 2020; Harrison 2017) – and the ways in which "cultural heritage" is not outside of but integrally a part of it (e.g. DeSilvey 2017; Harrison 2015). It also connected both conceptually and empirically with contemporary anthropological engagements with "futures" (e.g. Appadurai 2013; Salazar et al. 2017) and with current creative academic engagements with global climatological and environmental change (e.g. Haraway 2016; Tsing 2015; Tsing et al. 2017).

Conceptually, the research programme drew on and expanded previous work by the authors and others on the application of assemblage and actor network theory to the critical investigation of heritage and museums (Macdonald 2009a, 2009b; Harrison 2013a; Bennett et al. 2017), applying these perspectives to a range of other collections and institutional contexts. It was strongly influenced by the comparative perspectives of the "Endangerment and Its Consequences" project (Dias and Vidal 2016) in exploring a range of different cultural and natural heritage conservation practices collectively, drawing on the perspectives of histories of science and science and technology studies more generally in doing so. It was also influenced by the "ontological turn" in the social sciences, in particular Karen Barad's (2007) agential realism and various aspects of science and technology studies in seeing heritage practices of various kinds as enacting new realities through contingent practices of assembling and reassembling bodies, techniques, technologies, materials, values, temporalities and spaces in particular ways (see further discussion in Harrison et al. 2020).

Towards an ecology of heritage as future making practices

If we are to see heritage practices of various kinds as enacting new realities through contingent practices of assembling and reassembling bodies, techniques, technologies, materials, values, temporalities and spaces in particular ways, what does it mean to speak of "futures", "realities" and "worlds" in the plural?

> This is how I produced what I would call my first step towards an ecology of practice, the demand that no practice be defined as "like any other", just as no living species is like any other. Approaching a practice then means approaching it as it diverges, that is, feeling its borders, experimenting with the questions which practitioners may accept as relevant, even if they are not their own questions, rather than posing insulting questions that would lead them to mobilise and transform the border into a defence against their outside.
>
> *(Stengers 2005, 184)*

Invoking Isabelle Stengers' notion of ecologies of practices, I draw attention to the relative autonomy of different domains of heritage practices, with each of these domains specifying particular objects of conservation and specific accompanying methods of management. Examples of such domains include the fields of biodiversity conservation, built heritage conservation and endangered language preservation, each of which identifies a specific risk (respectively, loss of biological diversity, loss of cultural patrimony and loss of language and "culture") and an endangered object ("biodiversity", "built heritage" and "language diversity"). Each of these domains applies its own specific techniques for identifying, collecting, conserving and managing the endangered object and the factors that are perceived to threaten it (see Harrison 2015; Harrison et al. 2016; see also Dias and Vidal 2016). In thinking of how these domains exist contiguously and yet discretely from one another, I am influenced by Peter Sloterdijk's (2016) characterisation of the contemporary condition as one in which humans increasingly occupy a "foam" or complex ocean of fragmentary yet contiguous spheres. Insofar as heritage is generally tasked with preserving its endangered object for the "future" and each of these domains is concerned with establishing its respective conservation targets as both objects of knowledge and fields of intervention, these different heritage domains can be said to be actively engaged in the work of assembling and caring for future worlds. Although these domains of practice may sometimes come into relation with one another and may be sustained by discourses which arise from others, they often operate in relative isolation. Central here is a plural notion of heritage ontologies – understood as the world-making, future-assembling capacities of heritage practices of different kinds and the ways in which different heritage practices might be seen to enact different realities and hence to assemble radically different futures (Harrison 2015, 2017, 2018; see also Holtorf and Högberg 2015a, 2015b).

At the core of this plural notion of heritage ontologies is the idea that futures are not simply emergent but that futures are designed. What I mean by this is that futures are built and assembled as a result of actions in the present which are formed out of particular constellations of things, persons, places and practices and their coming together – in conflict or collaborations – at a particular moment in time. Here I understand conservation as the maintenance of plants, animals, languages, practices, ideas, persons, things, traces, residues and materials from the past, in the present, for the future (and we often hear this claim made for heritage as being something from the past which is conserved for "future generations"). But most importantly, I suggest that different forms of conservation practices often work towards assembling quite different futures and quite divergent worlds – and in doing so, they have the potential to undermine or come into conflict with one another.

Legacies: heritage and waste

I want to illustrate something of the distinctiveness of certain forms of heritage practices and, indeed, the value of comparative approaches which seek to work across them by working through some of the ways in which the comparison between heritage and waste might generate new insights for heritage. Obviously I'm not the first person to bring together the concepts of heritage management and waste management. Aspects of this comparison appear in the work of Mary Douglas (e.g. 1966) and in Michael Thompson's well-known book *Rubbish Theory* (1979), while more recently, the geologist Marcos Buser has written on what he terms the "heritage of toxic waste" (2016) to draw attention to the durability and persistence of chemotoxic and radioactive waste materials. Geographer Anna Storm (2014, and chapter 3) considers the landscape scars of industrial processes including nuclear power generation as forms of heritage, and Cornelius Holtorf and Anders Hogberg have for some time been exploring questions of communicating the dangers of buried nuclear waste with future generations and the broader issues this raises for thinking through heritage (e.g. 2015a, 2015b). Staffan Appelgren and Anna Bohlin (e.g. 2015a, 2015b) explore the relationship between heritage and recycling in their Re:Heritage project. Bjørnar Olsen and Þóra Pétursdóttir's important "Unruly Heritage" project also addresses similar issues (e.g. see Olsen and Pétursdóttir 2016; Pétursdóttir 2017; Pétursdóttir and Olsen 2018), as does Breithoff's (2020) work on the heritage of the Chaco war. And there are also resonances of Timothy Morton's (2013) work on hyperobjects in the arguments I deploy.

But rather than start with any of these, I want to begin with the logics of bringing together questions of heritage and waste. Heritage is a term that simultaneously identifies an object, place or practice as something which has value and as something which is considered threatened or at risk. That risk might simply be a result of the inherently perceived threat of time itself – which implies processes of forgetting, decaying, eroding or becoming worn with age. Sometimes the threat is a more active one of demolition or destruction – the flattening of a building, the bulldozing of a tree, the destruction of a tract of landscape by mining, perhaps, or even more seriously, the extinction of a plant or animal species or the genocide of a group of people during times of war. The element of potential or real threat to heritage – of destruction, loss or decay – links heritage historically and politically with a broader endangerment sensibility (I borrow this concept from my colleagues Nelia Dias and Fernando Vidal [2016]), which connects practices of cultural and natural heritage preservation across a range of different domains. This endangerment sensibility is articulated by way of practices of designation and listing (Harrison 2013a, 2016; Rico 2014, 2015). Even where a building, species or object is under no immediate threat of destruction, its listing on a heritage register – as something which is rare, unique or valuable – assumes a potential threat at some time in the future, from which it is being protected by legislation or designation. Waste, on the other hand, is a term which is clearly negatively valued – it designates a redundant object or useless by-product of some other operation. It is neither valuable nor at risk (although it might pose a risk to others). Yet both heritage and waste emerge from the same process of redundancy – these are both terms that denote superfluous objects which are no longer useful for the purpose for which they were originally produced. The museum and the rubbish dump might then be seen as two potential spatial end points for such redundant objects.

Containment 1: rubbish dumps and repositories

Or perhaps not so much end points as points in a cycle of consumption and re-use, as categories of spatial and discursive placement. Hetherington (2004) argues that disposal is not just about questions of waste and rubbish but is implicated more broadly in the ways in which people manage absence within social relations. Further, he explains that disposal is never final, as is implied by the notion of "rubbish", but that "disposal is a continual practice of engaging with making and holding things in a state of absence" (Hetherington 2004, 159). While Hetherington's concern here is with practices of consumption more broadly, I think his focus on the work of maintaining absence draws out another similarity between heritage and waste, which is that both are defined by practices of maintenance and containment. Intangible heritage practices require performance to maintain and preserve; nature and biodiversity are actively managed by practices of ecosystem service repair and management; endangered languages are recorded and archived, become extinct and might possibly be rediscovered and live again. But the important point here is that heritage and waste occupy a series of "other" spaces, outside of the realm of everyday life, set apart, sometimes hidden away in vaults, archives, banks, museums – repositories of different kinds.

I want to introduce two of these "other places" in which myself and members of the Heritage Futures team undertook research which I want to return to from time to time through the rest of this chapter. The first of these is the long-term nuclear waste and spent fuel repository site in Forsmark, Sweden, which is currently being constructed by SKB, the Swedish Nuclear Waste Management Company (Figure 2.1). And the second is the Svalbard Global Seed Vault, currently the world's largest secure seed storage facility, located on the Norwegian island of Spitsbergen near Longyearbyen in the remote Arctic Svalbard archipelago, about 1,300 kilometres (810 mi) from the North Pole (Figure 2.2). I bring them together here not only to highlight the strikingly similar physical architectures of the two repositories but also to think more broadly about the shared characteristics of such repositories which are built to contain and manage both heritage and waste.

SKB's facility at Forsmark currently includes the Final Repository for Short-Lived Radioactive Waste, SFR, as well as the facility where the Final Spent Fuel Repository for the most harmful radioactive wastes is planned. Since the mid-1980s, Swedish nuclear waste has been

FIGURE 2.1 Inside the SFR at Forsmark, Sweden

Photo: SKB.

FIGURE 2.2 The Svalbard Tube, inside the SGSV, Norway

Photo: R. Harrison.

managed and cared for at the Final Repository for Short-Lived Radioactive Waste and at the Central Interim Storage Facility for Spent Nuclear Fuel (Clab) in Oskarshamn. The planned long-term Spent Fuel Repository at Forsmark will ultimately involve the deposition of approximately 12,000 tonnes of spent nuclear fuel, which will remain radioactive and dangerous to living organisms for 100,000 years. A long ramp will descend to a depth of about 500 metres, where a system of tunnels will then be constructed in the bedrock. When fully developed – sometime in the 2080s – the repository will comprise 60 kilometres of tunnels with room for more than 6,000 copper canisters of spent nuclear fuel, packed inside bentonite casings. As each section is filled, it will be backfilled with concrete to completely encase the waste in concrete and bedrock. The total volume of space required underground in the rock will be approximately 4 cubic kilometres (SKB 2018).

In 2016, the Heritage Futures research programme ran one of a series of cross-research programme knowledge exchange workshops with SKB, which constituted a kind of extended field-based thought experiment (see further discussion in Harrison et al. 2020). What would it mean to plan for the management of the natural and cultural heritage materials and other conserved objects in the repositories in which various of our different project partners work for 100,000 years into the future? How might the management practices of archives, frozen zoos, seed and other biobanks, herbaria, historic sites, natural landscapes and museums change if we were to plan for the management of their respective conservation targets over such long durations? How would the objects of conservation themselves transform and change over those time periods? What emerged clearly from this workshop is that the futures which are planned for in heritage management are of much shorter duration – of one generation perhaps and generally little more – despite their rhetoric of fulfilling the needs of more distant futures (see also Högberg et al. 2017).

The popular understanding of the Svalbard Global Seed Vault as the "Doomsday Seed Vault" seems to echo these concerns with long-term potentially catastrophic futures. Situated on the remote island of Spitsbergen in the Norwegian Svalbard archipelago, high in the Arctic north, it received its first deposits of seeds in 2008. Nordgen, which is responsible for the day-to-day operations of the facility and maintains its public database of samples, reports that it holds approximately 933,000 "accessions" and over 50 million seeds from 234 countries and 733 depositor institutes in its frozen repository (NordGen 2016). Each accession represents an individual crop phenotype and is usually made up of approximately 500 individual seeds. The seed accessions are dried by depositing institutions to limit their moisture content to 5–6% and are then sealed inside an individual airtight aluminium bag. These bags are packed into standard-sized crates and stacked on shelving racks within one of the three separate, identical storage vaults, each measuring approximately 9.5 × 27 meters, which are refrigerated to maintain a constant temperature of −18 degrees Celsius. These vaults have been excavated approximately 120 metres into the side of a sandstone mountain at a height of 130 meters above sea level; entry to the vaults is via a 100-metre entrance tunnel. Equal parts bunker and frozen "ark", its dramatic façade includes a commissioned artwork, Perpetual Repercussion, which "renders the building visible from far off both day and night, using highly reflective stainless steel triangles of various sizes" (Government of Norway 2015). Cold climate and permafrost ensure that even if power is lost, the storage vaults would remain frozen for a significant period of time, even taking into account the possible effects of climate and sea level changes. "Designed for [a] virtually infinite lifetime", it is perceived to be "robustly secured against external hazards and climate change effects" (see further discussion in Harrison 2017).

Containment 2: ghosts, absences and unmanaged disposals

If the robust security of biobanks and nuclear waste repositories tells us something of the similarities between heritage and waste in their managed depositions in these "other places", what can we say about the absences and gaps in the landscape which are created by the gathering together of heritage and wastes and their storage in such a manner? And what happens when these affective materials leak from their repositories into the surrounding environment?

Drawing on the theme of cultural and historical debt in Derrida's *Spectres of Marx* (1994), Hetherington (2002, 2005, 2007) suggests that ghosts in the urban landscape represent the traces of unfinished or unmanaged disposal. We might think here of the obsessive conservation of the remaining traces of the Berlin Wall (Figure 2.3), for example, or of the almost fetishistic attempts to conserve the empty niche of the Great Buddha at Bamiyan in Afghanistan following its destruction by the Taliban in 2001 (see discussion of absent heritage in Harrison 2013a). Similarly, the early twenty-first century is haunted by other ghosts of unmanaged disposals, like the vast tracts of plastic wastes which converge within oceanic gyres (Figure 2.4; Morton 2013). But perhaps even more problematic and more authentically haunting are the invisible anthropogenic waste products: carbon monoxide as an artefact of agriculture and industrial processes, chemotoxic wastes within waterways, the radionuclides within the Earth's geology, which persist as atomic traces of nuclear energy production and warfare. These pollutants haunt as much because they constitute matter "out of place" (c.f. Douglas 1966), and

FIGURE 2.3 Traces of the former Berlin Wall conserved as "heritage" near Potsdamer Platz, Berlin, Germany

Photo: Aridd/CC BY-SA 4.0. https://creativecommons.org/licenses/by-sa/4.0/.

FIGURE 2.4 Plastic and rubbish floating in the ocean. Such images have themselves become a meme of the Anthropocene

Photo: Jens Cederskjold/CC BY 3.0. https://creativecommons.org/licenses/by/3.0/.

as such, they help shine a light more intensely on the ways in which these repositories create new realities through their collecting and ordering practices.

Transformation, toxicity and decay

One of the complex problems which the management of nuclear and other forms of hazardous waste presents is that the materials transform themselves over time. Marcos Buser (2016) points out that certain organic pollutants are transformed into what are termed "metabolites" by bacterial or chemical decomposition, forming new materials which are often far more toxic than their parent materials. In the same vein, we might think of the half-lives of nuclear materials and the ways in which they actively decay. On the other hand, decay contributes both positively and negatively to the aesthetic values of heritage (e.g. DeSilvey 2017). The affective power of heritage sites, both positive and negative, also changes, sometimes less predictably, but transforms nonetheless (e.g. Macdonald 2013). Conservation and heritage work is often framed as a process of slowing or managing such change and decay (DeSilvey 2017).

However, it is also clear that certain heritage sites also become more or less toxic according to other contextual shifts in their social, material, political, economic or ecological environment. I've written elsewhere (Harrison 2013a) about the ways in which what might otherwise be valorised and protected as "intangible heritage" might also form templates for identifying ethnic minorities and targeting them for expulsion, violence and genocides by majoritarian groups. I draw on Arjun Appadurai's *Fear of Small Numbers* (2006), in which he considers the connection between globalisation and extreme culturally motivated ethnic violence in the genocides that occurred in the 1990s in eastern Europe, Rwanda and India and subsequently in the 2000s in what has been termed the "war on terror". Each of these arose under circumstances in which "intangible" cultural differences amongst minorities became the focus for identifying specific groups for violence and genocide. Relatedly, in *Collecting, Ordering, Governing* (Bennett et al. 2017), we have shown how the culture concept has been utilised in the forms in which it has been articulated in museum and heritage sites to form templates for articulating difference in human populations for the purposes of the application of differentiated programmes of social governance. The ways in which these programmes of differentiated social governance can be administered through liberal and illiberal governmental means can shift according to changing political contexts.

Waste and wilderness as zones of exclusion

Another area where waste studies illuminates heritage is in thinking about waste and heritage as both contained within and defined by zones of exclusion. I draw here on the work of historian of science Peter Galison (2015), who notes this resonance across waste and wilderness in his work. Both heritage and waste are articulated through practices of boundary maintenance and the selective exclusion of humans. Wilderness is defined precisely by the absence of human traces and the boundaries managed to exclude such influences. IUCN Category Ib: Wilderness Areas are defined as "Protected areas that are usually large unmodified or slightly modified areas, retaining their natural character and influence, without permanent or significant human habitation, which are protected and managed so as to preserve their natural condition". They are "large … untouched areas where ecosystem processes, including evolution, can continue unhindered by human(s)" and should be managed in such a way so as to limit any human visitation. But in many other less extreme ways, we see heritage as defined by its controlled access and restriction of human intervention: the ubiquitous glass case of the museum, the roped barrier which keeps visitors from private rooms in country houses, the extreme isolation of the Svalbard global seed vault, for example.

This physical and discursive distance also produces problems of communication which resonate across waste and heritage management. We have become used to the idea that the values of objects, places and practices managed as heritage are not necessarily legible to "ordinary people", that "experts" must be engaged to "interpret" the values of such places to the public (e.g. Tilden 1957). Their values are mediated by interpretive signage, by costumed guides, by maps and plans, through audio guides. Like heritage, nuclear waste management is dominated by questions relating to the development of long-term warning messages which will communicate the danger of buried nuclear waste with the life forms which will occupy and inherit this human heritage hundreds of thousands of years in the future (e.g. see US Department of Energy 2004).

FIGURE 2.5 Interpretive heritage signage at the Ongarue Spiral Trail, Oregon, USA

Photo: Johnragla/CC BY-SA 3.0. https://creativecommons.org/licenses/by-sa/3.0/.

FIGURE 2.6 ISO nuclear radiation warning sign

Source: Public domain image.

Conservation, salvation and sacrifice

Another theme which emerged strongly from the empirical research we undertook within the Heritage Futures research programme is the ways in which the designation of certain forms of conservation landscape (for example, national parks and other forms of protected areas) could be argued to facilitate specific forms of environmentally damaging "non-conservation" activities outside of their borders. This is a sort of paradox in which conservation must be seen as integrally connected with and facilitative of its alter – landscape conservation emerges to compensate for practices of land degradation, biodiversity conservation compensates for practices which lead to species extinction, archival practices with the destruction of historical records and so on. This observation suggests that conservation landscapes in particular cannot be studied without looking at their broader landscape management effects – the "salvation" of certain designated areas and the "sacrifice" this facilitates for others. This invites a radical rethinking of conservation histories. We cannot understand the gazettal of protected buildings or landscapes without looking beyond the gazetted structure or reserved area to consider the negative actions which such conservation activity permits. The salvation of one endangered object could be said to be bound up with, and complicit in, the sacrifice of another (Harrison et al. 2020).

(Over)accumulation, remembering and forgetting

Returning to the Spitsbergen, the Svalbard Global Seed Vault is not a conventional seedbank but was conceived as part of a global system to facilitate the secure storage of a duplicate "backup" of seeds from national and regional repositories. These backed-up copies of seeds are stored free of charge and are held as part of an international agreement in which the seeds remain the property of the depositing institution and are available for withdrawal by the depositing institution (and only that depositing institution) at any time. It is thus not an active genebank but a literal "vault", a kind of Swiss bank containing a secure stock of duplicate seeds which can be used if seed stocks from the depositing institution become depleted or lost. The requirement for such a facility seemed to be clearly demonstrated when, in September 2015, scientists from the International Centre for Agricultural Research in Dry Areas (ICARDA) who had lost access to their genebank facility in Aleppo, Syria, requested the return of duplicate samples of seed which had been sent to the SGSV to reconstruct their collection in a new facility in Lebanon. This first withdrawal of seed samples from the SGSV as a result of the ongoing conflict in Syria was reported widely in the media and seemed to indicate clearly that the SGSV was already fulfilling a purpose which it had previously been assumed would arise in a more distant future, justifying the significant investment in this global "insurance policy". The manager of the new genebank facility in Terbol, Bekaa, was reported to have said of the withdrawal of seed samples,

> It [SGSV] was not expected to be opened for 150 or 200 years. … It would only open in the case of major crises but then we soon discovered that, with this crisis at a country level, we needed to open it.
>
> *(Alabaster 2015)*

Thus, in conjunction with ongoing processes of in situ crop diversity maintenance, themselves subject to continuing processes of natural and cultural selection which alter contemporary global crop diversity, the vault's collection reverses what is understood to be a natural entropic process of diversity decay (but one which is accelerated by anthropogenic activity) by increasing crop genetic diversity (Harrison 2017). In this sense, the values of its collection also increase with time – we can think about the SGSV as a bank for the accumulation of biodiversity as a specific formulation of bio-capital which relates to newly emergent late capitalist bioeconomies (see also Breithoff and Harrison 2020a, 2020b; van Dooren 2007, 2009).

However, questions of over-accumulation haunt both heritage and waste too. The concrete forms which the various apparatuses which have been produced to identify and manage forms of heritage at risk – the registers which form a record of endangered species and ecosystems, the List of World Heritage in Danger and so on – often don't include processes by which these objects may be removed from such registers (Harrison 2013b). The overburdened museum store room or the domestic spaces of the cluttered attic become the emblematic symbols of these processes of over-accumulation of heritage (see Macdonald and Morgan 2017, 2018; Morgan and Macdonald 2018). The over-accumulation of "stuff", the durability of plastic wastes and their accumulation in alarming quantities in landfill, the problem of managing digital objects, e-wastes, noise and light pollution and indeed the massive growth in human population itself all speak to this same anxiety of enumerating growth in things as forms of over-accumulation.

An obsession with memory practices also cuts across these fields. I've already referred to the work in nuclear semiotics which has at its core the study of intergenerational memory practices. Andreas Huyssen (2003) points to the emergence of memory and its materialisation through memorials, museums and other cultural institution as one of the key cultural and political phenomena of late twentieth-century modernity. Derrida's *Archive Fever* (1996) discusses the obsessive replication of such memory practices across many different domains of practice. I take this term "memory practices" from Geoffrey Bowker (2005) who studies the proliferation of archival and other classificatory systems across various fields of natural science. The archive stands in complicated relation here with the future; it orders and makes new worlds in its structuring of reality (see discussion in Bennett et al. 2017).

But these acts of deposition, preservation and interpretation are as much practices of forgetting as they are of remembering, as archives are actively selected, selectively retained and impartially interpreted. These selective practices of remembering are also hierarchical practices of valuing. We are reminded here of how the sites for the deposition of chemotoxic and radioactive wastes have also been selected in ways which reflect differentiation in the value of human lives between those who benefit from the consumption of the end products from which wastes are produced and those who are forced to live amongst and are most impacted by the presence of those wastes in the environment.

On life amongst the ruins: contamination, collaboration and contagion

I want to conclude with some notes on the light which these comparisons between heritage and waste might shed on living with the legacies of both. I do so with four further keywords, collusion, contagion, contamination and collaboration. I take this bringing together of two of these four terms from Anna Tsing, who notes that

> we are contaminated by our encounters, they change who we are as we make way for others. As contamination changes world-making projects, mutual worlds, and new directions, may emerge. Everyone carries a history of contamination; purity is not an option.
> *(Tsing 2015, 27)*

and further, "Collaboration means working across difference, which leads to contamination" (Tsing 2015, 28). While I'm interested in the ways in which she develops this line of thinking in relation to concepts of diversity and difference, which are also key heritage concepts, I want to highlight the ways in which the idea of contamination by chemical and radioactive wastes might help us to look with fresh eyes at the embodied experience or affect of heritage whilst also helping us to think through the ways in which a sense of collaborating or colluding with material and discursive legacies helps us to begin to look at the possible future-making projects which are implied across these various different domains of practice.

My understanding of the potentially affective, embodied experience of heritage has perhaps emerged most strongly from my work with Aboriginal people in Australia, for whom the management of the landscape as an aspect of heritage has a direct influence on both collective social wellbeing and individual physical wellbeing (see e.g. Harrison 2004). Take this statement from a colleague (now deceased), Arthur Hooper, a senior Muruwari man (then in his 70s) and honorary national park ranger with whom I worked recording the remains of a historic Aboriginal village in northwestern NSW in 2001. He said,

> ever since I've been coming out here, doing a little bit of work for people, I've been feeling really great. I'm really happy to see the old place again. And my feelings – inside me it's a very glad feeling, I have no worries about anything else. No aches and pains, I just walk around the place for hours and hours without getting tired.
>
> *(interview, 18 November 2001)*

June Barker, his contemporary, went on to say,

> We always liked to go out to Dennawan. We'd look around there ... it's hard to describe. Even yesterday I was at Dennawan and the little bit of a [house] frame is still standing there and I got a bit emotional. I don't know whether you understand it but it's within you ... and then yesterday I was out there again with Arthur Hooper and we went over and he said "I think this is the place here now, this is where you fellas used to live" and when I walked and stood I said "Yes Arthur, this is the place". You don't feel that just anywhere. You only feel that in special places and Dennawan is a special place. It will always be set in your heart.
>
> *(interview, 11 April 2002)*

Here, heritage is explained as a corporeal experience, an affective, embodied, emotional coming together of people and place.

We perceive chemical and radioactive wastes as invasive, contaminating waterways, invading our bodies and transforming us in the process. What separates this sense of the affective experience of heritage is not a qualitative difference in the nature of the embodied experience or transformative power of waste and heritage but practices of categorising and valuing. The separation between the rubbish dump and the museum begins to erode in the comparison.

What does it mean to present such comparisons now, in an epoch defined by the infiltration of radionuclides into the earth's very geology, in the Anthropocene, when any sense of the separation or isolation of humans from that thing which once we called nature can no longer be maintained? In my book *Heritage: Critical Approaches* (2013a), I asked rhetorically what it means to live amongst the ruins and spectral traces of the past, the heterogeneous piling up of historic materials in the present. And I think perhaps now I'm ready to venture an answer. Living with heritage and waste means collaborating with both, engaging thoughtfully in world-making practices which acknowledge the interconnectedness of each. The traces and residues we inherit are largely unchanged by these practices of categorising and valuing which relegate them to one or the other space but are the materials we will work with and through and which in turn will work with and through us. These traces equally constitute the legacies with which we will assemble future worlds, futures which are not predetermined but which remain open, emergent, unformed and multiple.

Acknowledgments

Heritage Futures was funded by an Arts and Humanities Research Council (AHRC) "Care for the Future: Thinking Forward through the Past" Theme Large Grant (AH/M004376/1), awarded to Rodney Harrison (as principal investigator), Caitlin DeSilvey, Cornelius Holtorf, Sharon Macdonald (as co-investigators), Antony Lyons (as senior creative fellow), Martha Fleming (as senior postdoctoral researcher), Nadia Bartolini, Sarah May, Jennie Morgan and Sefryn Penrose (as named postdoctoral researchers). Three PhD students were additionally funded as in-kind support for the research programme by their respective host universities – Kyle Lee-Crossett (University College London), Bryony Prestidge (University of York) and Robyn Raxworthy (University of Exeter). Martha Fleming left to focus on other responsibilities during the research programme's first year; the team of researchers was subsequently joined by Esther Breithoff (as postdoctoral researcher) and by Hannah Williams as administrative assistant and events coordinator, a role which was in turn later filled by Kyle Lee-Crossett and in its final year by Harald Fredheim (as postdoctoral researcher). Anders Högberg (Linnaeus University) was an associate researcher to the research programme. In addition, the research programme was supported by the work of a series of separate creative commissions with artists Shelley Castle (Encounters Arts), Nancy Campbell and Karen Guthrie and through collaborations with Henry McGhie, who acted as the curator of the Heritage Futures exhibition, which opened at the Manchester Museum in December 2018. See www.heritage-futures.org for further information. Work on this chapter was also supported by the AHRC Heritage Priority Area Leadership Fellowship Grant (AH/P009719/1). See www.heritage-research.org for further information.

Early versions of this chapter were presented at the international conference "On the Trace: Passing, Presence and the Persistence of the Past" held at the Saxo Institute at the University of Copenhagen, 22–23 September 2016, and as part of my inaugural lecture "Heritage Futures and Futures Heritages" delivered at University College London on 27 June 2017. I thank the organizers and participants in the "On the Trace" conference in particular for inviting me to present the paper and for comments which helped me to improve it.

References

Alabaster, O. 2015. "Syrian Civil War: Svalbard 'Doomsday' Seeds Transferred to Lebanon to Preserve Syria's Crop Heritage." *The Independent*, 10 October. Accessed 15 August 2016. www.independent.co.uk/news/world/middle-east/syrian-civil-war-svalbard-doomsday-seeds-transferred-to-lebanonto-preserve-syrias-crop-heritage-a6689421.html.

Appadurai, A. 2006. *Fear of Small Numbers: An Essay on the Geography of Anger*. Durham, NC: Duke University Press.

Appadurai, A. 2013. *The Future as a Cultural Fact: Essays on the Global Condition*. London: Verso.

Appelgren, S. and Bohlin, A. 2015a. "Introduction: Circulating Stuff Through Second-Hand, Vintage and Retro Markets." *Culture Unbound* 7: 1–11.

Appelgren, S. and Bohlin, A. 2015b. "Growing in Motion: The Circulation of Used Things on Second-Hand Markets." *Culture Unbound* 7: 143–168.

Barad, K. 2007. *Meeting the Universe Halfway: Quantum Physics and the Entanglement of Matter and Meaning*. Durham, NC: Duke University Press.

Bennett, T., Cameron, F., Dias, N., Dibley, B., Jacknis, I., Harrison, R. and McCarthy, C. 2017. *Collecting, Ordering, Governing: Anthropology, Museums and Liberal Government*. Durham, NC: Duke University Press.

Benson, E. 2010. *Wired Wilderness: Technologies of Tracking and the Making of Modern Wildlife*. Baltimore, MD: Johns Hopkins University Press.

Bowker, G. C. 2005. *Memory Practices in the Sciences*. Cambridge, MA: MIT Press.

Breithoff, E. 2020. *Conflict, Heritage and World-Making in the Chaco: War at the End of the Worlds?* London: UCL Press.

Breithoff, E. and Harrison, R. 2020a. "From Ark to Bank: Extinction, Proxies and Biocapital in Ex-situ Biodiversity Conservation Practices." *International Journal of Heritage Studies* 26(1): 37–55.

Breithoff, E. and Harrison, R. 2020b. "Making Futures in End Times: Nature Conservation in the Anthropocene." In *Deterritorializing the Future: Heritage in, of and after the Anthropocene*, edited by R. Harrison and C. Sterling, 155–187. London: Open Humanities Press.

Buser, M. 2016. *Rubbish Theory: The Heritage of Toxic Waste*. Amsterdam: Reinwardt Academy, Amsterdam University of the Arts.

Chakrabarty, D. 2009. "The Climate of History: Four Theses." *Critical Inquiry* 35: 197–222.

Derrida, J. 1994. *Specters of Marx: The State of the Debt, the Work of Mourning, and the New International*. Translated by P. Kamuf. London: Routledge.

Derrida, J. 1996. *Archive Fever: A Freudian Impression*. Chicago, IL: The University of Chicago Press.

DeSilvey, C. 2017. *Curated Decay: Heritage Beyond Saving*. Minneapolis, MN: Minnesota University Press.

Dias, N. and Vidal, F., eds. 2016. *Endangerment, Biodiversity and Culture*. Abingdon: Routledge.

Douglas, M. 1966. *Purity and Danger: An Analysis of Concepts of Pollution and Taboo*. London: Routledge & Kegan Paul.

Galison, P. 2015. "On the Building, Crashing, and Thinking of Technologies & Selfhood: Peter Galison in Conversation with Etienne Turpin." In *Art in the Anthropocene: Encounters Among Aesthetics, Politics, Environments, and Epistemologies*, edited by E. Turpin and H. Davis, 181–190. London: Open Humanities Press.

Ghosh, A. 2016. *The Great Derangement: Climate Change and the Unthinkable*. Chicago, IL: The University of Chicago Press.

Government of Norway. 2015. *Svalbard Global Seed Vault: More about the Physical Plant*. Last updated 23 February 2015. Accessed 8 August 2018. www.regjeringen.no/en/topics/food-fisheries-and-agriculture/landbruk/svalbard-global-seed-vault/mer-om-det-fysiskeanlegget/id2365142/.

Haraway, D. 2016 *Staying with the Trouble: Making Kin in the Chthulucene*. Durham, NC: Duke University Press.

Harrison, R. 2004. *Shared Landscapes: Archaeologies of Attachment and the Pastoral Industry in New South Wales*. Sydney: University of New South Wales Press.

Harrison, R. 2013a. *Heritage: Critical Approaches*. Abingdon: Routledge.

Harrison, R. 2013b. "Forgetting to Remember, Remembering to Forget: Late Modern Heritage Practices, Sustainability and the 'Crisis' of Accumulation of the Past." *International Journal of Heritage Studies* 19(6): 579–595.

Harrison, R. 2015. "Beyond 'Natural' and 'Cultural' Heritage: Toward an Ontological Politics of Heritage in the Age of Anthropocene." *Heritage & Society* 8(1): 24–42.

Harrison, R. 2016. "World Heritage Listing and the Globalization of the Endangerment Sensibility." In *Endangerment, Biodiversity and Culture*, edited by F. Vidal and N. Dias, 195–217. Abingdon: Routledge.

Harrison, R. 2017. "Freezing Seeds and Making Futures: Endangerment, Hope, Security, and Time in Agrobiodiversity Conservation Practices." *Culture, Agriculture, Food & Environment* 39(2): 80–89.

Harrison, R. 2018. "On Heritage Ontologies: Rethinking the Material Worlds of Heritage." *Anthropological Quarterly* 91(4): 1365–1384.

Harrison, R., Appelgren, S. and Bohlin, A. 2018. "Belonging and Belongings: On Migrant and Nomadic Heritages in and for the Anthropocene." In *The New Nomadic Age – Archaeologies of Forced and Undocumented Migration*, edited by Y. Hamilakis, 209–220. Sheffield: Equinox Publishing.

Harrison, R., DeSilvey, C., Holtorf, C., Macdonald, S., Bartolini, N., Breithoff, E., Fredheim, H., Lyons, A., May, S., Morgan, J. and Penrose, S. 2020. *Heritage Futures: Comparative Approaches to Natural and Cultural Heritage Practices*. London: UCL Press.

Harrison, R., Bartolini, N., DeSilvey, C., Holtorf, C., Lyons, A., Macdonald, S., May, S., Morgan, J. and Penrose, S. 2016. "Heritage Futures." *Archaeology International* 19: 68–72.

Harrison, R. and Sterling, C., eds. 2020. *Deterritorializing the Future: Heritage in, of and After the Anthropocene*. London: Open Humanities Press.

Heise, U. K. 2016. *Imagining Extinction: The Cultural Meanings of Endangered Species*. Chicago, IL: The University of Chicago Press.

Hetherington, K. 2002. "Phantasmagoria/Phantasm Agora: Materiality, Spatiality and Ghosts." *Space and Culture* 1(11/12): 24–41.

Hetherington, K. 2004. "Secondhandedness: Consumption, Disposal, and Absent Presence." *Environment and Planning D: Society and Space* 22: 157–173.

Hetherington, K. 2005. "Memories of Capitalism: Cities, Phantasmagoria and Arcades." *International Journal of Urban and Regional Research* 29(1): 187–200.

Hetherington, K. 2007. *Capitalism's Eye: Cultural Spaces of the Commodity*. New York: Routledge.

Högberg, A., Holtorf, C., May, S. and Wollentz, G. 2017. "No Future in Archaeological Heritage Management?" *World Archaeology* 49(5): 639–647.

Holtorf, C. and Högberg, A. 2015a. "Archaeology and the Future: Managing Nuclear Waste as a Living Heritage." In *OECD/NEA Radioactive Waste Management and Constructing Memory for Future Generations*, Nuclear Energy Agency no 7259, 97–101. Paris: OECD Publishing.

Holtorf, C. and Högberg, A. 2015b. "Contemporary Heritage and the Future." In *The Palgrave Handbook of Contemporary Heritage Research*, edited by E. Waterton and S. Watson, 509–523. London: Palgrave.

Huyssen, A. 2003. *Present Pasts: Urban Palimpsests and the Politics of Memory*. Stanford, CA: Stanford University Press.

Macdonald, S. 2009a. "Reassembling Nuremburg, Reassembling Heritage." *Journal of Cultural Economy* 2(1): 117–134.

Macdonald, S. 2009b. *Difficult Heritage: Negotiating the Nazi Past in Nuremburg and Beyond*. Abingdon: Routledge.

Macdonald, S. 2013. *Memorylands: Heritage and Identity in Europe Today*. London: Routledge.

Macdonald, S. and Morgan, J. 2017. *Perspectives on Disposal: Decluttering*. Accessed 8 August 2018. https://collectionstrust.org.uk/blog/perspectives-on-disposal-decluttering.

Macdonald, S. and Morgan, J. 2018. "What Not to Collect: Post-Coinnoisseurial Dystopia and the Profusion of Things." In *Curatopia: Museums and the Future of Curatorship*, edited by P. Schorch and C. McCarthy. Manchester: Manchester University Press.

Morgan, J. and Macdonald, S. 2018. "De-Growing Museum Collections for New Heritage Futures." *International Journal of Heritage Studies* 26(1): 56–70.

Morton, T. 2013. *Hyperobjects: Philosophy and Ecology after the End of the World*. Minneapolis, MN: Minnesota University Press.

NordGen. 2016. *Seed Portal of the Svalbard Global Seed Vault*. Accessed 15 August 2016. www.NordGen.org/sgsv/index.php?page=sgsv_information_sharing.

Olsen, B. and Pétursdóttir, Þ. 2016. "Unruly Heritage: Tracing Legacies in the Anthropocene." *Arkæologisk Forum* 35: 38–45.
Pétursdóttir, Þ. 2017. "Climate Change: Archaeology and Anthropocene." *Archaeological Dialogues* 24(2): 175–205.
Pétursdóttir, Þ. and Olsen, B. 2018. "Theory Adrift: The Matter of Archaeological Theorizing." *Journal of Social Archaeology* 18(1): 97–117.
Rico, T. 2014. "The Limits of a Heritage at Risk Framework: The Construction of Post Disaster Cultural Heritage in Banda Aceh, Indonesia." *Journal of Social Archaeology* 14(2): 157–176.
Rico, T. 2015. "Heritage at Risk: The Authority and Autonomy of a Dominant Preservation Framework." In *Heritage Keywords: Rhetoric and Redescription in Cultural Heritage*, edited by K. Lafrenz Samuels and T. Rico, 147–162. Boulder, CO: University Press of Colorado.
Salazar, J. F., Pink, S., Irving, A. and Sjöberg, J., eds. 2017. *Anthropologies and Futures: Researching Emerging and Uncertain Worlds*. London: Bloomsbury.
SKB. 2018. *Future Projects: The Spent Fuel Repository*. Accessed 8 August 2018. www.skb.com/future-projects/the-spent-fuel-repository/.
Sloterdijk, P. 2016. *Foams*. Cambridge, MA: MIT Press.
Stengers, I. 2005. "Introductory Notes on an Ecology of Practices." *Cultural Studies Review* 11(1): 183–196.
Storm, A. 2014. *Post Industrial Landscape Scars*. Cambridge, MA: MIT Press.
Thompson, M. 1979. *Rubbish Theory: The Creation and Destruction of Value*. Oxford: Oxford University Press.
Tilden, F. 1957. *Interpreting Our Heritage*. Chapel Hill, NC: University of North Carolina Press.
Tsing, A. L. 2015. *The Mushroom at the End of the World: On the Possibility of Life in Capitalist Ruins*. Princeton, NJ: Princeton University Press.
Tsing, A. L., Swanson, H., Gan, E. and Bubandt, N., eds. 2017. *Arts of Living on a Damaged Planet*. London: University of Minnesota Press.
United States Department of Energy. 2004. *Permanent Markers Implementation Plan, US Department of Energy Waste Isolation Pilot Plant*. Carlsbad, NM: Carlsbad Field Office.
van der Veer, P. 2016. *The Value of Comparison*. Durham, NC: Duke University Press.
Van Dooren, T. 2007. "Terminated Seed: Death, Proprietary Kinship and the Production of (Bio) Wealth." *Science as Culture* 16(1): 71–94.
Van Dooren, T. 2009. "Banking Seed: Use and Value in the Conservation of Agricultural Diversity." *Science as Culture* 18(4): 373–395.

3

SCARS

Living with ambiguous pasts

Anna Storm

A significant scar in classical literature is found on Odysseus' leg and originates from a childhood hunting accident. When the adult Odysseus returns in disguise from his long journey, this scar is what his old nurse finally recognizes. At the moment she is about to reveal his identity, the story of the scar's origin is told, but – a bit surprisingly to a present-day reader – not as a personal recollection, but as an "independent and exclusive present" (Auerbach 2003 [1946], 7). The scar works as a point of crisis, that is, as the unveiling of his disguise, and as a trigger to tell the reader about his early years. However, it does *not* work as a key to understanding Odysseus' individual biographical or psychological development. Erich Auerbach showed how the Homer heroes "appear to be of an age fixed from the very first" (Auerbach 2003 [1946], 17) and how everything in the epos is externalized, with events taking place in the foreground, on the surface, in no need of interpretation. Odysseus' scar allows the past to become part of the present, as a separate episode of importance in itself; the scar works as a physical bodily mark which affects the future course of events. It performs a relational function by connecting experience to identity while also constituting a basis for identification. Yet, unlike many present-day narrative uses of wounds and scars, it lacks the metaphorical capacity to articulate, for instance, the physical and mental process between wound and scar and its personal and political implications concerning both lived experience and living environments.

This metaphorical possibility was later employed frequently, among others, by the environmentalist Aldo Leopold, who famously characterized the fate of an ecologically conscious person as living in a "world of wounds" (Leopold 1970, 197). In a similar manner, scholars in the field of environmental humanities today often describe some of the dark visions of the Anthropocene as "struggles to survive in scarred, damaged lands" (Emmett and Nye 2017, 93; cf. Angel 2014; LeCain 2009; Tsing et al. 2017). In parallel, cultural and social processes linked to the notion of heritage are widely assumed to provide means of recovery from traumatic "wounds" and "scars", where post-conflict heritage discourse, particularly in relation to experiences of war and abuse (Chen 2002; Fitzpatrick 2017; Hochschild 2018; Lyde Cullen 2006; Miles 2014; Tachau 2008), abounds with "healing terms" (Giblin 2014, 503). A telling example is the winning proposal for a memorial to the victims of the 2011 massacre at Utøya,

Norway, entitled "Memory Wound", which suggested cutting through the island to form an artificial sound where the shootings occurred (Lindahl 2014, see also Jordheim, chapter 11).

The examples given are just a few among many similar metaphorical uses relating to the interconnected conditions of wounds, scabs, new tissues and scars. In particular, living with environmental harm, on the one hand, and living with difficult pasts or difficult heritage, on the other, seem to represent two areas where this range of metaphors is employed especially frequently. For the most part, these illustrations bear the character of everyday language and thus afford recognizable and catchy expressions rather than actively attempting to deploy any specific organic qualities of what was experienced. But could there actually be common analytical ground to be explored here? Could the metaphorical process from wound to scar imply a more conscious and literal conception and thus be developed into a means of examining and dealing with difficult pasts and environmental harm? In this chapter, I will follow the bodily changes from wound to scar and scrutinize their metaphorical potential in relation to "heritage ecologies".

Wounds: ambiguity and attention

A wound is the prerequisite, the starting point, for a scar, unless it is just a scratch or a scrape that soon disappears without a trace. A wound can be a bleeding cut in the skin or an open, acute sore that must be healed, either by the body itself or with the intervention of medical treatment. It may be accidental or deliberate, an expression of power abuse or assistance in an emergency or even a step towards status and beauty. For example, wounds incurred during a duel may be both expected and desired, as such a scar is a mark of class and honor, perhaps in imitation of war veterans. A dueling wound may also be seen as part of the wider category of "scarification", a practice where you self-inflict a pattern of wounds on the skin in order to create a permanent work of art or ritual mark on the body. A wound caused by surgery, such as a Caesarean section, has other connotations. It is often planned and controlled and necessary for the survival of, in this case, the mother and the child. A majority of wounds are certainly not pleasant or desired, but the existence of the previous examples brings some ambiguity into the picture; although the wound may be painful, it may still carry positive or productive meanings for the wounded bearer.

In an environmental context, when a landscape is seen as wounded, there is little if any room for ambiguity; it evokes primarily negative connotations. Any ambiguous interpretations may be based instead on the benefits created by the wounding activity, benefits which, in today's globalized world, are most often enjoyed elsewhere by people not directly exposed to the damaged land. Accordingly, the deliberate destruction of physical landscapes for the purpose of, for example, resource extraction has been conceptualized as the creation of "sacrifice zones". Here, a sacrifice is made in order to provide a "'common goods' of security or development or simply the private interest of short-term profit" (Holifield and Day 2017, 269), based on the perspective of a wider geography of production and consumption. As summarized by Robert S. Emmet and David E. Nye, the sacrifice zone logic, in short, is that "wealthy countries outsourced their pollution" (Emmett and Nye 2017, 18). Mining activities recur as expressions of environmental sacrifice zones, especially when they take the form of mountaintop removal (Fox 1999), open cut mining (Cottle 2013) or frac sand mining (Holifield and Day 2017), as are areas set aside to house toxic leftovers, for example, high-level nuclear waste (Endres 2012; Houston 2013; Masco 2006, 18).

FIGURE 3.1 Open cut mining is a technique of extracting rock or minerals from the surface rather than via underground tunnels. Over time, the result may be seen as a wounded landscape, harmful to most living organisms. The Chino copper mine, New Mexico, USA

Photo: Eric Guinther (Marshman)/CC BY-SA 3.0. https://creativecommons.org/licenses/by-sa/3.0/.

The "environmental injury" (Houston 2013, 420) highlighted in studies of sacrifice zones is inextricably intertwined with injustice. That is, a neglect of "the health and way of life of communities" (Holifield and Day 2017, 269) by the economic and social deprivation of people who are often poor to begin with and not seldom composed of indigenous groups (Emmett and Nye 2017, 17). To describe and give voice to experiences of environmental devastation and its consequences, not only scholars but also activists and local residents use organic metaphors. For example, Holifield and Day (2017, 275) quote local opponents to frac sand mining who state that it is "like selling off our own body's essential organs to make a quick buck" and that "we go to the woods to heal the wounds of life … we are losing our place of home". Their experiences are not unique. Worldwide, Albrecht et al. suggest that there is an "increase in ecosystem distress syndromes matched by a corresponding increase in human distress syndromes" (Albrecht et al. 2007, 95). While Donna Houston emphasized how sacrifice zones are not only environmentally and socially damaged but often also invisible in a wider public consciousness, I would argue that the use of organic metaphors here may work as a verbal protest against such relegation of places and landscapes to the "shadows" (Foote 2003; Plumwood 2008). A wound, as well as a scar, calls for public attention, to be seen, acknowledged and healed.

A call for public attention is also a key component for things and practices labeled "heritage". A heritage process involves the contemporary articulation and valuation of past events and their materialities, to be understood and recognized by a large public. However, in contrast to the pain and problems of "sacrifice zones", heritage is "generally invoked as a positive quality" (Harrison 2013, 7), something to enjoy and cherish and which is to smoothly foster affirmative human identities and belonging in a present-day context. Nevertheless, a rather limited exchange between heritage perspectives and contemporary social, political and ecological issues (cf. Harrison 2015, 28; Harrison 2013, 4) sometimes challenges these notions of enjoyment and smoothness. For example, Sharon Macdonald shows how the problematic past of World War II atrocities in Germany became "a lurking presence – a virus in the system – that would lead to sickness in society" (Macdonald 2015, 14). Here, heritage as a contemporary concern was not a source of positive identity and belonging but quite the opposite.

Without question, heritage scholarship and practice not only engage with assumed positive heritage qualities but also acknowledge and analyze difficult, dark and dissonant pasts and how these are manifested in today's society (cf. Logan and Reeves 2009; Macdonald 2009; Soyez 2009; Tunbridge and Ashworth 1996). Often, this engagement has not shaken the traditional understanding of what characterizes heritage. Within the basic logic of regarding heritage as a way to shape affirmative and, first and foremost, national identities, it has been rather easy to incorporate some difficult experiences by "bolstering senses of national togetherness and pride" (Macdonald 2015, 9) and by limiting or "containing" more troublesome aspects.

FIGURE 3.2 War atrocities of the past often become wounds of a community, not least when evidence is discovered long after the events. Mass grave, originating from the Spanish Civil War, Estépar, Spain

Photo: Mario Modesto Mata/CC BY-SA 4.0. https://creativecommons.org/licenses/by-sa/4.0/.

Yet, with Germany as a pioneer, triggered by the "lurking presence" mentioned previously, another approach has developed which calls instead for addressing this past politically by means of transparency, apology and morals. Nevertheless, in a wider heritage context, the efforts to face difficult pasts outside the general frame of a "positive quality" are both recent and relatively rare (Macdonald 2015, 7).

A wound is thus an epicenter of pain and a focus of attention, and it can be used metaphorically to describe both environmental damage and difficult pasts. The cause and connotations of a wound may vary, but its function is to signify and materialize an important event or change. In environmental studies, the wound as a metaphor calls for temporal and geographical awareness as well as for public attention to devastated lands, which otherwise might elude recognition by being shadowed as sacrifice zones. In heritage scholarship and practices, a greater concern for past and present wounds in the landscape and in society could challenge the overarching emphasis on heritage as an apolitical source of positive but often shallow identities. A "reorientation" towards contemporary concerns, as Rodney Harrison and others have requested (Harrison 2013, 4), may thus include the recognition that not all pasts and presents are equally important to account for but that the wounded and ambiguous elements of heritage unquestionably deserve a place of significance.

Scabs and new tissue: retrospective and prospective approaches

If the bleeding stops and the wound begins to heal, a contracting, protective coating forms. A vulnerable shield, the scab, covers the ongoing repair of tissue underneath. The new tissue that will later form a scar differs from normal unwounded tissue in that it contains more fibers. Moreover, these are aligned in a single direction rather than randomly and supported by relatively few blood vessels. As a consequence, the three primary layers making up normal skin tissue are replaced by a fibrous tissue of another texture and quality. Sweat glands, hair glands, muscle cells and nerves are seldom restored but replaced by the more dense tissue. For humans, most wounds lead to some degree of scarring, and the tissue is said to "repair". However, for some species, the regrowth of tissue can happen without any scar formation, which is then labeled "regeneration" and thus signifies something different from "repair". This is the case for lower vertebrates, such as salamanders, and for invertebrates.[1] Yet from a human bodily perspective, the process after wounding most often involves some kind of "repair", and what takes shape is not a complete and identical reconstruction or "regeneration" of what was before but the creation of something new and different.

Ecological restoration work has traditionally been directed towards reinstating a pre-disturbed status: that is, an imagined original, pristine natural environment (Kowarik 2005, 17; cf. Cronon 1995). Species that have disappeared from an area they previously inhabited may be reintroduced (Jørgensen 2015, 484), and "invasive" or "alien" species may be cleared away to allow "native" species, both flora and fauna, to prosper. In this view, (pristine) nature is seen as an "inherent good and humans (and their technologies) as inherently destructive to that good" (Jørgensen 2017, 849 with reference to Gandy 2002). Within such a dichotomy between nature and culture, the goal might even be to "erase human history and involvement with the land and flora and fauna" (Jørgensen 2015, 487) in order to make space for the original "nature". As one consequence of this view, the understanding of a "wilderness" without humans has emerged as a powerful tool to repress human practices that have long been intertwined with a landscape (Cronon 1995; Woods 2018).

Lately, however, another position has been articulated, emphasizing more hybrid landscapes as possessing ecological as well as social value, in what is sometimes labeled a "natureculture" approach. Dolly Jørgensen highlights how "[n]atureculture as an epistemological, ethical, and aesthetic category has been widely accepted in the humanities and social sciences as a condition of the Anthropocene" (Jørgensen 2017, 849) including by some, but certainly not all, ecological restorationists based in the natural sciences. From this point of departure, it is, for example, possible to put forward an "inverted value system" (DeSilvey 2017, 112) where species that would have been labeled invasive in a traditional restoration context are instead given precedence, as they represent a specific "natureculture" hybrid. As "hybrid" in the strict sense denotes an infertile offspring of two different species, Joseph Masco suggests that "mutant" might be a more precise and productive concept for examining the complexity of natureculturess. He highlights how the (basically evolutionary) concept of mutation assumes change, but it does not from the outset judge either the temporal scale or the type of change that will take place. It also marks a transformation that is reproduced generationally, making the mutation a specific kind of break with the past that reinvents the future (Masco 2006, 301).

One kind of hybrid, or perhaps mutant, landscape that has drawn much scholarly attention is urban-industrial wastelands (cf. Gandy 2013; Keenan and Jorgensen 2012; Storm 2014). Here, the spontaneous overgrowth of abandoned urban or industrial sites creates enclaves of a "new wilderness" (Kowarik 2005, 22), where so-called ruderal species are often the first to establish. Ruderals are pioneers that can prosper from the specific combination of sometimes toxic substances remaining in the ground from whatever human activity was abandoned there. They are fast-growing plants with short life cycles, often producing great numbers of seeds and typically dominating a disturbed area for a few years before gradually losing out in competition with other species. From an ecological point of view, Ingo Kowarik suggests that if the vegetation's capacity for process and growth is what is valued and measured, urban-industrial wastelands actually represent the most "natural" nature (Kowarik 2005, 19). He calls this approach "prospective", in contrast to the traditional "retrospective" ideal of ecological restoration. While traditional restoration ecologists might not appreciate urban-industrial wastelands, urban dwellers both relish and dismiss them. Some find this "postpastoral nature" (DeSilvey 2017, 112) exciting and beautiful, forming rare recreational spots in a densely built environment, while others regard them as ugly and dangerous, useless areas merely awaiting new development (Gandy 2013, 1302).

The retrospective and prospective approaches to ecological work and their ideals resonate closely with much heritage thinking, and not least with building conservation. In this field, a traditional ideal is to restore a damaged or decaying building to its "original shape", whatever this is decided to be, and to preserve a building's "authenticity" by using original construction techniques and materials. In general, building conservation practice has often tried to "freeze" structures in time as "testimonies of specific moments in the past" (Weiler 2017, xxi, xvii) and in a search for "genuineness, originality, or truth" (Weiler 2017, xviii) – that is, a retrospective approach very similar to the ecological restoration ideal described previously. The retrospective heritage ideal generally dominates building conservation practice, at least in the Western context, but has also been broadly questioned for being nostalgic, for focusing too much on "heritage as a source of enjoyment [and not as] a social and spiritual asset to help bring about

FIGURE 3.3 Abandoned urban and industrial sites quickly become overgrown by pioneer species, like St. John's wort in this picture. These "wastelands" make up a contested category of ecological and social urban space: on the one hand, flourishing and inclusive, on the other, decaying and useless. Overall, they are marked by an ambiguous state of becoming, the state of a scab. Lowestoft Station, Lowestoft, UK

Photo: Mark Watson.

a better future" (Lowenthal and Jenkins 2011, 36). That is, the retrospective ideal is questioned by more prospective heritage approaches.

One such prospective approach was suggested by the architect Lebbeus Woods, who designed several projects for places marked by different kinds of crisis, for example, cities after war bombings or natural catastrophes. He calls his work "radical reconstruction", featuring proposals named "scabs", "tissues" and "scars". In many ways, the new architectural elements also resemble organic texture, formed through interplay with existing, damaged buildings and sites, by means of contrast, mirroring and outgrowth. In an essay, Woods denotes a scab as "a first layer of reconstruction, shielding an exposed interior space or void, protecting it during its transformation" (Woods 1997, 16).

The philosopher Elizabeth V. Spelman engages with Woods's work, highlighting how the architectural proposals intend "to face and explore the damage, the brokenness, not find ways (if any are possible) of undoing it" (Spelman 2007, 135). "Facing" the brokenness, she suggests, means to "look into the damage and be attentive to what it can tell us or suggest to us" (Spelman 2007, 132), which resounds with Caitlin DeSilvey's appeal to "understand change

FIGURE 3.4 The architect Lebbeus Woods submitted several design proposals for places marked by crisis, such as war bombings or natural catastrophes. This particular proposal suggests a first protective coating on a damaged building and is entitled "War and Architecture: SCAB Construction", 1993. Lebbeus Woods, Architect, American (1940–2012)

Source: © Estate of Lebbeus Woods.

not as loss but as a release into other states, unpredictable and open" (DeSilvey 2017, 3). At the same time, Spelman acknowledges that one has to be "aware of the moral and political dangers entailed in facile celebration of the bright possibilities to be found in rubble or brokenness" (Spelman 2007, 135f; see also Millington 2013 on "ruin porn"; Lepawsky et al. 2017, 59–60, on "repair-scapes"; and Macdonald 2015, 17, on "washing away" difficulties through acts of recognition). So, a prospective heritage approach may imply a close involvement with the repair as it unfolds into something new, without falling into the trap of enchantment of damaged livelihoods.

Both the retrospective and the prospective approaches bear promises of respect, beauty and meaning, as well as risks of neglect and violation. However, the metaphor of a scab – the vulnerable stage in between the wound and the scar – which understands repair as the creation of a scab and later scar tissue definitely resonates primarily with the prospective approach. Something new is becoming and taking shape, in relation to what was there before and in relation to its surroundings. A scab does not grow in isolation but as an integrated part of the skin and of the body as a whole, yet it stands out as transformation and as something new. Attention inwards is important during this process, for example, to watch out for possible infection. The basic potential of prospective thinking in the metaphorical sense is, however, that there is something to discover, imagine and learn if one engages with what is damaged but on its way to repair, or perhaps mutate. Through this potentiality, prospective thinking clearly contrasts with the retrospective idea of attempting to reinstate exactly what was before. The search for "truth" through retrospective restoration into an "original" state or shape – either in an ecological or in a heritage setting – can therefore be contradicted or at least complemented by the "truth" of a prospective approach acknowledging the sometimes painful traces of historical change.

Scars: integration, healing and concealing

A scar is the result of the repair, a new mark on the body. It is often discolored and depigmented and without hair. Scars are less resistant to ultraviolet radiation, and the new tissue is at least 20% weaker than non-scarred tissue. The different layers of unwounded skin have been replaced by an integrated fibrous tissue, and its strength is mainly tensile, that is, it resists being pulled apart. Because of its high density, the scar is less flexible and sensitive and may thus be discomforting, although the scarred area itself tends to be less susceptible to pain. A scar can chafe and itch, and it does not go away. Like its forerunners, the wound and the scab, it draws attention; the bearer may be either proud or ashamed of its existence and try to show or to hide it accordingly. It forms an integrated part of the living body as it embodies an important experience.

To be able to repair from a damaged status, to form something new which performs similar functions as before, can be termed "resilience". In ecological work, resilience is closely related to what was described previously as a prospective approach, in that it rejects the idea of an ecosystem as a "balanced arrangement that has persisted for millennia" (Emmett and Nye 2017, 20). That is, it rejects the notion of a "pristine" nature in terms of an ideal status quo or equilibrium. Resilience thinking instead emphasizes ecosystems as open and dynamic and with the potential to "attain a different configuration of stability" (Emmett and Nye 2017, 20) after disruption or damage. As Libby Robin states in a commentary to the formative 1973 text on resilience by C. S. Holling: "[s]ystems had a history, and this history was a force in the evolutionary pathways of biota" (Robin 2013, 257).

The ability of a resilient system to incorporate change has made the concept attractive to a range of disciplines also outside ecology and the natural sciences, for example, sociology, archaeology and geography. Along this conceptual journey, an entangled understanding of "social-ecological resilience" has been widely explored to describe, for example, community responses to change (cf. Trell et al. 2018, 7), that is, an application of ecological thinking to the social sphere. This has also resulted in a frequent orientation of resilience scholarship and practice toward policy work such as resource management and adaptive governance (Walker et al. 2004; Robin 2013, 257), resembling work in the field of "sustainability", along with all of the difficulties its implementation entails. These difficulties have also rightfully resulted in a broad critique of policies driven by too much faith in general "technological fixes" (cf. Emmett and Nye 2017, 88; Viitanen and Kingston 2014) and universalized "technologies of recovery" (Easthope 2018) that do not easily meet local post-disaster "situated realities" (Easthope 2018; cf. Guarnacci 2016).

Resilience perspectives also focus on interdependence and connectivity, such as when a single animal or plant is understood to exist in an inextricable relation to soil and climate, to human activities and so on. On different scales, living organisms and abiotic elements are seen as integrated and co-constitutive of each other. An overall social-ecological resilience approach, thus, not only acknowledges the temporal aspects of disturbance and change but also the interlinkages and dependencies between human and non-human species and between organic and inorganic components. From this perspective, categories like

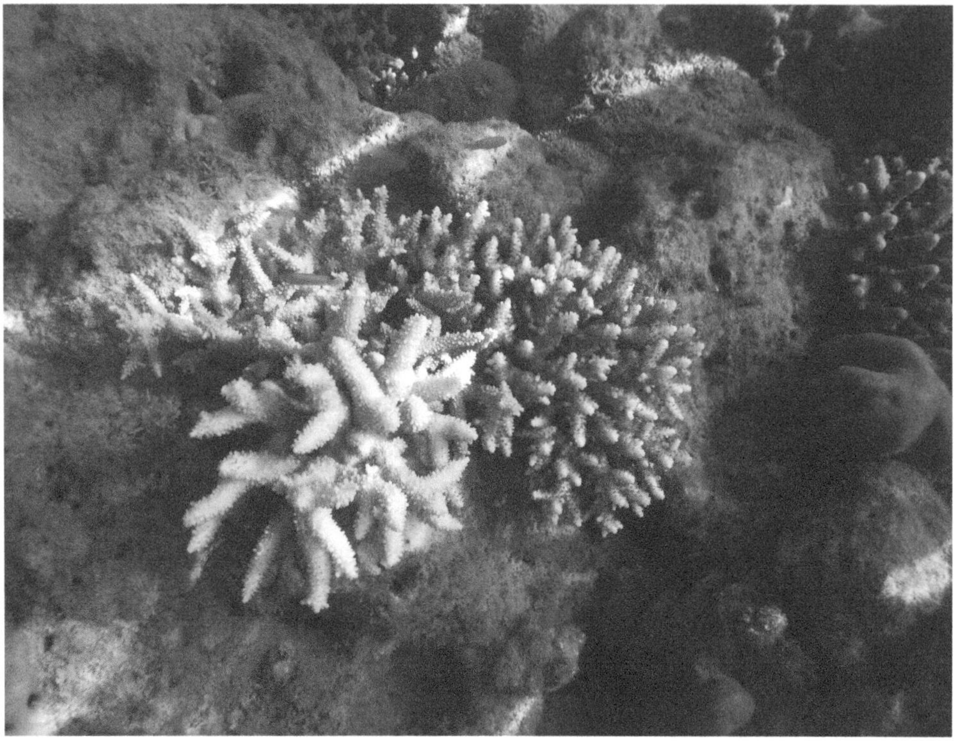

FIGURE 3.5 Ecosystems, too, have a history of change, of slow or sudden transformations. Sometimes the new configuration is able to perform the same functions as before, as a scar on the body. Sometimes, remnants of, for example, toxicity, are merely superficially concealed by new verdure. Coral reef, Andaman Islands, India

"wilderness" and "heritage" suddenly become blurred, often seeming interchangeable or even simultaneous. For example, Robert S. Emmet and David E. Nye suggest that "the mix of reverence, awareness of long human inhabitation, and national symbolism [in the Fuji-Hakone-Izu National Park, Japan] makes 'heritage' a more apt category than 'wilderness'" (Emmett and Nye 2017, 42). In a similar vein, Whitney E. Woods reminds us how "wilderness" in fact forms a critical part of what in many contexts has long been understood as heritage (Woods 2018, 1; cf. Cronon 1995).

In a corresponding manner, heritage scholarship has articulated questions about the relation between human and non-human, as well as about permanence versus process. For some time, the field has pushed to abandon a retrospective understanding of heritage as a physical, tangible and static "thing" from the past to account for in the present and instead put forward heritage as intangible (Ruggles and Silverman 2009; Graham and Howard 2008, 4) and as an ongoing contemporary cultural and social process of experiencing and of ascribing meaning (Smith 2006). Brian J. Graham and Peter Howard, for example, state that "heritage is less about tangible material artefacts … of the past than about the meanings placed upon them and the representations which are created from them", and furthermore, that it is "meaning that gives value … to heritage" (Graham and Howard 2008, 2). Lately, this constructivist effort has been complemented by studies aiming to bring materiality back into the picture (Pétursdóttir and Olsen 2014, 14), not, however, in the "old" fashion but in ways that blur precisely the dividing line between animate and inanimate matter. These approaches might be called new materialism, post-humanism or object-oriented ontology (Bangstad and Pétursdóttir, chapter 1). One area that has drawn particular attention is the different forms and affordances of material processes such as decay and ruination (DeSilvey 2017; Pétursdóttir and Olsen 2014).

Caitlin DeSilvey, for example, engages with objects as "processual events" (DeSilvey 2017, 29) and with an understanding of subjectivity in which "materiality is not a static field of reference that awaits inscription from an active mind but is itself constitutive of … human selfhood, as distributed through intimate relations with other entities – plants, stones, dust" (DeSilvey 2017, 14), echoing what Rodney Harrison has termed "connectivity ontologies" (Harrison 2015, 27). DeSilvey points out that, if we accept that any system can unfold in many different ways, this entails that "what may appear as erasure on one register may be generative of new information on another" (DeSilvey 2017, 13), and therefore, she suggests, there is a potential to "uncouple memory work from material stability" (DeSilvey 2017, 15).

FIGURE 3.6 Overgrown trenches from World War I battlegrounds may speak of reconciliation, and in that sense, they can be understood as a scar. From a heritage perspective, we are to consider the risks and possibilities of their continuous material changes. Perhaps a respectful heritage process might well also include processes of material decay? Verdun, France

Photo: Bjoertvedt/CC BY-SA 4.0. https://creativecommons.org/licenses/by-sa/4.0/.

To let heritage practice embrace material change in terms of neglect, decay and, eventually, destruction is certainly to challenge a retrospective conservationist approach with a prospective one. DeSilvey argues that there is an increasing openness to policies based on no intervention (DeSilvey 2017, 49), although this is certainly not a position commonly held by heritage scholars and practitioners.

The risk of engaging with damage and decay in ways close to celebration and enchantment, as noted previously in the section on scabs and new tissue, now resurfaces as a more explicit question of power relations. The resilient new ecological "configuration", or the new generative heritage "register" following change, undoubtedly works as means of "assembling futures" (Harrison 2015, 24) and is thus highly political. To use the metaphors of wounds, scabs and scars also imply a process of healing that might appear linear, predetermined and unproblematic. However, this is definitely not the case. The physical bodily wound can remain as a chronic wound, the scab can become infected and the scar can reopen to become a wound again. In the same way, the metaphorical process of healing can bring difficult pasts to the fore (Bohlin 2012) or lead to less desirable outcomes. For example, the deliberate overgrowing of former mining landscapes (Bailey and Hooey 2017) and abandoned industrial structures (Braae 2015) as a way to shape new futures in crises struck areas or, in other words, "to heal the scars left by industry with a soothing layer of greenery" (DeSilvey 2017, 106), has also been seen as way to obscure, conceal or excuse an uncomfortable past and even to work as a "substitution story" (Storm 2014, 122ff) for difficult heritages. What is seen as or anticipated to infer a process of healing and, ultimately, reconciliation may therefore turn out to be a process of concealing past and present injustice and harm, sometimes resembling the persistence of phantom pain after an amputation.

The connectivity and integration pinpointed by the scar metaphor, between past and present, and between human and non-human, resonates with both social-ecological resilience thinking and the recently suggested "connectivity ontologies" within heritage scholarship. As the normal skin layers are replaced with integrated fibrous scar tissue, the scar metaphor also challenges a layered, or palimpsestual, understanding of history and heritage per se. A palimpsest is a handwritten manuscript on parchment or papyrus on which one text has been scraped away and replaced by another (cf. Storm 2014, 3f). This recycling practice of Antiquity and the Middle Ages was often used repeatedly in order to re-use valuable parchment, even though it was sometimes possible to discern the erased text behind the new addition. In postmodern contexts, the concept has been used metaphorically to describe the different layers of significance that make up heritage (cf. Smith 2006; Huyssen 2003). New layers are added continuously over time, like new texts inscribed on a scraped, cleaned parchment.

However, this metaphorical use of the palimpsest is misleading if one wishes to emphasize the interconnectedness, the linked relevance, between the different layers. If heritage consists of a multitude of parallel or successive stories and perspectives, all contributing to one another and to an overall picture, the palimpsest is a poor choice of metaphor, since the palimpsest in the original sense implies that new texts are added without regard or connection to those previously erased, and thus the end result is a picture made up of fragments and additions without any mutuality or interrelated content. A palimpsestual approach, "in which the present is merely the sum of past episodes" (Harvey 2013, 154), like in the Odyssey, is therefore an expression of a "retrospective memory", differing from a "prospective memory", which David Harvey denotes as an "unfolding and ongoing relationship between past, present and future" (Harvey 2013, 155; cf. Holtorf and Williams 2006, 237f).

So, instead of trying to uncover and acknowledge the manifold layered stories of the past, the scar metaphor suggests that heritage is an integrated and crucial part of human living with all its inherent contradictions and ambiguities. It is a prospective approach that attempts to discern wholeness out of complexity, divergences and change. This is not to say that ambiguity is always and inherently good or liberating but rather that, in the words of Shiloh R. Krupar, "ambiguity is what must be explored to contest routinized recitations of evidence and established truths" (Krupar 2013, 17). Even if the scar itself could be understood as dead, less sensory skin, where the normal layers have been replaced by a new dense scar tissue, it forms a part of the living body. The scar is not a palimpsest.

Living with ambiguous pasts

A well-known scar in modern children's literature emblazons the forehead of the young wizard Harry Potter (Rowling 1997). One can argue that this scar in fact makes up the key material entity in the stories; trumping Harry's broomstick, invisibility cloak and Marauder's Map. The scar is shaped like a bolt of lightning and marks a wound he incurred as a baby when the evil Lord Voldemort attacked his family. While both his parents died, Harry miraculously survived, protected by his mother's love. As a boy and teenager, Harry's scar works as a sign of his legendary survival, although he himself prefers to hide it behind his bangs. The scar also forms a constant connection to both the good and bad of his past, epitomized by his parents, his own survival and Lord Voldemort. Through the partly failed attack, Voldemort involuntarily shared some of his powers with Harry, who struggles not to let the dark side within him win, aided by the memory of his parents. Occasionally Harry feels how the scar is itching and painfully burning, and after some time, he learns that this happens when Voldemort experiences strong feelings of success or anger. Unlike Odysseus, Harry lives with his scar as a constant node of remembrance and of present challenges and obligations. The past is not only a separate episode told to the reader but integrated in Harry's evolving lived experience. The scar is also a trigger of remaining difficulties, injustices and ambiguities, making up part of Harry's body as an entity with distinct and special functions.

In this chapter, I have suggested that the metaphorical process from wound to scar closely mirrors existing scholarly discussions and professional practices related to ecology and heritage, especially to the area in between, inhabited by various "natureculture" approaches. The bodily conditions of wound, scab and scar have proved able to highlight common analytical standpoints of the different traditions, such as a call for public attention to harm and abuse (wound), a prospective idea about how to respond to damage (scab), the inherent political risks of reconciliation processes and, finally, an integrative "connectivity ontology" including temporal as well as human – non-human aspects (scar). In fact, the set of metaphors has shown how the space where ecology turns social and temporal, and where heritage turns post-human and recognizes value in disappearance, is in many respects a shared intellectual space between the different practices and scholarly traditions of ecology and heritage. While the explicit exchange hitherto has been limited and fragmented, it does have capacity for growth.

In some more detail, I would argue that in response to shadowed environmental and social sacrifice zones, or to dominant positive-affirmative understandings of heritage, the wound metaphor normatively calls for attention to past and present harm and difficulties, with ambiguous components to be acknowledged and attended to. To use the wound metaphor is to say that what it represents is significant and of importance, although its meaning might not be clear cut or easily defined. The scab, for its part, highlights the difference between retrospective and prospective approaches, corresponding to the bodily processes of complete and exact "regeneration" of unwounded tissue versus "repair" into new, transformed tissue. This is translated, first, into the traditional ideal of ecological restoration toward "pristine" natures versus a valuation of an ecological "new wilderness" in urban-industrial settings. Second, it is translated into a traditional building conservation ideal toward "original shape" as opposed to regarding material and immaterial change as a generative heritage process revealing information on new "registers", that is, in the words of Caitlin DeSilvey, memory work "uncoupled" from material stability.

The scar, finally, emphasizes that the promise of healing, or, metaphorically, of reconciliation with past and present difficulties, comes with a risk of hiding or obscuring injustice or enduring harm under a seemingly well-functioning surface. For example, the technological "fixes" for achieving social-ecological resilience, or the strategic overgrowing of an abandoned industrial structure, might well, as Rodney Harrison writes, "assemble futures" that are fair and legitimate in that they acknowledge inherent ambiguities. However, they might also involve political exploitation for certain exclusive benefits or merely constitute negligence of complexity and pain. The line between healing and concealing is thin, and the scar offers a way to conceptualize the process of navigating toward a better future in a defined direction, without being deterministic in terms of normative "positive qualities" or ready-made "technologies of recovery".

In summary, the wound, the scab and the scar joined into a bodily process of potential healing metaphorically articulate the constant presence of past difficulties in an evolving contemporaneity and suggest an emphasis on temporal integration rather than a conception of separate, palimpsestual, time layers. By letting the wound and the scab, in consequence, form part of the scar, the scar becomes embodied experience, something you live with. As is widely acknowledged, not everything can be remembered, and not all pasts can be incorporated in the present. What the scar metaphor does is to steer our concern, to say that among those pasts and presents that are most important and that need to be attended to, it is not least the difficult and ambiguous ones that stand out and crave our attention. With Lebbeus Woods' telling and essentially hopeful wording, I will close this chapter. He writes that a scar is a

> deeper level of reconstruction that fuses the new and the old … a mark of pride and of honor, both for what has been lost and what has been gained. It cannot be erased. … To accept a scar is to accept existence.
>
> *(Woods 1997, 16)*

Acknowledgments

The author thanks Anders Houltz, Anna Bohlin, Arn Keeling, Kenneth Knoespel, Lars Kaijser, Magnus Rodell, Martin Schmid, Mattias Frihammar and Staffan Appelgren for their help and valuable comments on earlier versions of this chapter.

Note

1 All medical information comes from the Wikipedia entries "Human skin", "Wound healing", "Skin repair" and "Scar," accessed 20 July 2018.

References

Albrecht, G., Sartore, G.-M., Connor, L., Higginbotham, N., Freeman, S., Kelly, B., Stain, H., Tonna, A. and Pollard, G. 2007. "Solastalgia: The Distress Caused by Environmental Change." *Australasian Psychiatry* 15: 95–98.

Angel, T. 2014. *The Scars of Project 459: The Environmental Story of the Lake of the Ozarks*. Fayetteville, AK: University of Arkansas Press.

Auerbach, E. 2003 [1946]. *Mimesis: The Representation of Reality in Western Literature*. Princeton, NJ: Princeton University Press.

Bailey, T. and Hooey, C. 2017. "'From Wasteland to Wonderland': New Uses for Mined Land in Rural Southeast Kansas." *The Midwest Quarterly* 58(3): 252–263.

Bohlin, A. 2012. "Land Restitution and Reconciliation: A Lost Opportunity? Emotion, Land and Heritage in Post-Apartheid South Africa." Cited in J. D. Giblin. 2013. "Post-conflict Heritage: Symbolic Healing and Cultural Renewal." *International Journal of Heritage Studies*: 13f.

Braae, E. 2015. *Beauty Redeemed: Recycling Post-Industrial Landscapes*. Risskov: IKAROS Press.

Chen, Y.-F. 2002. "Review of *Scars of War: The Impact of Warfare on Modern China*, by D. Lary and S. MacKinnon." *The Journal of Asian Studies* 61(3): 1033–1035.

Cottle, D. 2013. "Land, Life and Labour in the Sacrifice Zone: The Socio-Economic Dynamics of Open-cut Coal Mining in the Upper Hunter Valley, New South Wales." *Rural Society* 3: 208.

Cronon, W. 1995. "The Trouble with Wilderness; or, Getting Back to the Wrong Nature." In *Uncommon Ground: Rethinking the Human Place in Nature*, edited by W. Cronon, 69–90. New York: W. W. Norton & Co.

DeSilvey, C. 2017. *Curated Decay: Heritage Beyond Saving*. Minneapolis, MN: University of Minnesota Press.

Easthope, L. 2018. *The Recovery Myth: The Plans and Situated Realities of Post-Disaster Response*. Cham: Palgrave Macmillan.

Emmett, R. S. and Nye, D. E. 2017. *The Environmental Humanities: A Critical Introduction*. Cambridge, MA: MIT Press.

Endres, D. 2012. "Sacred Land or National Sacrifice Zone: The Role of Values in the Yucca Mountain Participation Process." *Environmental Communication* 6(3): 328–345.

Fitzpatrick, M. 2017. *Invisible Scars: Mental Trauma and the Korean War*. Studies in Canadian Military History. Vancouver: UBC Press.

Foote, K. E. 2003. *Shadowed Ground: America's Landscapes of Violence and Tragedy* (rev. edn). Austin, TX: University of Texas Press.

Fox, J. 1999. "Mountaintop Removal in West Virginia: An Environmental Sacrifice Zone." *Organization & Environment* 12: 163–183.

Gandy, M. 2002. *Concrete and Clay: Reworking Nature in New York City*. Cambridge, MA: MIT Press.

Gandy, M. 2013. "Marginalia: Aesthetics, Ecology, and Urban Wastelands." *Annals of the Association of American Geographers* 103(6): 1301–1316.

Giblin, J. D. 2014. "Post-Conflict Heritage: Symbolic Healing and Cultural Renewal." *International Journal of Heritage Studies* 20(5): 500–518.

Graham, B. J. and Howard, P. 2008. "Introduction: Heritage and Identity." In *The Ashgate Research Companion to Heritage and Identity*, edited by B. J. Graham and P. Howard, 1–15. Aldershot: Ashgate.

Guarnacci, U. 2016. "Joining the Dots: Social Networks and Community Resilience in Post-conflict, Post-disaster Indonesia." *International Journal of Disaster Risk Reduction* 16: 180–191.

Harrison, R. 2013. *Heritage: Critical Approaches*. Abingdon: Routledge.

Harrison, R. 2015. "Beyond 'Natural' and 'Cultural' Heritage: Toward an Ontological Politics of Heritage in the Age of Anthropocene." *Heritage & Society* 8(1): 24–42.

Harvey, D. 2013. "Emerging Landscapes of Heritage." In *The Routledge Companion to Landscape Studies*, edited by P. Howard, I. Thompson and E. Waterton, 152–165. Abingdon: Routledge.

Hochschild, A. 2018. "Scars of the Gulag." *The New York Times Book Review*: 16.

Holifield, R. and Day, M. 2017. "A Framework for a Critical Physical Geography of 'Sacrifice Zones': Physical Landscapes and Discursive Spaces of Frac Sand Mining in Western Wisconsin." *Geoforum* 85: 269–279.

Holtorf, C. and Williams, H. 2006. "Landscapes and Memories." In *The Cambridge Companion to Historical Archaeology*, edited by D. Hicks and M. C. Beaudry, 235–254. Cambridge: Cambridge University Press.

Houston, D. 2013. "Environmental Justice Storytelling: Angels and Isotopes at Yucca Mountain, Nevada." *Antipode* 45(2): 417–435.

Huyssen, A. 2003. *Present Pasts: Urban Palimpsests and the Politics of Memory*. Stanford, CA: Stanford University Press.

Jørgensen, D. 2015. "Rethinking Rewilding." *Geoforum* 65: 482–488.

Jørgensen, D. 2017. "Competing Ideas of 'Natural' in a Dam Removal Controversy." *Water Alternatives* 10(3): 840–852.

Keenan, R. and Jorgensen, A. 2012. *Urban Wildscapes*. Abingdon: Routledge.

Kowarik, I. 2005. "Wild Urban Woodlands: Towards a Conceptual Framework." In *Wild Urban Woodlands: New Perspectives for Urban Forestry*, edited by I. Kowarik and S. Körner, 1–32. Berlin: Springer-Verlag Berlin Heidelberg.

Krupar, S. R. 2013. *Hot Spotter's Report: Military Fables of Toxic Waste*. Minneapolis, MN: University of Minnesota Press.

LeCain, T. J. 2009. *Mass Destruction: The Men and Giant Mines That Wired America and Scarred the Planet*. New Brunswick, NJ: Rutgers University Press.

Leopold, A. 1970. *A Sand County Almanac: With Essays on Conservation from Round River*. New York: Ballantine Books.

Lepawsky, J., Liboiron, M., Keeling, A. and Mather, C. 2017. "Repair-Scapes." *Continent* 6(1): 56–61.

Lindahl, B. 2014. "Dramatiskt verk till minne av Utøya." *Svenska Dagbladet*, 27 February.

Logan, W. S. and Reeves, K., eds. 2009. *Places of Pain and Shame: Dealing with "Difficult Heritage"*. Abingdon: Routledge.

Lowenthal, D. and Jenkins, S. 2011. "Prizing the Past for the Present and the Future." *British Academy Review* 18: 34–40.

Lyde Cullen, S. 2006. "Review of *Battle Scars: Gender and Sexuality in the American Civil War*, edited by C. Clinton and N. Silber." *The North Carolina Historical Review* 4: 493–494.

Macdonald, S. 2009. *Difficult Heritage: Negotiating the Nazi Past in Nuremberg and Beyond*. London: Routledge.

Macdonald, S. 2015. "Is 'Difficult Heritage' Still 'Difficult'?" *Museum International* 67(1–4): 6–22.

Masco, J. 2006. *The Nuclear Borderlands: The Manhattan Project in Post-Cold War New Mexico*. Princeton, NJ: Princeton University Press.

Miles, W. F. S. 2014. *Scars of Partition. Postcolonial Legacies in French and British Borderlands*. Lincoln, NE: Nebraska University Press.

Millington, N. 2013. "Post-Industrial Imaginaries: Nature, Representation and Ruin in Detroit, Michigan." *International Journal of Urban and Regional Research* 37(1): 279–296.

Pétursdóttir, Þ. and Olsen, B. 2014. "An Archaeology of Ruins." In *Ruin Memories: Materialities, Aesthetics and the Archaeology of the Recent Past*, edited by B. Olsen and Þ. Pétursdóttir, 3–29. Abingdon: Routledge.

Plumwood, V. 2008. "Shadow Places and the Politics of Dwelling." *Australian Humanities Review* 44: 139–150.

Robin, L. 2013. "Commentary: C. S. Holling, Resilience and Stability of Ecological Systems (1973)." In *The Future of Nature: Documents of Global Change*, edited by L. Robin, S. Sörlin and P. Warde, 257–259. New Haven, CT: Yale University Press.

Rowling, J. K. 1997. *Harry Potter and the Philosopher's Stone*. London: Bloomsbury (the following books in the series were published 1998, 1999, 2000, 2003, 2005 and 2007).

Ruggles, D. F. and Silverman, H. 2009. "From Tangible to Intangible Heritage." In *Intangible Heritage Embodied*, edited by H. Silverman and D. F. Ruggles, 1–14. New York: Springer.

Smith, L. 2006. *Uses of Heritage*. New York: Routledge.

Soyez, D. 2009. "Europeanizing Industrial Heritage in Europe: Addressing Its Transboundary and Dark Sides." *Geographische Zeitschrift* 97(1): 43–55.

Spelman, E. V. 2007. "Embracing and Resisting the Restorative Impulse." In *Healing Natures, Repairing Relationships: New Perspectives on Restoring Ecological Spaces and Consciousness*, edited by R. L. France, 127–138. Faringdon: Libri Publishing.

Storm, A. 2014. *Post-Industrial Landscape Scars*. New York: Palgrave Macmillan.

Tachau, F. 2008. "Scars of War, Wounds of Peace: The Israeli-Arab Tragedy." *Shofar* 26(3): 183–186.

Trell, E.-M., Restemeyer, B., Bakema, M. M. and van Hoven, B. 2018. "Governing for Resilience in Vulnerable Places: An Introduction." In *Governing for Resilience in Vulnerable Places*, edited by E.-M. Trell, B. Restemeyer, M. M. Bakema and B. van Hoven. Abingdon: Routledge.

Tsing, A. L., Swanson, H., Bubandt, N. and Gan, E., eds. 2017. *Arts of Living on a Damaged Planet: Ghosts of the Anthropocene*. Minneapolis, MN: University of Minnesota Press.

Tunbridge, J. E. and Ashworth, G. J. 1996. *Dissonant Heritage: The Management of the Past as a Resource in Conflict*. Chichester: Wiley.

Viitanen, J. and Kingston, R. 2014. "Smart Cities and Green Growth: Outsourcing Democratic and Environmental Resilience to the Global Technology Sector." *Environment and Planning A* 46(4): 803–819.

Walker, B., Holling, C. S., Carpenter, S. R. and Kinzig, A. P. 2004. "Resilience, Adaptability and Transformability in Social-Ecological Systems." *Ecology and Society* 9(2).

Weiler, K. 2017. "Introduction." In *Authenticity in Architectural Heritage Conservation: Discourses, Opinions, Experiences in Europe, South and East Asia*, edited by K. Weiler and N. Gutschow. Transcultural Research – Heidelberg Studies on Asia and Europe in a Global Context, xvii–xxiv. Cham: Springer.

Woods, L. 1997. *Radical Reconstruction*. New York: Princeton Architectural Press.

Woods, W. E. 2018. "Framing Wilderness as Heritage: A Study of Negotiating Heritage in Environmental Conflict." *Heritage & Society*: 1–24.

4

WILDERNESS HERITAGE

For an ontology of the Anthropocene

Levi R. Bryant

Correlationism, the Anthropocene, and golems

The Anthropocene is both the culmination and undoing of correlationism. Quentin Meillassoux defines correlationism as "the idea according to which we only ever have access to the correlation between thinking and being, and never to either term considered apart from the other" (Meillassoux 2008, 5). In its classical form, correlationism lends itself to forms of idealism and social constructivism, exploring the ways in which mind, society, and language construct reality. What reality might be apart from us, we cannot know, the correlationist claims, for reality is always filtered through our linguistic and conceptual schemes. With the Anthropocene, we get a very different form of correlationism, where physical reality itself has been transformed by thought. While debates might rage as to when we should mark the advent of the Anthropocene, the Anthropocene nonetheless is that epoch in which the nature of the planet has been transformed by humanity at a geological, ecological, and climatic scale.

In this regard, the earth, in its physical reality, has been transformed through thought and the practice that accompanies thought, becoming an embodied reflection of human thought. However, what a strange reflection this is! Unlike a faithful reflection in a mirror or picture in a photograph, the earth is akin to the painting in Oscar Wilde's *The Picture of Dorian Gray*, where it endlessly ages and reflects the ravages of how he has lived his life, while Mr. Gray remains eternally young. The correlate of our thought, the earth, increasingly presents itself as a ruined wasteland transformed by our thought and practice, while humanity still regards itself through the distorted lens of the bloom of innocent youth. In this regard, humanity does not recognize itself in its own painting.

The Anthropocene thus announces a strange sort of correlationism. Yes, the earth has been literally transformed through thought and the practice that accompanies thought, but far from mastering the earth like the prosthetic gods imagined by Freud, this earth transfigured by thought is something akin to the figure of the golem in Jewish folklore. In Jewish folklore, the sorcerer constructs the golem from the earth to do his bidding, but in these stories, something often goes wrong, and the golem goes on rampage, destroying the village. It is in this sense that the Anthropocene is *also* the undoing of correlationism. In forming matter according to thought, we thought we had mastered the earth. We have conceived ourselves as akin to

Plato's demiurge, taking the chaos of matter and fashioning it according to our image. The earth is transformed by thought in the Anthropocene, but that transformed earth takes on an uncontrollable life of its own, exceeding the capacities of economic, calculative, and instrumental reason to master it. In the Anthropocene, we discover an earth beyond thought, a force of matter beyond matter that we have formed, and an earth and matter beyond our mastery; we discover a heritage that is ours because it is the result of our thought and practice but do not recognize ourselves in this heritage.[1]

The ontic and the ontological wilderness

In the Anthropocene, we discover the wilderness as that which has been lost *and* that is all that ever was. Here I wish to maintain that the first conception of the wilderness was a sort of illusion that we would do well to abandon, but that we must simultaneously preserve, though under erasure as the wilderness. I will have more to say about this later. I call this first conception of the wilderness the "ontic wilderness".[2] At the ontic level, we discover that the wilderness is lost insofar as there is today no region of the earth that is untouched by humanity. In ordinary language, the wilderness is that which is untamed and outside of civilization. It is a realm of pure nature untouched by humans. Under this ontic conception of the wilderness, the wilderness is conceived *spatially* as a *place* to which one can journey. There is, on the one hand, civilization where the material world has been domesticated, tamed, and given form and structure according to human intentions, and then, on the other hand, there is the untamed wilderness, where nature is governed by its own internal principles.

This first ontic conception of the wilderness embodies the logic of the raw and the cooked described by Lévi-Strauss (1983). We can call this the ontology of calculative and instrumental reason. The civilized is the cooked, and all those things that are given form by human agency whether it be food, well-tended lawns, concrete roads, or built homes. All of these things fall into the category of the cooked. By analogy to Plato's demiurge, cooking consists in giving form to unformed matter through the thought and the agency that arises from thought. Here, form is brought to bear on matter from the external agency of humans. Things are not *naturally* this way. The marble is not naturally a statue of Apollo. The raw, by contrast, is that which takes on form according to its own internal nature, as in the case of every Aristotlean's favorite example of the acorn becoming an oak tree. Unlike the cooked example of an oak log becoming a statue of Zeus through the agency of the artist, the raw case of the acorn contains the oak tree as its destiny within itself of its own accord. The ontic distinction between the raw and the cooked, the wilderness and civilization, governs every aspect of our thought, ranging from our scientific categories, to how we think about the relationship between society and nature, to how we sort between civilized humans and so-called "savages" and differences between men and women. The cooked, civilized man has had his animal nature transformed through culture and education, rising above both nature and his animal nature, while the raw man and woman are said to be governed by their animal nature and are those who live close to nature. Often it is suggested that the raw man cannot be cooked – this tacit or unconscious belief is at the foundation of racism and sexism – though the cooked man of culture has often feared being cooked by the raw man or witch. This last iteration of the cooked/raw distinction has often led to monstrous consequences and abuses of other populations. On our planet, at least, this ontic wilderness is now gone, as there is no region of the globe that has not been impacted by our species in some way. There is a sense, both literal and figurative, in which everything is cooked today.

At the ontological level, the Anthropocene leads us to discover that the wilderness is all that there ever was, insofar as there is no humanity outside of nature, and that there never was a humanity outside of nature. The ontological *concept* of the wilderness places the ontic, raw/cooked distinction between the wilderness and civilization under erasure. Henceforth, there will *only* be the wilderness, and the spatial conception of the wilderness as that which is *outside* civilization will be abandoned. There is nothing outside the wilderness. In this regard, the wilderness is another name for being or what is. Even when you are in the center of Paris, Hong Kong, or New York, you are in the wilderness. Moreover, you are no more in the wilderness when you are in a remote region of Alaska or Svaerholt, Norway, than when you are in the suburbs of Los Angeles. This might appear to be a perverse use of the ordinary language signification of "wilderness"; however, it is the Anthropocene that occasions this rethinking of nature and the wilderness. The Anthropocene provokes us to think our technological and industrial inheritance but also leads us to rediscover the manner in which we are dependent upon a natural world that is also our heritage. The ontological situation of the Anthropocene calls for a rethinking of heritage and what we inherit both at the level of the inherited distinctions that govern our thinking of nature and our relation to nature and what we inherit from both culture and nature.

Wilderness heritage and ~~wilderness~~ heritage

In what sense are we authorized in claiming that in the Anthropocene, there has only ever been the wilderness? In the Anthropocene, we discover two unexpected forms of heritage. In the common concept of heritage, "heritage" tends to largely be conceptualized as *cultural* heritage and those forms of natural heritage we *value* because they are singular, rare, and irreplaceable. As defined by UNESCO, "[h]eritage is our legacy from the past, what we live with today, and what we pass on to future generations" (UNESCO 2018). For example, UNESCO goes on to cite the Great Barrier Reef and the Egyptian pyramids as examples of natural and cultural heritage. Further, despite nods to natural heritage, in the photo collage of heritage sites on the website, we are exclusively shown pictures of *cultural* monuments in the form of architectural achievements from around the world. UNESCO praises this definition of heritage for its *universal* application. However, between the explicit definition of heritage it proposes and the *examples* of heritage it proposes, we can detect a tension. As Olsen and Pétursdóttir point out, heritage is here conceptualized in social constructivist terms "as staged and negotiated in response to contemporary human interests and conflicts" (Olsen and Pétursdóttir 2018). At the level of cultural heritage, heritage is that past we would like to preserve because it either speaks to who we are and where we came from or speaks to those forms of the cultural past that continue to trouble us today, such as the continuing legacy of the Civil War in the American South. With regard to natural heritage, heritage refers to those singular and irreplaceable places on the planet, such as the redwood forests of California, that we would like to preserve for future generations.

However, if we take UNESCO's definition of heritage *literally* or "to the letter" – and here resides the tension – we should not restrict the concept of heritage to those things we wish to preserve but should instead include all of those things that we inherit without necessarily wishing to inherit them. It is precisely this undesired heritage that we discover in the Anthropocene. Heritage is no longer merely the delightful heritage of Greek monuments and those lost sites of significance as a result of the Three Gorges Dam in China. Heritage is *also* industrial waste, polluted air and water, floating islands of plastic in the ocean, radioactive waste, landfills, genetically modified plants and organisms that perhaps fundamentally transform ecosystems, and all the rest. All of these things that we generally classify as waste or garbage form a sort of "Lacanian" heritage in that they follow a logic similar to the *objet a*. In Lacanian psychoanalysis, the *objet a* is, among other things, a sort of ineradicable excess or remainder produced in our pursuit of the objects of our desire. These remainders plague us in all sorts of ways and generate a compulsion to repeat as we attempt to empty them from our psychic system. This other, dark heritage of waste is the ineradicable remainder produced by civilization. We would like literally and figuratively to throw it away and believe that we can throw it away, but, like the return of the repressed, the undead, or the compulsion to repeat encountered in the clinic, this waste returns to haunt us, challenging the very possibility of civilization's continued existence, at least in civilization's current form.

Modernity thought we were prosthetic gods, that we had finally conquered nature and fully cooked it through our scientific and technological mastery. Or, at least, it conceived itself as being on the way to fully cooking nature. Among the many meanings of "nature" in antiquity is that of the difficult-to-translate term "*τύχη*", which is ordinarily rendered as "fate" or "luck" but which might equally be rendered as "chance" or "chaos". As Serres reminds us, farmers and sailors know nature is something unpredictable and incalculable (Serres 2011, 28–29). Modernity thought that it had overcome the incalculable. Yet in the Anthropocene, the very products of our technological and scientific mastery return as a remainder, an excess, that undermines that mastery such that the very existence of civilization is called into question. In this connection, Serres never tires of reminding us that the French word *temp* simultaneously refers to *both* time *and* weather (Serres 2011, 27). Just as the weather is aleatory, there is an aleatory dimension to heritage in the Anthropocene that undermines our ability to calculate what we are passing on. In the Anthropocene, we discover an unruly heritage.

This leads us to discover a second natural heritage. With the Anthropocene, we discover that the wilderness itself, nature itself, is a part of heritage. Under the ontic conception of the wilderness, we thought that civilization was outside of nature, that we had somehow escaped nature or the wilderness in the spatial sense of the terms. Yet insofar as the sustainability of the biosphere is now threatened, we discover that this was always an illusion. We discover that the existence of civilization was always already dependent upon the biosphere and that civilization was therefore never outside of nature. In this regard, it is not simply those remarkable and singular places like the Great Barrier Reef that constitute natural heritage. It is the wilderness itself, in the ontological sense, that is a form of heritage. The wilderness must be passed on for civilization to exist.

We are now in a position to understand why there has never been anything but the wilderness. Speaking ironically, the ontic distinction between the wilderness and civilization has always been an artificial distinction. If such a claim is ironic, then this is because the very distinction between the artificial and the natural is itself a distinction premised on the ontic distinction between the wilderness and civilization. If the ontic distinction between the wilderness and civilization is an artificial distinction, then this is because civilization has always-already been a part of the wilderness in the ontological sense. In ordinary language, when we speak of the wilderness, we refer to it as that in which a person becomes lost. The wilderness is that in which we are lost and where we cannot find our way home. We also speak of the wilderness as that which is unexplored, unknown, and filled with mysteries. Like the logic of the veil, the wilderness is thought as that which must be unveiled, tamed, domesticated, and mastered.[3] In the Anthropocene, we discover that we have always-already been in the wilderness and that we are lost every bit as much in "civilization" as we can be lost in the wilderness. Further, we discover that the civilization that we thought we knew and that represented our mastery is, in fact, the unknown and undiscovered. We are unaware of the conditions for civilization.

The situation of the Anthropocene is both a danger and an opportunity. It is a danger, of course, insofar as it threatens our continued existence. It is an opportunity insofar as it invites us to think the being of being otherwise under the ontological conception of the wilderness. It was a certain way of thinking, a certain form of ontological thinking, that has placed us in the predicament of the Anthropocene. If we are to move beyond this predicament, we must both uncover the ontological assumptions that put us here and seek forms of ontological thought that move beyond these ways of thinking. Only in this way can we hope to develop forms of practice adequate to the situation in which we find ourselves. Part of this consists of coming to understand the wilderness not as a place outside of civilization but that it is all that there is, such that civilization itself is a part of the wilderness. The earth has been dismembered and lacerated by distinctions both at the level of thought and at the level of physical reality. We must uncover the ways in which the world has been dismembered and lacerated by thought leading to forms of practice that have led to the Anthropocene, for if the wilderness is all that is, then ways of thinking are also part of our wilderness heritage.

Aletheia and the distinctions that haunt calculative reason

However, this physical dismemberment and laceration of the earth has been rendered possible based on a series of distinctions that haunt the unconscious of thought as a sort of spontaneous ontology of the world. In one form or another, all of these distinctions are variations of the wilderness/civilization distinction, or the distinction between the raw and the cooked. Let us call this form of thought the ontology of calculative or instrumental reason. Wilderness ontology aims to replace the ontology of calculative or instrumental reason with a different way of thinking our relationship to nature. What we do is, in part – and we must emphasize, only in part – a function of the conceptual distinctions we deploy in approaching the world. These distinctions are the filter through which we apprehend the world. In the transcendental deduction in the *Critique of Pure Reason*, Kant observes that the conditions for the possibility of experience are also the conditions for the possibility of the *objects* of experience (Kant 1998, A111). Speaking crudely, in Kant's correlationist framework, the conditions for the possibility of experience are the transcendental forms of intuition or space and time, coupled with the pure a priori categories of the understanding. While I will not here go all the way in claiming with Kant that the conditions for the possibility of experience are the conditions for the objects of experience, I do claim that the conditions for experience – whatever they happen to be – are the conditions for the possibility of the *manifestation* of the objects encountered in experience.[4] Whatever the conditions turn out to be, these conditions would be the conditions under which the objects of the world manifest or show themselves. Following Foucault's earlier work such as *The Order of Things*, these conditions would be variable throughout history and in different societies.[5] They would also be variable depending on the form of practice one is engaged in. Put differently and in Heideggerian terms, they are the conditions under which objects manifest, disclose, or reveal themselves. These conditions are what lead objects to stand in relief, in the foreground, in a particular way against a background. In disclosedness or manifestation, things are pulled into relief in a particular way, while other things are veiled.

Heidegger called this sort of manifestation *aletheia*, or truth as disclosedness: the way in which things show themselves (Heidegger 1962, 257–273). In his discussion of space and distance in *Being and Time*, Heidegger claims there is a sense in which the glasses on a person's face are further from them than the object across the room. One sees *through* the glasses; one does not see the *glasses*. This is how it is with the distinctions that govern the manifestation of objects in the world about us and is the reason they form a sort of spontaneous, unconscious ontology structuring thought and practice.[6] We encounter the world through these distinctions and are often unaware of the distinctions that structure our thought and practice. I would like to argue that the Anthropocene has placed these distinctions into question and calls for them to be reconceptualized or even abandoned. While there are many distinctions that structure the form of thinking that contributed to the age of the Anthropocene and that constitute part of our wilderness heritage, there are two that I will focus on here: the ecological and the social and the agent and the patient. These distinctions form the unconscious, spontaneous ontology of calculative or instrumental rationality and form a sort of episteme for the thought and practice that led to the Anthropocene.

First distinction: the ecological and the social

Among the fundamental distinctions that govern the onto-epistemology of instrumental rationality is the distinction between the ecological and the social. In many respects – and perhaps entirely – the ecological/social distinction reflects the ontic wilderness/civilization distinction. In other words, ecology leads us to discover the interdependence of various plants and animals forming something like a hyper-organism, as discussed by Lynn Margulis under the title of the Gaia hypothesis (Margulis 1999). When we think of ecology, we often immediately think of *nature*, and some of the first things that come to mind are forests, deserts, and coral reefs. I hasten to add that this is not wrong, only that it indicates a deep-rooted cognitive tendency to see ecology as about something *other* than *society*. In doing so, we effectively place the social outside of the ecological and the ecological outside of the social.

By contrast, when we think of society, we think of people, language, signs, and norms. In his magnificent masterpiece, *Theory of Society*, Niklas Luhmann, reviewing the various theories of society in very broad strokes, says that society is either conceived as an assemblage of people and their relations, that it is a product of consensus, or that societies are defined by territories (Luhmann 2012, 6). For his part, Luhmann perhaps advocates the most radical position of all, arguing that people belong to the *environment* of society and that society consists *entirely* and *solely* of communications (Luhmann 2012, 42–43).[7] Investigation of society therefore means investigation of something other than nature. Society is understood as that which is not nature. This holds for politics as well. Politics is conceived as being about rights, identities, and all things that pertain to people. Although the situation is improving, it is notable that climate issues often take a back seat to economic and civil rights struggles. Those who are concerned with climate, it is said, are concerned with the fate of spotted owls and other endangered species and are therefore concerned with frivolous things.

While these distinctions are perhaps necessary for isolating objects of research, they become reified and ontologized, informing how we view these two domains treating what is continuous as discrete. Society comes to be understood solely in terms of what pertains to people and their relations, language, signs, or, in Luhmann's case, communications. What is missed here is that societies exist in an interface with the broader natural world upon which they are dependent both for food and various forms of energy but also all of the nonhuman things such as roads, fiber-optic cables, satellites, buildings, and all the rest necessary for social relations to exist at all. In *Theory of Society*, Luhmann makes progress over his earlier work *Social Systems* (Luhmann 1995)[8] in that he begins to recognize that mediums of communication such as writing make a decisive difference in the form communication takes for a society and that technologies such as the internet fundamentally transform social relations (Luhmann 2012, Chapter 2), yet he fails to recognize that recognition of these elements, so similar to thinkers like Latour and McLuhan, would require a complete transformation of his concept of society. The point here is that the bracketing of materiality and nature that takes place in our social theory broadly construed distorts our understanding of society. Or, following Timothy LeCain, can we really understand the "culture" of the American West without taking into account soil conditions of Montana, the grasses of the region, and the behavior and needs of cattle (LeCain 2017)? Were not the rhythms of the cattle a part of how the cowboys organized their social relations and how they lived their lives? The cattleman had to wake up early to milk the cows and attend to how they wandered about the land. All of this would contribute to how they lived their lives and related to other people. All of this disappears, becomes veiled, or becomes invisible when we restrict society to communication, language, people, or signs treating everything else as outside society.

On the other end of the spectrum, ecological thought, in its focus on organisms of all kinds, misses its fundamental insight. Ecological thought believes that its task is to investigate nature, which is something outside of society. Yet the fundamental insight of ecology is not that *organisms* form a system of dependencies, feedback loops, and relations. The fundamental insight of ecology is that *being* – what I am here calling "the wilderness" – is composed of dependencies, relations, and feedback loops. There is no less an ecology of an office or corporation than there is an ecology of an Amazonian rain forest. A corporation is not ecological because it involves micro-organisms, perhaps rodents running about the building, and viruses, though it is ecological for that reason as well, because a corporation involves these things as well. A corporation is ecological because of how the different people in the corporation relate to one another in all sorts of dependencies and feedback relations and because of how the corporation as an entity in its own right relates to other corporations. How, for example, are we to explain the behavior of banks in the 2000s when giving bad loans to people for home loans? We can, of course, give individualistic explanations. "The loan officers were greedy!" "The loan officers were ignorant of what they were doing and did not realize they were giving bad loans that people would not be able to pay back". However, what if we gave an ecological explanation similar to the role that wolves might play in a broader ecosystem? What if the banks and bankers knew full well what they were doing, that they were giving bad loans? What if the reason they gave these loans was not because they were ignorant or greedy but because they knew that if *they* did not do it, another bank would, and *their* bank would lose stock value, making it go bankrupt? What if it was the entire *eco*system of relations among banks that was the problem? That would lead us to think of the problem of investment in a very different way. We would no longer seek culpability in this or that banker or bank but would instead look at how a set of ecological relations is toxic, tending towards positive feedback relations that run amok. The point here is that the discipline of ecology deserves a far broader scope than the one that it currently enjoys – indeed, the term "ecology" is another name for wilderness – and, that in thinking a wilderness ontology beyond calculative reason, we should see ecology as all that there is, seeing society as continuous with nature and as merely a local ecology like a particular coral reef. For wilderness ontology, everything is ecological through and through. Ecology is not something outside of society, nor is society something that is not ecological.

Working within the framework of the ontic conception of the ~~wilderness~~, the distinction between the ecological and the social can lead us to miss or overlook our unruly heritage, causing us to both distort the complexity of social assemblages and pose problems and questions poorly. Take the example of the Islamic State group, Boko Haram. Boko Haram has wrought profound suffering throughout Northeastern Nigeria, as well as Cameroon, Chad, and Niger (Taub 2017). If we are thinking in terms of the ontic conception of the ~~wilderness~~, we draw a strong distinction between the social/cultural and the ecological. On the side of the social and cultural will fall meaning, beliefs, ideologies, signs, signifiers, groups, and institutions. On the side of the ecological will fall all that belongs to ~~nature~~. The ecological is largely to be ignored because ~~nature~~ falls outside of society. As a consequence, we might think that addressing the humanitarian crisis in this region solely revolves around defeating the form of Islamic fundamentalism advocated by Boko Haram either militarily or by persuading people to abandon these beliefs and this ideology. In short, we see these beliefs and this ideology as the *cause* of the conflict in the region.

While a situation as complex as the crisis in this region of Africa cannot be reduced to one explanation,[9] such an approach does tend to overlook questions of what might contribute to making such a form of belief and ideology appealing in the first place. Here it is important to emphasize that the answer to questions such as this will differ from social assemblage to social assemblage, as all ecologies are unique. What might apply in one case will not apply in another. In the case of Boko Haram, that explanation premised on the ontic conception of the wilderness overlooks the material conditions of the region or the unruly heritage of farming in that part of the world. Niger, Nigeria, Chad, and Cameroon all border Lake Chad. For centuries, Lake Chad has been the key to life in this region, providing fish, water for farming, and a variety of other forms of sustenance. Since 1963, the size of Lake Chad has shrunk dramatically as a result of climate change, "lack of rain, [and] also the development of modern irrigation systems for industrial agriculture" (Skretteberg 2018). Once again, we here encounter the figure of Golem or a heritage of our own making that has exceeded our control. Climate change and unwise irrigation practices have created profound food shortages and have contributed to mass migrations as people struggle to survive in a growing wasteland. We can speculate that in a situation such as this, struggles over resources emerge that might render such ideologies and groups appealing. If this is true, then part of addressing the situation in the region would consist not simply of seeking to defeat Islamic fundamentalism but also the material problems of the region. Dire situations such as this make a strong case for abandoning the strong distinction between the social and the ecological, instead seeing the social as necessarily embedded in ~~nature~~ as a part of its heritage and the elements of nature as members of social assemblages.

Second distinction: the agent and the patient

A second core distinction at the heart of the ontology of calculative and instrumental rationality is the distinction between the agent and the patient. On the one side, there is the domain of agency, while on the other, there is the patient upon which agency is exercised. The agent is the side of activity, while the patient is the side of passivity. The agent is the seat of action, while the patient is the recipient of action. Agency is restricted to the domain of the living and has degrees ranging from the humble bacteria all the way up to humans, who are said to possess the highest degree of agency. If humans are said to have the highest degree of agency, then this is because they have the capacity of reason that grants them autonomy or the capacity to direct themselves through themselves through reason. Thus, for example, Kant will contrast autonomy and heteronomy (Kant 1996, 83). A determination of the will is heteronomous when it arises from pathological causes such as an instinct, appetite, emotion, or outside force. In this case, the will is not determining itself but is determined by something else. A will is autonomous or has agency when it determines itself through itself. For Kant, reason meets this criteria. The categorical imperative is a law formulated by reason and determines the will to action through reason. Rather than action being determined heteronomously through something outside itself such as our animal nature or an external force, reason calls the shots. For this reason, only a being with reason is fit to govern, because only a being with reason governs itself and is capable of producing its own law. Most importantly, insofar as autonomous beings are capable of formulating their own law to determine their action, only autonomous beings have the characteristic of *dignity*.[10]

Insofar as the patient does not have reason, it is right for the autonomous to rule over it. In the first instance, the agent is composed of two parts, the body and the soul. The soul rules over the body. Rational agents then rule over those who are heteronomous, and in their turn, men rule over beasts and the material world (Aristotle 1984b, Book I, Section 5). Insofar as the heteronomous is not defined by reason, it lacks dignity and therefore need not be treated with dignity. Instead, the world of matter and "beasts" (Aristotle's word) is there to be used as we see fit.

Questions of agency are incredibly complex and controversial, so I will not be able to do them justice here. Indeed, if there is one criticism that the object-oriented ontologies and new materialisms have repeatedly elicited, it is that in extending agency to nonhumans, they denigrate humans and place the dignity of a virus, for example, on equal footing with that of human beings. I cannot address these ethical and political criticisms here, but the aim of the object-oriented ontologies and new materialisms is certainly not to place everything on an equal ethical footing in our social assemblages. This is especially true of the new materialisms. Within the ontological framework of instrumental and calculative reason, the agent is conceived as a sovereign over the material and organic world such that it can be exploited as we see fit. For example, we merely see the mountaintop as something standing in the way of the coal and ore beneath, thereby thinking nothing of removing it. Insofar as we see the mountaintop as a patient without agency, we do not consider the consequences of this removal. No dignity is granted to things themselves. This denial of agency to things has contributed, in part, to the situation of the Anthropocene.

If we are to overcome this way of thinking, we must, as Jane Bennett has argued, see agency as *distributed* (Bennett 2010, 20–24). No agent ever acts alone, but rather, whether in reasoning or otherwise, it is always assemblages of agents that act. In other words, agency is itself ecological. If we do not notice the distributed nature of agency, then this is because the things we rely upon to act are often so reliable that they fall into the background and it seems to be we alone who are acting. On 20 July 1969, Apollo 11 landed, placing men on the moon for the first time in history. If you watch the film footage from that first landing, you notice something remarkable. The astronauts cannot *walk* because the Moon's mass is about 1.2% that of Earth, making for very different gravitational conditions. As they try to walk, they fall over and spin about on the surface and are unable to move about until they develop a sort of hopping gait. Examples such as this reveal that something as mundane as walking is ecological and distributed. We don't walk *on* the earth but *with* the earth, insofar as walking is rendered possible not just by our bodies and will but by the gravitational features of our planet. We can thus think of movements like walking as a sort of wave. While waves are distinct individuals, they are not separate from the ocean. They fold, pleat, or internalize wind, currents, and the ocean bed into themselves in order to form themselves. As these things vary, the nature of the wave varies. This is how it is with walking. We pleat the gravity of the earth into ourselves, giving rise to the activity we call walking. We do not do this alone but with the earth in a way that makes walking a distributed activity.

There are all sorts of ways in which our agency does not merely arise from us but rather depends on features of the natural environment as well as the tools and technologies we use. Moreover, these things are not simply patients that we use according to our own will as we see fit, but they modify what it is that we do, contributing something of their own. In *Entangled*, Ian Hodder discusses the archaeological discovery that the interior walls of dwellings in Çatalhöyük in Turkey had many layers of plaster. He explores many reasons they plastered their walls, ranging from the aesthetic; to protecting against mosquitoes; to absorbing the smells of cooking, people, and the decaying bodies of ancestors they buried in the foundations of their dwellings. Yet why the multiple levels of plaster? He notes that the clay of the plaster very quickly would begin to chip and darken from the smoke of the fire and cooking. It had to be repeatedly maintained to bring it back to its original state (Hodder 2012, 71). In Hodder's example, humans find themselves entangled in another series of activities distinct from the reasons they plaster their wall by virtue of the features and qualities of things. In *In Defense of Things*, Bjørnar Olsen gives a striking example of the distributed nature of agency with respect to art. Drawing on the work of Hesjedal and Bjerck, he notes how red cave paintings in Northern Norway mingle with red iron deposits. Olsen asks whether there was a sense in which the paintings were already slumbering in the *stone*, and whether the role of the artist was to bring forth what already slumbered there (Olsen 2010, 85). In this regard, the matter or cave wall would not simply be a blank slate upon which the artist projects meanings and forms that are already in their mind but would contribute something to the work itself.

Reason, too, would be distributed, relying on all sorts of nonhuman things outside the mind. This is evident in Andy Clark's extended mind hypothesis that sees mind not merely as something between the ears but rather as something that emerges from the relationship between brain, body, and things of the world. In one striking example, Clark points out that most of us cannot easily mentally multiply large numbers like 796547 × 4179645. Rather, we rely on paper and pencil and a series of operations that simplify the problem. Unless we are a savant of some sort, when we try to do problems like this in our mind, we encounter a confusing mess. Instead, we literally offload memory onto the paper, allowing it to remember on our behalf so we can focus on simpler operations (Clark 2003, 74). The paper and pencil contribute something to our process of reasoning, and it is unlikely that things like complex mathematics would have been possible without the development of writing. Indeed, the very numbering system we use – Arabic numerals – contributes to our ability to do complex mathematics. Imagine trying to multiply those numbers using Roman numerals! For Clark, the crucial role that things like this play in our cognition entails that we should treat them as a part of the mind. He does not intend this in the idealist sense in which we say that the mind structures reality but rather that if we are to understand cognition and our cognitive capacities, we must take the role that materiality plays into account as a scaffolding for our thought. If agency is distributed, then we are always entangled with a broader ecology of things that afford and constrain what we do. By emancipating agency from its imprisonment in the sovereign individual and uncovering its distributed nature, we also cultivate greater attentiveness to our ecological embeddedness in the broader wilderness. Life on Mars, for instance, no longer looks so appealing when we recognize that pregnancy is not something that simply takes place in the body but in conjunction with the gravity of the earth and that insofar as Mars is half the mass of the earth, there is a question of whether a fetus can even properly develop in that environment. The claim that things have agency is not the absurd claim that they have intentions, desires, and wishes, only that they contribute to our agency and modify it in a variety of different ways.

The distinction between the agent and the patient premised on the ontic conception of the ~~wilderness~~ can lead us to overlook the unruly heritage at work in social assemblages. Take the example of road layouts in downtown Dallas. As you approach Dallas from the north side, you hit Harry Hines Boulevard and enter a special sort of hell. It is difficult to escape the impression that the city planners – if there were any – suffered from a perverse madness. The driver is confronted with a maze of curving and intersecting roads with forks, bends, and side streets in the midst of constant construction, poor signage, difficult-to-discern traffic lights, and the ever-present necessity of being in just the right lane to reach your destination. Not only is this area dangerous for driver and pedestrian alike, but it is clearly not adequate to the volume of traffic entering the city daily. Dallas is like a heart with respect to the human body. Every day, the city draws in tens of thousands of people from the surrounding suburbs from miles away who work there and then pumps them back out at the end of the day. The problems endemic to this area are exacerbated by the fact that Dallas has very poor public transit connecting it to its surrounding areas.

If we are thinking in terms of the ontic conception of the ~~wilderness~~ that sees agency as residing solely in persons, we will see this city design solely as the result of incompetent city planners. However, this approach fails to take into account the materiality of the history of urban design in the city of Dallas. Our tendency is to think the past as spiritual, residing only in memory. Yet there is a materiality of the past. Drawing on Serres' conception of time not as a linear line but topologically as something like a crumpled piece of paper, Christopher Witmore argues that the past persists in the present and that things are, in part, assemblages of percolating time (Witmore 2006, 278–281). Time is something that is profoundly material. We see this in the city. We cannot say that the city is merely present, but rather past design features of the city are present in the present. In this regard, the monstrosity of Dallas urban design is not merely the result of thoughtlessness but arises from having to navigate the past design of the city with new design decisions in response to an unforeseen future that has arrived. In this regard, the Golem appears again. We become patients of agents that we ourselves created but that have taken on a life of our own, generating all sorts of activity that was not of our own choosing.

The wasteland and the posthuman

The heritage we inherit from the ontology of calculative and instrumental reason that underlies the Anthropocene is the Wasteland. On the one hand, there is the material wasteland of the earth. We are now told that in the decades to come, regions of the earth will be uninhabitable due to extreme temperatures. Other, once-green, regions are being swallowed up in deserts. Fukushima has potentially been transformed for centuries as a result of meltdown of their nuclear power plant, while there are regions of the ocean that have become uninhabitable due to floating islands of discarded plastic and the run-off of fertilizers that have caused algae blooms that exhaust oxygen in the ocean. At the material level, we everywhere encounter wastelands encroaching upon us. These are the golems of the Anthropocene.

However, there is also the discursive level of the wasteland as well. Under the ontology of instrumental and calculative reason, all values are reduced to means, and all things, including human life, come to be valued only as means. In the *Nichomachean Ethics*, Aristotle distinguishes between two types of values. There are, he says, those things that are desired for their own sake and those things that are desired for the sake of something else (Aristotle 1984a, 1094^{a}18–20). Although it has become unfashionable to speak in these ways, we can call the first sort of ends "absolute ends" and the second sort of ends "relative ends". Work and money, for example, are relative ends. Work is not valuable for its own sake but is valuable for the sake of what it produces. Likewise, money is not valuable for its own sake but is valuable for the sake of the security it produces and the goods that it can buy. The person that hoards their money, that does not spend their money – the miser – suffers from a sort of sickness, confusing relative values with absolute values. For Aristotle, the relative values are always subordinate in dignity and value to the absolute values. For example, the science of strategy is not valuable for its own sake but is only valuable for the sake of victory. It is subordinate to the end of victory in war (Aristotle 1984a, 1094^{a}6–8). And victory, in its turn, is only valuable for the peace and security that it brings. The relative ends or values are valuable by virtue of their utility or instrumentality, while the absolute values are not valuable for the sake of anything else. Indeed, the splendor of absolute values is precisely that they are useless. They are not *for* anything. They are desired for their own sake, and it is this that makes them absolute. Examples of absolute values or ends in themselves might be health, friendship, beauty, love, knowledge, and many other things besides. These things are not desirable for any utility or instrumentality they might possess – it would be a strange person who valued their friend because of the ways in which their friend is useful, and we would harshly regard the person who abandons a friend who has ceased to be useful – they are desired for their own sake. When Kant, in the *Critique of the Power of Judgment*, describes our judgments of beauty as a sort of disinterested satisfaction, his point is that the beautiful is appreciated for its own sake and is not appreciated for the sake of some utility the beautiful object might possess (Kant 2000, 96). And indeed, look at all the waste we happily engage in in the name of the beautiful, expending money beautifying our homes despite the fact that there is no utility in doing so and setting aside public lands that are rich in resources from which profits could be drawn simply because they are beautiful.

The essence of the Wasteland is that absolute ends or values undergo their twilight and disappearance, becoming more and more difficult to discern, and are replaced entirely by relative values. All beings, including human beings, are reduced to their utility or instrumentality, and there are no longer any superior values to which relative values are subordinate. The Wasteland is the twilight of ends, which is to say that the Wasteland is the end of ends, such that there is only the bad infinity of utility and instrumentality that is for the sake of nothing ultimate. In the Wasteland, the relative ends reign supreme, and everything is only valued in terms of its utility and profit that it can confer. For example, education is no longer valuable for its own sake but rather is only valuable for the jobs that it can get us, and increasingly labor comes to be valued for its own sake and not for the sake of final end such as a good life. In reducing everything to relative ends or values, we thus paradoxically encounter the end of all value or the twilight of value. And in a world where value has taken flight, all things are devalued and are valuable only as sources of profit and utility. Everything becomes expendable if it does not serve these ends, and everything is to be exploited. The discursive wasteland of the social and intellectual world, of a world where there are only relative ends, thus produces the wasteland of the earth where even life itself, in all of its diversity, becomes subordinate to the pursuit of utility. Golem becomes the destroyer of worlds and the destroyer of the world. We gnash our teeth and express regret that the irreplaceable Great Barrier Reef is dead, but what is to be done? We must pursue profit, and people must work. It becomes difficult to even raise the question of whether profit and work should be limited to preserve places like the Great Barrier Reef, and this is because, in our non-ecological thinking, we see the reef as something other than us that is not our concern anyway.

Wilderness ontology aims to combat the twilight of absolute values, to open a space where they might shine and manifest themselves and become governing principles in our thought and decision-making. We must avoid that form of ecological thought that says we should preserve the rain forests because perhaps they contain the cure for cancer. This form of reasoning repeats the calculative reasoning that produced the golem of the Anthropocene in that it only sees the value of things in their utility. Instead, we must strive to make a post-human turn in which we cultivate a sensibility in which we can see certain things as valuable for their own sake. In a remarkable essay, Þóra Pétursdóttir calls for a way of relating to things otherwise than in terms of representation and utility, instead drawing on Heidegger's concept of *Gelassenheit*, letting them be and retain something of their aura of mystery (Pétursdóttir 2014, 348). There is, says Pétursdóttir, an "ownness" to things. In adopting this sensibility, we discover a certainty dignity of things, challenging that gaze that approaches objects only as resources to be exploited. We must come to see that it is not just persons that possess dignity but that there is a certain dignity to things and the other creatures of the world. It is only in this way that we can hope to cultivate a different way of relating to the world that might allow us to escape the wasteland.

Notes

1 In a brilliant article, Christopher Witmore has beautifully captured the logic of the Anthropocene as golem with his proposal that we should refer to the Anthropocene as the "hypanthropocene". The "hyp" of hypanthropos, Witmore contends, refers to the Greek *uper* and *upo*, capturing the sense in which the Anthropocene is something that both exceeds the human and under which humanity finds itself (Witmore 2019).

2 For an excellent discussion of the conceptual semiotics of what I am here calling the "ontic wilderness", cf. Rivers (2015).

3 As Stacy Alaimo notes, there is an entire logic to the language of domestication that reflects gendered language of the relationship between men and women and the distinction between the public and the private, the social and the domestic. We would do well to question and complicate this logic, exploring how these gendered logics and ways of conceiving the relation between the domestic as the private and the social or public have contributed to the situation of the Anthropocene (cf. Alaimo 2016).

4 The point here is subtle. Kant's transcendental idealism leads to the sceptical conclusion that we can never know whether the objects of experience exist in their own right independent of our experience of them. Clearly this position is untenable for any realism and materialism committed to the existence of beings independent of mind, signs, and social constructions. The concept of manifestation or

disclosedness allows for the recognition that entities are revealed to mind and society under certain conditions and in certain ways while maintaining a space for their independent existence. For an account of what authorizes realism with respect to objects, cf. Bryant (2011, 39–57).

5 Foucault's concept of the episteme outlines how different discursive structures lead things to be manifested or shown in different ways, bringing some into relief, pushing others into hiddenness. Cf. Foucault (1994).

6 In discussing the way in which objects manifest themselves, one might object that I have fallen prey to correlationism. However, elsewhere I have argued that we simultaneously need to attend to the richness of phenomenological experience, the symbolic structuration of the world, *and* the materiality of the world. We need to think the interplay of all three registers together (cf. Bryant 2016). Here it must be recalled that an object is not merely something but also that each object encounters the world in its own particular way. This calls for a heterophenomenology observing how other beings encounter the world about them. For example, in what way does a bat or a dog phenomenologically encounter the world?

7 Luhmann's sociological theory is based on the distinction between system and environment. To say that something is in the environment is a system is to claim that it is *outside* the system. In claiming that humans are outside the system of society, he is not claiming that humans do not exist, only that communications are what compose society.

8 In his great work, *Social Systems*, Luhmann is far more dogmatic, restricting social systems entirely to communications while largely ignoring the role that materiality of various forms plays in the form social relations take.

9 Elsewhere I have argued that we need a form of theory that is supple enough to think the complex interplay of the material, symbolic, and experience of agents (Bryant 2016; Bryant 2011).

10 It is notable that some variation of this conception of agency has been historically used to grant a subordinate place to women and indigenous populations. Insofar as they are said to be governed heteronomously by appetites and emotions, it is concluded they are not fit to govern themselves or others (cf. Aristotle 1984B, $1254^{b}13$–14).

References

Alaimo, S. 2016. *Exposed: Environmental Politics and Pleasures in Posthuman Times*. Minneapolis, MN: University of Minnesota Press.

Aristotle. 1984a. "Nichomachean Ethics." In *The Complete Works of Aristotle: Volume 2*, edited by J. Barnes. Princeton, NJ: Princeton University Press.

Aristotle. 1984b. "Politics." In *The Complete Works of Aristotle: Volume 2*, edited by J. Barnes. Princeton, NJ: Princeton University Press.

Bennett, J. 2010. *Vibrant Matter: A Political Ecology of Things*. Durham, NC: Duke University Press.

Bryant, L. R. 2011. *The Democracy of Objects*. Ann Arbor, MI: Open Humanities Press.

Bryant, L. R. 2016. "Knots: Notes for a Daemonic Materialis." *Symplokē* 24(1–2): 27–45.

Clark, A. 2003. *Natural-Born Cyborgs: Minds, Technologies, and the Future of Human Intelligence*. Oxford: Oxford University Press.

Foucault, M. 1994. *The Order of Things: An Archaeology of Human Sciences*. New York: Vintage Books.

Heidegger, M. 1962. *Being and Time*. Translated by J. Macquarrie and E. Robinson. San Francisco, CA: Harper San Francisco.

Hodder, I. 2012. *Entangled: An Archaeology of the Relationship Between Humans and Things*. Oxford: Wiley-Blackwell.

Kant, I. 1996. "Groundwork for the Metaphysics of Morals." In *Practical Philosophy*, edited by M. J. Gregor, 37–108. Cambridge: Cambridge University Press.

Kant, I. 1998. *Critique of Pure Reason*. Translated by P. Guyer and A. W. Wood. Cambridge: Cambridge University Press.

Kant, I. 2000. *Critique of the Power of Judgment*. Translated by P. Guyer and E. Matthews. Cambridge: Cambridge University Press.

LeCain, T. 2017. *The Matter of History: How Things Create the Past*. Cambridge: Cambridge University Press.

Lévi-Strauss, C. 1983. *The Raw and the Cooked, Mythologiques: Volume One*. Translated by J. and D. Weightman. Chicago, IL: The University of Chicago Press.

Luhmann, N. 1995. *Social Systems*. Stanford, CA: Stanford University Press.

Luhmann, N. 2012. *Theory of Society: Vol.1*. Stanford, CA: Stanford University Press.

Margulis, L. 1999. *Symbiotic Planet: A New Look at Evolution*. New York: Basic Books.

Meillassoux, M. 2008. *After Finitude: An Essay on the Necessity of Contingency*. New York: Continuum.

Olsen, B. 2010. *In Defense of Things: Archaeology and the Ontology of Objects*. Lanham, MD: AltaMira Press.

Olsen, B. and Pétursdóttir, Þ. 2018. "About." *Unruly Heritage: An Archaeology of the Anthropocene*. Accessed 9 October 2018. https://unrulyheritage.com/about/.

Pétursdóttir, Þ. 2014. "Things-Out-of-Hand: The Aesthetics of Abandonment." In *Ruin Memories: Materialities, Aesthetics and the Archaeology of the Recent Past*, edited by B. Olsen and Þ. Pétursdóttir, 335–364. New York: Routledge.

Rivers, N. A. 2015. "Deep Ambivalence and Wild Objects: Toward a Strange Environmental Rhetoric." *Rhetoric Society Quarterly* 45(5): 420–440.

Serres, M. 2011. *The Natural Contract*. Translated by E. MacArthur and W. Paulson. Ann Arbor, MI: The University of Michigan Press.

Skretteberg, R. 2018. "Why You Should Care About the Lake Chad Crisis." *Norwegian Research Council*. Accessed 3 September 2018. www.nrc.no/news/2018/september/why-should-you-care-about-the-lake-chad-crisis/.

Taub, B. 2017. "Lake Chad: The World's Most Complex Humanitarian Disaster." *The New Yorker*. Accessed 4 December 2018. www.newyorker.com/magazine/2017/12/04/lake-chad-the-worlds-most-complex-humanitarian-disaster.

UNESCO. 2018. *World Heritage*. Accessed 9 October 2018. https://whc.unesco.org/en/about/.

Witmore, C. 2006. "Vision, Media, Noise and the Percolation of Time: Symmetrical Approaches to the Mediation of the Material World." *Journal of Material Culture* 11(3): 267–292.

Witmore, C. 2019. "Hypanthropos: On Apprehending and Approaching That Which Is in Excess of Monstrosity, with Special Consideration Given to the Photography of Edward Burtynsky." *Journal of Contemporary Archaeology* 6(1): 136–153.

5

CULTURAL HERITAGE AND MEMORY OF THE ECUMENE IN THE AGE OF THE ANTHROPOCENE

Laurent Olivier

The following anecdote may not be true, but it hardly matters; what counts is the moral of the story. On the morning of 15 July 1789, when Louis XVI was informed that the Bastille had been overrun the day before, he asked the Duke de la Rochefoucauld, who had brought him the news, "There's a revolt on?" "No, sire", replied the duke, "a revolution" (Robert and Cougny 1891). The New Climate Regime (Latour 2015) which we have now entered is not simply a matter of "global warming" that we can hold in check by moderating the amount of "greenhouse gases" released into the atmosphere, just as the fall of the Bastille was not simply an uprising that could be quelled with calming words and the free distribution of bread. This is a new era in the history of Humanity. It is a new age, a new planet-wide archaeological period that it behooves us to deal with as such (Olivier 2013a).

In dealing with the upheavals engendered by the expansion of the Anthropocene, which have changed our experience of the world on a scale that Latour has characterized as "cosmo-colossal" (2015, 9–10) and Witmore (2014) as "hypanthropic", we have clearly found it difficult to change our ways of thinking. There has been the very real temptation to keep applying thought patterns that are basically inoperative, to continue using old conceptual tools that are wholly unsuited to what is taking place – in a word, to try to make something new out of something old. And yet, in the final analysis, it is not a good idea. Had Louis XVI not been so blind to change, he could have told us: the old world, from which we originate, now lies moribund. And the new one on the horizon is, in the strict sense of the term, inconceivable.

This is why we archaeologists find ourselves flailing about in a paradox that has stymied our ability to conceptualize the situation. On the one hand, we are increasingly interested in subtle, discreet forms of cultural heritage; we now believe we have to take into account intangible forms of cultural production like concepts and ideas, and even tastes and feelings (Ruggles and Silverman 2009; Smith and Akagawa 2009). At the same time, cultural heritage in all its various forms is daily being destroyed on a gigantic scale throughout the world (González-Ruibal 2008). It is not just the rural cultural heritage of the distant past that is disappearing as land development wreaks havoc on the subsoil. Nor is it limited to the distant heritage of "primitive" peoples snuffed out under the impact of urbanization and industrialization, which have obliterated traditional ways of life. It is rather all the ways in which the land has been used, everything that is involved in the way we inhabit the world, here and elsewhere, that is being attacked and rent asunder, along with all the memory it holds. And the phenomenon has been *accelerating* since the middle of the twentieth century.

Something is happening

We have witnessed all about us the collapse of the barriers that long separated themes and the boundaries that kept disciplines distinct but without grasping what has led to this tectonic upheaval heretofore unseen in the history of ideas. Some refer to it as an "ontological turn" in the social sciences (Holbraad 2017). It involves nothing less than a shift in focus away from *anthropos:* human beings no longer occupy the central place in our understanding of cultures and societies. As contemporary anthropologists have noted, human actions and ideas are not the lone contributors to the creation of a civilization's structures and objects or the development of societal forms. Other kinds of "life", a variety of other non-human organisms, contribute to their creation as well. They most notably include, along with places, what we generally refer to as things: objects, constructions, and materials (Olsen 2010). In effect, they include all the organic and non-organic components of the world about us. These "beings", both animate and inanimate, are not simply present in the world: they give it its existence and life (Olivier 2011a).

This is not simply some new theoretical approach. It is rather the after-effect of a phenomenon that is all too real. The spectacular fashion in which climate conditions in the Holocene, which had remained largely stable for some 11,700 years, have been upended has shown us that human beings, their societies, and their activities are an integral part of a "terrestrial system". As climatologists have noted, anthropogenic factors do not constitute an external force capable of disturbing some intrinsically "natural" environmental system (Oldfield and Steffen 2004). They are part of that system and thus dependent upon it. Above all, as Latour pointed out, we are going to have to come to terms with the force with which the environment can act and the fact that it is not mere background against which human actions play out (Latour 2017, 56). In reacting to the pressures that human actions bring to bear upon it, the environment changes very quickly and, in that it does, it constitutes an historical entity inseparable from the history of mankind. The history of the world, heretofore basically "eco-social", now resembles, in the long-term perspective of geologists, a geo-history (Chakrabarty 2009) or, archaeologically, what some French scholars have called an "archaeo-geography" (Chouquer 2007; Chouquer 2008).

In the world of social sciences, the shift has indeed been "ontological". Anthropologists no longer reduce the formal links that so-called "primitive" societies establish between the human and the non-human to linguistic metaphors or symbolic representations. They now rather consider them other views of reality that are just as structured and pertinent as our own (Keck, Regehr and Walentowitz 2015, 34). It behooves us to understand "others" in their very otherness if we are to grasp from within the workings of their representations. We have to acknowledge the "ecologies of others" (Descola 2011) in all their intelligence and recognize "ontologies" of cultures different from our own (Descola 2005) in all their conceptual soundness (Kohn 2013). For the first time, we stand poised to admit not only that we may have something to learn from cultures different from our own but that our own history has a strangeness, an otherness of its own.

The definitive collapse of the long-standing dichotomy between nature and civilization to which this ontological shift has led is not without practical consequence for the very way in which we practice archaeology. It has effectively stripped humans of their privileged status as the lone beings endowed with knowledge and capable of action in an incognizant world subjected to their undertakings. But it seems to me that at a deeper level, the ontological turn that ushered in the demise of processual and post-processual archaeology needs to be seen as symptomatic of the transformations that post-industrial societies are currently undergoing. As Bruno Latour pointed out, the New Climate Regime that we entered with the period of the Anthropocene known as the Great Acceleration (Steffen et al. 2011, 848) now obliges us to rethink our own "ontology", which we inherited from the Moderns of the seventeenth and eighteenth centuries (Latour 1991), at the same time that it raises the question of the place history occupies in this enormous process that has so drastically altered living conditions on Earth (Latour 2015, 182).

We have entered a new age

Beginning with the end of World War II, human activity radically transformed the climate conditions of the period known as the Holocene, which dates back to the Ice Age that followed the end of the Pleistocene, more than 10,000 years ago. We have entered a new climate regime that is most notable for the rapid rise in temperatures and sea levels and the accelerated disappearance of species. In one or two centuries' time, the latter could affect 75% of living species, leading to a sixth mass extinction comparable in scope to the one which saw the end of the dinosaurs (Cerballos, Ehrlich and Dirzo 2017). The change became clearly visible with the onset of the Industrial Revolution in the nineteenth century (Crutzen and Steffen 2003). Since 1945, it has been accelerating at a tremendous pace (Zalasiewicz et al. 2015) and, the good intentions voiced by various governments notwithstanding, the rate at which carbon emissions are being released into the atmosphere shows no sign of abating (Angus 2018).

Until now, the climate conditions of the Holocene were remarkably stable, which coincided with the development of settled populations, the appearance and growth of agricultural communities, and, more generally, the increase in the number of humans. The preceding period of the Pleistocene was marked by a chaotic succession of vast climate changes that included abrupt, short-lived swings in temperature that apparently ranged from just one to three years (Zalasiewicz and Williams 2012, 205). Climatologists fear that with climate destabilization in the Holocene now at the tipping point, the state of the global system will be altered (which has probably already begun) and eventually catapult us into a chaotic period similar to that of the Pleistocene. But the Holocene, it would seem, is the only state in which the planet can sustain contemporary societies (Steffen et al. 2015). We thus now live under the threat of the collapse of human civilization as profound as the impact the post-industrial world has had upon the Earth system.

Viewed in this context, the erosion of archaeological heritage assumes new meaning. The devastation wrought by the man-generated Great Acceleration has not affected the atmosphere, species, and natural environments alone. It has attacked the Earth system as a whole, which includes the traditional anthropogenic milieus: the countryside of the distant rural world, the urban fabric of pre-industrial towns and cities, in effect, all "inhabited land". We have noted that this process of mass destruction became clearly evident with the onset of the Industrial Revolution and that the phenomenon began accelerating in the mid-twentieth century. I propose that we call the material impact that contemporary post-industrial civilization has had on the natural and cultural environment "the Age of Devastation".

The paradox of cultural heritage in the Age of Devastation

The Great Acceleration of the Anthropocene has forced us to rethink our fundamental relation to things, in particular what we understand to be our cultural heritage. As the geographer Augustin Berque noted, the Age of Modernity that is ours "is that of a civilization that has wrought the worldwide mass destruction of the historical ways in which land was inhabited" by standardizing the built environment everywhere on the planet and thereby stripping the land of its distinct cultural identities (Berque 2010, 79). It was because industrialization and urbanization inflicted massive, irreparable damage on the fabric of traditional habitation patterns that there arose our preoccupation with preserving and protecting the natural environment and/or our cultural heritage, the aim being to compensate for "the standardization and the loss of identity of the land". This preoccupation appeared earlier in Europe than elsewhere simply because Europe was first to modernize. To put it another way, the concern for our cultural heritage is a symptom of Western modernity's relation to the natural and/or the built environment.

But as Berque pointed out, this relation is fundamentally conflicted, for it appears in the form of two attitudes that are at once antithetical and consubstantial: on the one hand, we have painstakingly restored miniscule archaeological enclaves to "preserve" them, while, on the other, we have allowed for the "development" of immense tracts of land that eradicates every last trace of the material memory found there. In effect, as Berque put it, we have combined occasional "mummification" with the extensive practice of *tabula rasa* (Berque 2010, 83). In other words, what we take to be indispensable measures of safeguarding – rescue digs, the preservation and restoration of monuments and archaeological sites, the protective housing of archaeological and anthropological collections in museums – are merely the flip side of the frenzied destruction and eradication of our natural and cultural environment. As archaeologists, scholars, and curators, we are not on the "good side" of things because we attempt to preserve some of what is threatened. We are, in truth, just one aspect of the process of devastation into which modernity has thrust us.

The combination of global annihilation and local mummification invites two further remarks. First, the Age of Devastation into which we entered with the Great Acceleration strikes at what Berque called "urban composition" (Berque 2010, 83), that is, structures and the associated land development that over the course of history has slowly evolved from the local level, villages, to extend globally to territories. This is the hybrid composite, half cultural, half natural, that holds the memory of inhabited spaces, which Berque has labelled the ecumene, "the land occupied by humans, in other words, the veritable home of humans" (Berque 2000; Berque 2010, 125). But, as Berque has pointed out, the preservation of the ecumene would appear to be incompatible with anthropogenic modernity (Berque 2010, 86), and it is their incompatibility that has been exacerbated by the Great Acceleration. The Age of Devastation has been attacking the inhabited world that preceded it in its entirety in a continuous process that I propose we call the "Devourment" of the living world, both animate and inanimate.

The second comment deals with the paradoxical nature, under these circumstances, of the very notion of cultural heritage. On the one hand, implicit in our choice to preserve objects, structures, and landscapes is our recognition that they hold specific historical value and that their particularities render them different from today's objects, structures, and landscapes. On the other hand, we invoke their historical value as part of the cultural identity of mankind today; we argue for the need to preserve these remains on the grounds that they are part of our cultural heritage. And yet, to make that argument is to deny the existence of the distinction between past and present, which, within our own cultural system, Berque contends, is what "allows for the modern notions of monuments and cultural heritage" (Berque 2010, 86). Thus does the very notion of "cultural heritage objects" become highly problematic, for they are reduced to images or representations of something hollow. They become what Marx would have called fetishes, or more plainly put, objects of cultural consumption. Callously stated, within the process of massive destruction that targets memory above all, the only identity cultural heritage objects can have is that of some fictional creation. In effect, the more their environment is ravaged, the more we mummify and fetishize objects. That, in truth, is the nature of the work we have been called upon to perform.

The effects of the Great Acceleration

These profound changes, among them the need we now feel to preserve our cultural heritage, have directly affected the way in which we practice archaeology and have become increasingly manifest over the last ten or 20 years. The first effect that the Great Acceleration has had on disciplines like archaeology and history that study mankind's past is the emergence of what the historian François Hartog sees as "a new regime of historicity" and which he has called "Presentism", a term that has not been without generating some controversy (Hartog 2003). In Hartog's view, the continuity between past and present has been severed, and the past no longer exists as part of the present, as demonstrated by the aforementioned "devourment" of the material memory of the past. In other words, the present has become the lone horizon for history and archaeology, the lone entry point from which we can gain access to the past (Olivier 2013b).

This new status enjoyed by the present has stimulated work on recent periods of contemporary history, such as the collective twentieth-century traumas generated by the two world wars and their consequences (Schnitzler and Landolt 2013; Bernbeck 2017; Andreassen, Bjerck and Olsen 2010). It has also led to research in what now must be called the "archaeology of the present" (Olivier 2017a), that is, the study of remains from the present day, with pioneering work having been done by Rob Bonnichsen in his article "Millie's Camp" (Bonnichsen 1973) and William Rathje in "Garbage Project" (Rathje 1974). It is striking to note that interest in the archaeology of the recent past appeared in the course of the second half of the twentieth century, that is, at the same time that the destabilizing effects of the Great Acceleration were beginning to be felt.

To date, little attention has been paid to a third effect that the Great Acceleration has had on archaeology, that of revealing its limits, which can be likened to the way in which the potentially unlimited expansion of the economic exploitation of the Earth system has revealed the planetary limits of "human enterprise" (Rockström and Klum 2015; Steffen et al. 2015). Much of the work done by archaeologists over the last 30 years has come in response to the Anthropocene. Because industrial and urban land development is threatening increasingly larger swaths of the ecumene with irreversible destruction, archaeologists have adopted a new approach, working on extensive sites and prioritizing strip methods. This new approach obliges us to investigate all periods and all types of archaeological occupation, and the resulting mass of information and remains that have to be recorded, processed, and preserved has increased to the point that it has already become impossible to manage. Larger and larger amounts continue to pour in from dig sites, and no one knows what to do about it.

The other limit that archaeology is now forced to face is that imposed by the material transformations of the present. It is accepted fact that the material imprint of human activity has never been as profound nor as long-lasting as it is today, and this is especially so if we understand nuclear accidents like Chernobyl and Fukushima to be major archaeological events in the Age of Devastation (Olivier 2011b; Olivier 2017b). Above all, contemporary material creations are of so vast a scope that it has become impossible for archaeologists to deal with them. No one knows how to carry out a dig on an airport, a nuclear facility, or even a modest suburb. They are far too massive and far too dehumanized to be approached in the way archaeologists studied the remains of previous periods of the history of mankind. The Great Acceleration is the point of zero visibility for archaeology, which deals only with man: with man-made things that individuals use in spaces inhabited by groups of humans. It is not equipped to handle the "non-places" of supermodernity (Augé 1992).

What can be done?

We are cornered. The Anthropocene is everywhere, and there is no escaping it. And more perniciously still, whatever ideas we come up with to free ourselves from the weight it bears upon us will be derived from the anthropocenic state in which we now live. For the misleading ideology of the Age of Devastation reduces the world to pure object while ennobling individual persons and bodies as the lone subjects in the world, whereas, in truth, that world has been dehumanized and human beings are prey to their "fetishes", which they take to be interactive entities rather than the hollow shells they are. It is indeed a cyborg world (Berque 2010, 132). Is it our lot to accept it, to resign ourselves to finding the least uncomfortable position within the Great Acceleration in which we now find ourselves trapped?

If we can neither eliminate the Anthropocene nor pretend that it does not exist, we can at least try to resist it to the extent that our feeble means allow. After all, nothing forces us to condone through our silence and inertia the recent mutations that archaeology has undergone and which, driven by a market-based economy, have transformed an unalienable collective treasure – by which I mean our archaeological heritage – into marketable items. What goes by the name of "commercial" archaeology is a business construct intended to free developers from archaeological constraints, as the technocrats call them, by "purging" the land to be developed of its archaeological memory, sweeping it clean (Olivier 2013c; Olivier 2016). Nor does anything force us to go along with the prevailing trend to "objectify" our archaeological heritage while lauding the sentiments and commitment of individuals. At bottom, these constitute little more than the respectable face of the morbid dismantling of the "humanly inhabitable" world, its academic justification, as it were.

For the expansion of the Anthropocene does not pose a threat to natural and anthropized environments alone. The Great Acceleration also directly threatens archaeology, both as a discipline and in its ethical core. But in that archaeology is the discipline of the material memory of the inhabited world, it represents a formidable means of resistance. Confronted with the growth of the Anthropocene, we need to defend the memory of the world, of the human and animal, cultural and natural, *livable* world. The Anthropocene did not appear out of nowhere. It is the product of the capitalist economy of industrialized countries, epitomized by today's

America, from the elder Bush through to Trump. The former said, "Our way of life is not negotiable", the latter, "the concept of global warming was created by and for the Chinese in order to make US manufacturing non-competitive". The hellfire kindled by the nineteenth-century industries that harnessed the energy of fossil fuels has grown into the blaze that is the Anthropocene and continues to feed the flames (Malm 2017). And it is the profits made by these predatory economies that provide our salaries and finance our research. Are they not thereby offering us a magnificent opportunity to combat the Anthropocene?

So what now?

Fine, you say, you have presented the situation clearly, but what can we do about it, and more specifically, in what ways has the Anthropocene already changed the way we practice archaeology and what are the challenges that we shall soon have to face? What we call, for ease of reference, the Anthropocene actually corresponds to two phenomena that are often conflated. The Anthropocene is foremost defined as a new era or age in the history of mankind that we fully entered in the course of the twentieth century. As such, it is an area of study that naturally draws the interest of archaeologists who work on the contemporary past. But because the Anthropocene wreaks physical damage to the ecumene, it also alters the archaeology of earlier periods, for it transforms remains in their very materiality. This is an aspect that has drawn little attention in the discipline, even though it concerns all archaeologists.

One of the first questions that arises with regard to the Anthropocene, understood as a new archaeological period of the contemporary era, is how to define it (González-Ruibal 2018). What, in archaeological terms, characterizes this new Age of Devastation? Obviously, there is its massive scale, which has produced heretofore unseen "mega-monstrosities" (Witmore 2019), but even more so, perhaps, there are the increasing number of structures and practices that have been generated by what we might call "mass technologies" and that have been devised to deal, with methodical efficiency, with huge quantities of human and non-human beings as well as both living and inert matter. The use of weapons of mass destruction and the death camps of the twentieth century quickly come to mind, along with, in the same category but related to animals, meat-processing plants and other such industrial-scale slaughterhouses. One also thinks of the infrastructures needed for the mass extraction of raw materials, first among them those of the oil industry, which were so superbly photographed by Edward Burtynski (2016), but also of those of industrialized agricultural landscapes of which North America presents striking examples. And then there are all those "non-places" – bedroom communities, superhighway networks, train stations, airports – that can be found everywhere in the world, all of them designed to hold or direct and dispatch human masses like so much "flowing" matter. How did all of this come to be, and how did it develop and spread?

The scholars who address these issues will naturally be led to look at the consequences that the Age of Devastation has had on the way we think and on the human condition generally. The changes that our physical environment has undergone have taken place so quickly over the last few decades and have been so profound that an examination of the status of the things and substances that the Great Acceleration has either introduced or transformed is all but unavoidable (Pétursdóttir 2017). Nor will it be possible to avoid dealing with the political frameworks fostered by the Age of Devastation and which sustain its growth (González-Ruibal 2019), from Donald Trump's reactionary America to the generalized cyber-surveillance of today's China. Conversely, we will have to attempt to determine what kinds of economic and social structures might be capable of, if not restoring what the Anthropocene has destroyed, at least arresting its destructive progress. Some people will join the struggle against this mad dash to catastrophe, while others, through their silence and passivity, will endorse it. On an even deeper level, we will surely have to examine the ways in which the Age of Devastation has severed our ties to the environments of the past that were so instrumental in forging humanity and thereby transformed our behaviour, our aspirations, and our very identity as human beings.

We need to change our approach to archaeological timeframes. The Age of Devastation will continue to confront us with paradoxes that undermine the notions of synchrony and contemporaneousness that are the basis for our approach to archaeological interpretation (Lucas 2015). Mass-scale urbanization in one place will have as its counterpart, elsewhere, deforestation and the desertion of longstanding rural living environments (González-Ruibal 2008), for the two phenomena are actually one and the same. In those outlying areas most forcefully subjected to the effects of a devoured ecumene, archaeological time will clearly reveal its fundamentally multi-temporal (Olivier 2011c) or "heterochronic" dimension. Even now we can observe indigenous peoples using cell-phones and iPhones and indigent farmers toiling the soil with wooden ploughs drawn by tractors (González-Ruibal 2018, 7, Figure 2). The mechanistic, globalizing schemas long in use will prove incapable of dealing with the diversity and, above all, the heterogeneity of today's world.

But the consequences of the Anthropocene go well beyond the present. In our fieldwork, we have already had to get used to working on terrains and sites disrupted and polluted by physical restructuring and chemicals. The way in which iron debris, pipelines, and metal constructions of all kinds have blurred geomagnetic maps is a distinctive sign of the Great Acceleration. Archaeological remains that have become chemically or physically unstable, especially those made of metal, are another. And the disruption of ecological milieus has also generated a rebound effect. Global warming has brought to the surface buried remains, such as the mummified body of Ötzi, which reappeared in 1991 after lying dead for some 3,000 years, or the giant viruses in the Arctic Circle that the glacier melt revitalized after they had lain latent for tens of thousands of years (Legendre et al. 2015). Elsewhere, environmental transformations have reactivated the effects of soil degradation that first appeared in prehistoric times (Bailey 1997) and discharged into waterways chemical substances, in particular the ones that contain lead or mercury, which had been trapped in sediment for centuries or even millennia (Hansson et al. 2017). These phenomena are bound to multiply in the decades ahead, especially those involving the highly contaminating substances that were dispersed into the environment in the twentieth century during the early stages of the Age of Devastation.

Scholars who deal with these phenomena will be forced to reassess the framework of their chronological and cultural areas of specialization. Their research will increasingly have to be carried out in a diachronic perspective forced upon them by the interpenetration of archaeological temporalities that has accompanied the Anthropocene. François Hartog was absolutely right in identifying, in the early twenty-first century, a change in the "regime of historicity" (Hartog 2003), but, as any conventional historian would, he disregarded the material, which is to say the archaeological dimension of the phenomenon. In his view, "Presentism" existed solely in the minds of historians, which is not the case. The "presentist" temporality of the Anthropocene is the result of the coexistence of all of the pasts that survive materially in the present. And the process that is the Age of Devastation brings these pasts to the fore by reactivating them. The Age of Devastation is thus, paradoxically, an age of memory, a present from which we best observe the phenomena of latency, leaps in time, and things being reactivated and brought back into play (Olivier 2019).

These two approaches, one historical, the other archaeological, one that interprets phenomena, the other that deals with their nature, do not stand in contradiction to each other; they are rather complementary. For the Age of Devastation has brought down the walls separating disciplines, in particular those that had isolated history, archaeology, sociology, and anthropology. These disciplines effectively began opening up to each other with the coming of the Great Acceleration. So what should we do? Well, we should do what we do best: observe. We have front-row seats from which to watch this phenomenon of unprecedented magnitude as it unfolds before our very eyes, even as we do not fully understand where it is leading us. We need to get out of our offices and laboratories and libraries and observe what is going on, for no one who comes after us will be able to see it. We need to give our full attention to this phenomenon; we can't allow ourselves to miss the slightest bit of it. There is a whole world out there that needs to be explored: ours.

References

Andreassen, E., Bjerck, H. B. and Olsen, B. 2010. *Persistent Memories: Pyramiden, a Soviet Mining Town in the High Arctic*. Trondheim: Tapir Academic Press.

Angus, I. 2018. *Face à l'Anthropocène. Le capitalisme fossile et la crise du système terrestre*. French translation by N. Calvé. Montréal: Écosociété.

Augé, M. 1992. *Non-lieux. Introduction à une anthropologie de la surmodernité*. Paris: Le Seuil.

Bailey, G. 1997. "The Klithi Project: History, Archaeology and Structures of Investigation." In *Klithi: Palaeolithic Settlement and Quaternary Landscapes in Northern Greece, Volume 1*, edited by G. Bailey. Cambridge: McDonald Institute for Archaeological Research.

Bernbeck, R. 2017. *Materielle Spuren des nationalsozialistischen Terrors. Zu einer Archäologie der Zeitgeschichte*. Bielefeld: Transcript.

Berque, A. 2000. *Écoumène. Introduction à l'étude des milieux humains*. Paris: Belin.

Berque, A. 2010. *Milieu et identité humaine. Pour un dépassement de la modernité*. Paris: Donner lieu.

Bonnichsen, R. 1973. "Millie's Camp: An Experiment in Archaeology." *World Archaeology* 4(3): 277–291.

Burtynsky, E. 2016. *Essential Elements*. London: Thames & Hudson.

Cerballos, G., Ehrlich, P. and Dirzo, R. 2017. "Biological Annihilation Via the Ongoing Sixth Mass Extinction Signaled by Vertebrate Population Losses and Declines." *Proceedings of the National Academy of Sciences of the United States of America*, 25 July. Accessed 6 October 2018. www.pnas.org/cgi/doi/10.1073/pnas 1704949114.

Chakrabarty, D. 2009. "The Climate of History: Four Theses." *Critical Enquiry* 35: 197–222.

Chouquer, G. 2007. *Quels scénarios pour l'histoire du paysage? Orientations de recherche pour l'archéogéographie*. Coimbra and Porto: CEAUCP.

Chouquer, G. 2008. *Traité d'archéogéographie. La crise des récits géohistoriques*. Paris: Errance.

Crutzen, P. J. and Steffen, W. 2003. "How Long Have We Been in the Anthropocene Era?" *Climatic Change* 61: 251–257.

Descola, P. 2005. *Par-delà nature et culture*. Paris: Gallimard.

Descola, P. 2011. *L'écologie des autres. L'anthropologie et la question de la nature*. Paris: Quae.

González-Ruibal, A. 2008. "Time to Destroy. An Archaeology of Supermodernity." *Current Anthropology* 49(2): 247–279.

González-Ruibal, A. 2018. "Beyond the Anthropocene. Defining the Age of Destruction." *Norwegian Archaeological Review* 51(1–2): 10–21.

González-Ruibal, A. 2019. *An Archaeology of the Contemporary Era*. London: Routledge.
Hansson, S., Claustres, A., Probst, A., De Vleeschouwer, F., Baron, S., Galop, D., Mazier, F. and Le Roux, G. 2017. "Atmospheric and Terrigenous Metal Accumulation Over 3000 Years in a French Mountain Catchment: Local vs. Distal Influences." *Anthropocene Review* 19(5): 45–54.
Hartog, F. 2003. *Régimes d'historicité. Présentisme et expériences du temps*. Paris: Le Seuil.
Holbraad, M. 2017. *The Ontological Turn: An Anthropological Exposition*. Cambridge: Cambridge University Press.
Keck, F., Regehr, U. and Walentowitz, S. 2015. "Anthropologie: le tournant ontologique en action." *Tsantsa* 20: 34–41.
Kohn, E. 2013. *How Forests Think: Toward an Anthropology Beyond the Human*. Berkeley, CA: University of California Press.
Latour, B. 1991. *Nous n'avons jamais été modernes. Essai d'anthropologie symétrique*. Paris: La Découverte.
Latour, B. 2015. *Face à Gaïa. Huit conférences sur le nouveau régime climatique*. Paris: La Découverte.
Latour, B. 2017. *Où atterrir? Comment s'orienter en politique*. Paris: La Découverte.
Legendre, M., Lartigue, A., Bertaux, L., Jeudy, S., Bartoli, J., Lescot, M., Alempic, J.-M., Ramus, C., Bruley, C., Labadie, K., Shmakova, L., Rivkina, E., Couté, Y., Abergel, C. and Claverie, J.-M. 2015. "In-Depth Study of *Mollivirus sibericum*, a New 30,000-y-Old Giant Virus Infecting *Acanthamoeba*." *Proceedings of the National Academy of Sciences of the United States of America*, September. Accessed 6 October 2018. www.pnas.org/cgi/doi/10.1073/pnas.1510795112.
Lucas, G. 2015. "Archaeology and Contemporaneity." *Archaeological Dialogues* 22(1): 1–15.
Malm, A. 2017. *Fossil Capital: The Rise of Steam Power and the Roots of Global Warming*. London: Verso.
Oldfield, F. and Steffen, W. 2004. "An Integrated Earth System." In *Global Change and the Earth System: A Planet Under Pressure*, edited by W. Steffen, A. Sanderson, P. D. Tyson, J. Jäger, P. A. Matson, B. Moore III, B. F. Oldfield, K. Richardson et al., 1–9. Berlin: Springer.
Olivier, L. 2011a. "Les choses aussi sont des êtres." *Libération*, 11 January.
Olivier, L. 2011b. "Quelque chose s'est ouvert que nous n'aurions pas dû voir." *Libération*, 25 April.
Olivier, L. 2011c. *The Dark Abyss of Time: Archaeology and Memory*. Landham, MD: AltaMira Press.
Olivier, L. 2013a. "Nous sommes à l'âge de la Dévastation." In *Destruction. Archaeological, Philological and Historical Perspectives*, edited by J. Driessen, 27–36. Louvain-la-Neuve: Presses universitaires de Louvain.
Olivier, L. 2013b. "The Business of Archaeology is the Present." In *Reclaiming Archaeology: Beyond the Tropes of Modernity*, edited by A. Gonzalez-Ruibal, 117–129. London: Routledge.
Olivier, L. 2013c. "Notre passé n'est pas à vendre. Nuestro pasado no esta en venta." *Complutum* 24(1): 29–39.
Olivier, L. 2016. "How I Learned the Law of the Market." In *Archaeology and Neoliberalism*, edited by P. A. Resco, 223–237. Madrid: JAS Arqueologia Editorial.
Olivier, L. 2017a. "I Can't Get No Satisfaction: For an Archaeology of the Contemporary Past." In *Clashes of Time: The Contemporary Past as a Challenge for Archaeology*, edited by J.-M. Blaising, J. Driessen, J.-P. Legendre and L. Olivier, 11–22. Louvain-la-Neuve: Presses universitaires de Louvain.
Olivier, L. 2017b. "Fukushima, un désastre qui continue à grandir." *La Nouvelle Quinzaine Littéraire*, n° 1172, 1 May: 24.
Olivier, L. 2019. "The Future of Archaeology in the Age of Presentism." *Journal of Contemporary Archaeology* 6(1): 16–31.
Olsen, B. 2010. *In Defense of Things: Archaeology and the Ontology of Objects*. Lanham, MD: AltaMira Press.
Pétursdóttir, Þ. 2017. "Climate Change? Archaeology and Anthropocene." *Archaeological Dialogues* 24(2): 175–205.
Rathje, W. 1974. "The Garbage Project." *Archaeology* 27: 236–241.
Robert, A. and Cougny, G. 1891. "La Rochefoucault (François-Alexandre-Frédéric de)." In *Dictionnaire des parlementaires français, Volume III*, 601–602. Paris: Edgar Bouloton.

Rockström, J. and Klum, M. 2015. *Big World, Small Planet: Abundance within Planetary Boundaries*. Stockholm: Max Strom.

Ruggles, D. F. and Silverman, H., eds. 2009. *Intangible Heritage Embodied*. Dordrecht: Springer.

Schnitzler, B. and Landolt, M. 2013. *À l'Est, du nouveau! Archéologie de la Grande Guerre en Alsace et en Lorraine*. Catalogue de l'exposition des Musées de la Ville de Strasbourg. Strasbourg: Musées de la Ville de Strasbourg.

Smith, L. and Akagawa, N. 2009. *Intangible Heritage*. Abingdon: Routledge.

Steffen, W., Grinevald, J., Crutzen, P. J. and McNeill, J. R. 2011. "The Anthropocene. Conceptual and Historical Perspectives." *Philosophical Transactions of the Royal Society A* 369(1938): 842–867.

Steffen, W., Richardson, K., Rockström, J., Cornell, S. E., Fetzer, I. and Bennet, E. M. 2015. "Planetary Boundaries: Guiding Human Development on a Changing Planet." *Science* 347(6223), 13 February: 736–747.

Witmore, C. 2014. "Archaeology, the Anthropocene and the Hypanthropocene." *Journal of Contemporary Archaeology* 1(1): 128–132.

Witmore, C. 2019. "Hypanthropos: On Apprehending and Approaching that which is Excess of Monstrosity, with Special Consideration Given to the Photography of Edward Burtynski." *Journal of Contemporary Archaeology* 6(1): 136–153.

Zalasiewicz, J., Waters, C. W., Barnosky, A. D., Cearreta, A., Crutzen, P., Ellis, E., Ellis, M. A., Fairchild, I. J., Grinevald, J., Haff, P. K., Hajdas, I., Leinfelder, R., McNeill, J. R., Odada, E. O., Poirier, C., Richter, D., Steffen, W., Summerhayes, C., Syvitski, J. P., Vidas, D., Wagreich, M., Wing, S. L., Wolfe, A. P., Zhisheng, A. and Oreskes, N. 2015. "When Did the Anthropocene Begin? A Mid Twentieth Century Boundary Level Is Stratigraphically Optimal." *Quaternary International* 383: 196–203.

Zalasiewicz, J. and Williams, M. 2012. *The Goldilocks Planet: The Four Billion Year Story of Earth's Climate*. Oxford: Oxford University Press.

6

OIL MATTERS

Esther Breithoff

The twentieth century was in many ways the "black century", marked by our obsession with oil, the "black gold" (More 2009) extracted from the veins of the earth. Today, as governments are committed to reducing carbon emissions globally as part of the Paris Agreement, many hope to move away from oil and embrace a carbon-free existence as a necessary step in desperate efforts to curb global warming and take action on mass extinctions and the climate emergency. To stop using oil, however, is easier said than done. Carbon-based economies and their by-products have come to infiltrate our lives and bodies and everything around us. In the words of media artist Marina Zurkow, we "are soaking in petroleum and wouldn't know how to live, feed, shelter, clothe, or express ourselves without oil-based products" (Plueker 2012). Oil is, so it seems, everywhere: not just in engines and energy production but in the fertilisers and pesticides which allow us to feed ourselves, in the petroleum plastics that make their way into the food chain, in the clothing we wear and in the emollients and lotions with which we nourish our skin. From the iconic oil platforms that, like majestic steel giants, rise from the rough waters of the North Sea and Gulf of Mexico, to climate and the carbon dioxide air pollutants that fill our lungs, carbon-based economies have impacted the planet as a whole, creating a deeply entangled, unruly and inescapable heritage that constitutes an ominous legacy of the Anthropocene. We can look away all we want, but this sticky heritage resists escape and persists in every part of the contemporary world. Indeed, rather than being optional and selected, the legacy of our addictive and dependent relationship with oil imposes itself on us in all its material and immaterial forms. Its physical and discursive imprints remain in the form of unforeseen monuments and unwanted material memories from a past that keeps haunting the present (and indeed the future).

Oil infiltrates everything in the Anthropocene. The legacies of oil are not optional but something all humans and non-humans have to live with (see Olsen and Pétursdóttir 2016). A massive globally distributed material phenomena, it constitutes what Timothy Morton (2013) defines as a "hyperobject" *par excellence* – oil's impact is tangible, viscous, inter-objective and often non-local and challenges conventional understandings of time and space. This visual essay captures the lingering and persistent physical legacies of "petrocultures" (e.g. Wilson, Carlson and Szeman 2017) – understood here as the manifold material-ecological-human assemblages created by oil – and their global spread and impact. Acknowledging that oil has been both the source of armed conflict over resources as well as a resource with which to fuel war, these images aim not only to critically uncover the haunting material traces of such petrocultures but also to highlight their involvement in contemporary processes of recycling, re-use, re-shaping and re-definition of the heritage ecologies of petroleum which point to new ways of living with these material legacies in the present and future. In doing so, I point to alternative ways of thinking and doing heritage which go beyond conventional narratives of endangerment and preservation (c.f. Vidal and Dias 2016) to explore instead the creative possibilities of decay (e.g. DeSilvey 2017), repurposing (c.f. Appelgren and Bohlin 2015), curating and adapting to such persistent and hyperobjective heritages in the present.

"Petromelancholia"

The move away from fossil fuels towards renewable energies has been accompanied by what Stephanie LeMenager has termed "petromelancholia" – a concept that she understands in the context of North American culture as the "the grieving of conventional oil resources and the pleasures they sustained" (2014, 18). "Humanity", she writes, is "at the mercy of collapsing naturecultures" (LeMenager 2014, 107). One of the ways in which this feeling of melancholia and mourning the loss of a "failed modernity" (González-Ruibal 2006) materialises in places around the United States is in the form of oil memorabilia. Here, the country's oil heritage seems to be worn like a badge of honour. It evokes a yearning for the ways of life which are imminently declining and desperately tries to hold on to a culturally constructed image of the carbon economies as a symbol of progress and wealth, even as it dissipates in the face of the urgency of the current global climate crisis.

The global within the local

A small abandoned house stands forgotten at the edge of a Mennonite village deep inside the remote Paraguayan Chaco region of South America. Dating from the early years of Russian-Canadian Mennonite migrations to the Chaco in the late 1920s, the owners have long gone. The little house now lies in ruins, weeds sprouting from the cracks and crevasses of its crumbled walls. On closer inspection, however, the picturesque scene reveals a troubling reality that has withstood the test of time. Otherwise built in the wattle-and-daub tradition set within a wooden frame, the house has an unusual roof that has survived years of heavy rain and drought. Glinting in the midday sun amongst sections of corrugated iron and bricks are panels of sheet-metal roofing made of flattened oil canisters, which it is said the original house owners found in the bush. In their attempt to build an existence in the Chaco scrub, the farmers thus unknowingly transformed objects used to fuel South America's first modern industrialised war into a structure that represents protection and family life. Equally troubling, it hints at a larger picture of the Chaco War (1932–35), fought between Paraguayan and Bolivian troops and involving local Mennonite newcomers to the region as well as indigenous people, in which global conflict is materialised in the local (see Breithoff 2020).

This point is further illustrated by two objects found in local Mennonite museums. The first one is a butter churn on display in the Koloniehaus Museum in Filadelfia, the main town of the Fernheim Mennonite colony. Butter churns were used by the early colonists to convert cream into butter and would have traditionally consisted of a wooden container and staff. In this case, however, the maker had substituted the wooden container with an old gas canister from the war with the seashell logo of the Royal Dutch Shell oil and gas company embossed on it. Similarly, the Geschichtliches Archiv Kolonie Menno in Loma Plata preserves an old baking tray fashioned out of a flattened oil canister with the Shell emblem still faintly visible. Although waged predominantly by indigenous and mestizo farmhands and miners in the distant thicket of South America, the implications of the Chaco War moved far beyond the physical boundaries of the Chaco scrubland. Fought with European and North American weaponry (Hughes 2005) and backed by powerful multi-national corporations such as British Royal Dutch Shell and Standard Oil in the United States (Chiavenato 1989), the Chaco War was very much a global event that is reflected in the local material culture. Similar to the American and Russian helmets found in post-conflict Ethiopia, the house roof, butter churn and baking tray become a "metaphor of globalization" (González-Ruibal 2006, 188) while simultaneously embodying the local, personal suffering and sacrifices of the soldiers, Mennonites and indigenous peoples of the Chaco.

Hiding in plain sight

Halfway across the world in the Laotian jungle lies the community of Long Tieng. During Laos' so-called Secret War (1964–1973), the village served as a clandestine air base for US forces (Kurlantzick 2017). Oil fuelled thousands of planes on their mission to drop bombs on neighbouring Vietnam, many of which ended up emptying their cargo over Laos, contributing to it gaining its tragic status of most heavily bombed country in history (Sisavath 2019). According to Legacies of War, a US-based organisation dedicated to raising awareness about the US bombing initiatives in Laos during the Vietnam War, over two million tons of ordinance were dropped over Laos in 580,000 bombing missions. That is a planeload of bombs every eight minutes, 24 hours a day, seven days a week, for nine long years (Legacies of War 2019). During the war, Long Tieng did not officially exist and did not appear on any contemporary maps. The material evidence, however, is firmly present, constituting an alternative archive of military waste (Sisavath 2019) that belies the lack of conventional historical archives which might document its function. The remnants of war and the oil that fed it are hiding in plain sight and have quietly infiltrated and blended in with the everyday: village houses and fences are partly built from petrol drums and airplane parts, forcing people and animals to forge an existence amongst the material legacies of oil-fuelled war.

Curating oil

The city of Stavanger, known as the Oil Capital of Norway, has grown in importance to the Norwegian economy and labour market since the development of the offshore oil industry during the 1960s. Norsk Oljemuseum (Norwegian Petroleum Museum) in Stavanger was built in 1999 with the aim of displaying the history of Norway's offshore petroleum industry in the North Sea as a symbol of the nation's industrial and economical success. Despite models of platforms attempting to convey the size of rigs, the material magnitude of these objects and their related machinery remains unfathomable inside the museum walls. How then curate and display such enormous things? How to make them accessible and sensible when their own legacies are so messy and senseless? Can the ugly and disordered ever be tamed and transformed into easily legible heritage?

While heritage and musealisation represent one conventional way of giving a "second life" to objects, places and practices which are defunct and no longer profitable or useful for their intended purpose (Kirshenblatt-Gimblett 1998), there are other ways of recirculating and giving new functions to things that have become apparently superfluous or redundant. Utilising elements recycled from the abandoned Frigg Oil Platform, progressively decommissioned after the closure of the oil field in 2004, as well as other recycled materials from the petroleum industry, Geopark is a designed "junk playground" located immediately adjacent to the Norsk Oljemuseum. Opened in 2008, the park was designed by architects Helen and Hard in consultation with the museum and local youth groups to represent the geology and contours of the "Troll" gas and oil field, reconstructed on a scale of 1:500 (Helen and Hard 2019). The formerly abandoned site and the recycled objects have been used to create a skate and bicycle park, a space for street art, structures for climbing and other adventure play areas.

"Dark ecology"

The 2010 Deepwater Horizon or BP oil spill in the Gulf of Mexico was the largest marine oil spill in the history of the United States' petroleum industry, with catastrophic consequences for local wildlife. It is now thought to have been one of the most devastating environmental disasters in US history. Monitoring of the effects of the spill itself, and subsequent "clean-up" efforts in which large quantities of oil coagulants and dispersants were used to deal with the spill, have shown significant impacts on the mortality rates of marine wildlife in the area, including dolphins, tuna and sea turtles (e.g. Colegrove et al. 2016). The concept of ecology is not just a romantic celebration of everything green and natural – and by extension beautiful. Timothy Morton's notion of *Dark Ecology* (2016) forces us to face the "mesh", the intrinsic interconnectedness of human and non-human entities, which is dark, ugly, grimy and messy. Oil is dark – both in its physical and philosophical properties. It is sticky, destructive, deadly, leaving a huge environmental imprint. Like the mud in the trenches of the First World War, which Sanatu Das says had "aggressive agency" and a "leech-like suction on the skin" (2005, 45), the viscous nature of oil blurs the categories of object and subject. The Deepwater Horizon disaster revealed this dark side of oil in all of its monstrous power, as an opaque black mass that appeared to smother everything in its path.

Sustainable futures?

Ever since their development in the 1930s, offshore drilling platforms have been a huge physical, financial and environmental presence globally. Over 6,000 such installations extract hydrocarbons from the oil and gas reserves of the world's oceans. Ever-present media reports of the environmental consequences of the oil industry are graphic, catastrophic and, above all, undeniable. They show coral bleaching caused by rising water temperatures, in large part a result of the rise in atmospheric carbon caused by fossil fuel burning, and oil spills suffocating birds and other wildlife in toxic goo.

These images appear to realise the central Anthropocene thesis that human intervention and the natural environment are inseparably entwined. But this interplay is not necessarily driven by human initiative. Hidden from view and unknown to many, new heritage ecologies develop. Underwater worlds flourish on and around the gigantic steel legs that run hundreds of meters down to the ocean floor. Fostering a unique habitat, the platforms have become vertical reefs for myriad different forms of marine life. Initiatives like Blue Latitudes, a marine environmental consulting firm that "designs sustainable, creative, and cost-effective solutions for environmental issues that surround the offshore industry", support Rigs-to-Reefs – the modification of suitable platforms that have come to end of their working lives as oil extraction facilities into officially recognised artificial reefs and in the process reducing the costs and environmental risks linked to decommissioning processes (Blue Latitudes 2019) .

These heterogeneous and geographically widespread examples speak not only to the ways in which oil has infiltrated every aspect of the contemporary material world but also to the necessity of finding new, innovative and creative ways to live with these persistent material heritages into the future. While heritage practices are conventionally understood to be pitted against processes of material decay which should be slowed or arrested (cf. DeSilvey 2017), the heritage of oil reverses this to ask how instead we might forge lives amongst these large, ubiquitous, widespread and sometimes toxic by-products and remnants of what will in the future represent extinct petrocultures but which will nonetheless persist and do not readily decay. These examples show why and how oil *matters*, both in Karen Barad's (2007) sense of materialising and making worlds but also its importance to contemporary heritage discourses, which have barely engaged with the long-term futures (cf. Harrison et al. 2020) into which such materials will endure.

FIGURE 6.1 Back wall of a restaurant covered in oil memorabilia in Cayucos, California, USA

Photo: E. Breithoff.

FIGURE 6.2 Ruined Mennonite house with roof partly made out of flattened oil canisters, Paraguay

Photo: E. Breithoff.

FIGURE 6.3 Butter churn on display in the Koloniehaus Museum, Fernheim, Paraguay

Photo: E. Breithoff.

FIGURE 6.4 Baking tray from the Geschichtliches Archiv Kolonie Menno, Loma Plata, Paraguay

Photo: E. Breithoff.

FIGURE 6.5 Village fence partly made of oil drums at Long Tieng, Central Laos

Photo: E. Breithoff.

FIGURE 6.6 Inside the Norsk Oljemuseum in Stavanger, Norway

Photo: E. Breithoff.

FIGURE 6.7 Geopark "junk playground" outside the Norsk Oljemuseum in Stavanger, Norway

Photo: E. Breithoff.

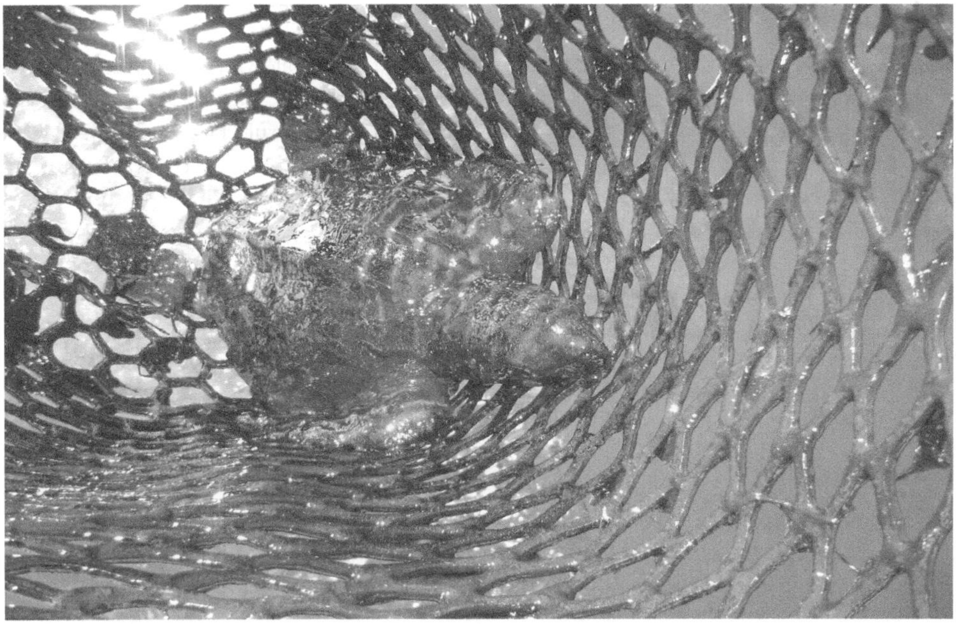

FIGURE 6.8 Oil-covered Kemp's ridley sea turtle rescued by the Louisiana Department of Wildlife and Fisheries, USA

Photo: Louisiana Department of Wildlife and Fisheries. CC-BY-SA-2.0 https://creativecommons.org/licenses/by-sa/2.0/.

FIGURE 6.9 Artificial reefs

Photo: Joseph Platko, courtesy of Blue Latitudes.

Acknowledgments

For the past ten years, the material afterlives of oil have been following me across countries and research projects. From my PhD research on conflict heritage in the Paraguayan Chaco (University of Bristol), to my postdoctoral work as part of the Unruly Heritage project (UiT The Arctic University of Norway), to my current research on the conflict legacies in Laos (UCL/Birkbeck, University of London), the material culture of oil has made both its physical as well as global presence known. This photo essay is the result of these chance and yet unavoidable encounters with the tangible remnants of carbon-based economies. I would like to thank the Fond National de la Recherche Luxembourg, the Unruly Heritage project at UiT and the UCL Institute of Archaeology for funding fieldwork on which this photo essay draws. Work on this chapter was completed as part of my UKRI Future Leaders Fellowship MR/S016236/1 on "Material Memories" at Birkbeck, University of London.

References

Appelgren, S. and Bohlin, A. 2015. "Growing in Motion: The Circulation of Used Things on Second-Hand Markets." *Culture Unbound* 7: 143–168.

Barad, K. 2007. *Meeting the Universe Halfway: Quantum Physics and the Entanglement of Matter and Meaning*. Durham, NC: Duke University Press.

Blue Latitudes. 2019. *Environmental Solutions for Offshore Industry*. Accessed 18 October 2019. www.rig2reefexploration.org/.

Breithoff, E. 2020. *Conflict, Heritage and World-Making in the Chaco: War at the End of the Worlds?* London: UCL Press.

Chiavenato, J. J. 1989. *La Guerra del Chaco. Petróleo*. Asunción: Carlos Schauman.

Colegrove, K. M., Venn-Watson, S., Litz, J., Kinsel, M. J., Terio, K. A., Fougeres, E., Ewing, R., Pabst, D. A., McLellan, W. A., Raverty, S., Saliki, J., Fire, S., Rappucci, G., Bowen-Stevens, S., Noble, L., Costidis, A., Barbieri, M., Field, C., Smith, S., Carmichael, R. H., Chevis, C., Hatchett, W., Shannon, D., Tumlin, M., Lovewell, G., McFee, W. and Rowles, T. K. 2016. "Fetal Distress and in Utero Pneumonia in Perinatal Dolphins During the Northern Gulf of Mexico Unusual Mortality Event." *Diseases of Aquatic Organisms* 119: 1–16.

Das, S. 2005. *Touch and Intimacy in First World War Literature*. Cambridge: Cambridge University Press.

DeSilvey, C. 2017. *Curated Decay: Heritage Beyond Saving*. Minneapolis, MN: Minnesota University Press.

González-Ruibal, A. 2006. "The Dream of Reason: An Archaeology of the Failures of Modernity in Ethiopia." *Journal of Social Archaeology* 6(2): 175–201.

Harrison, R., DeSilvey, C., Holtorf, C., Macdonald, S., Bartolini, N., Breithoff, Esther, Fredheim, H., Lyons, A., May, S., Morgan, J. and Penrose, S. 2020. *Heritage Futures: Comparative Approaches to Natural and Cultural Heritage Practices*. London: UCL Press.

Helen & Hard. 2019. *Geopark: A New Playground Using Resources from the Offshore Oil Industry*. Accessed 18 October 2019. www.helenhard.no/thinking/geopark.

Hughes, M. 2005. "Logistics and the Chaco War: Bolivia Versus Paraguay, 1932–1935." *The Journal of Military History* 69(2): 411–437.

Kirshenblatt-Gimblett, B. 1998. *Destination Culture: Tourism, Museums, and Heritage*. Berkeley, CA: University of California Press.

Kurlantzick, J. 2017. *A Great Place to Have a War: America in Laos and the Birth of a Military CIA*. New York: Simon and Schuster.

Legacies of War. 2019. *Legacies of War: Frequently Asked Questions*. Accessed 18 October 2019. http://legaciesofwar.org/about/faq/.

LeMenager, S. 2014. *Living Oil: Petroleum Culture in the American Century*. Oxford: Oxford University Press.

More, C. 2009. *Black Gold: Britain and Oil in the Twentieth Century*. London: Continuum.

Morton, T. 2013. *Hyperobjects: Philosophy and Ecology after the End of the World*. Minneapolis, MN: Minnesota University Press.

Morton, T. 2016. *Dark Ecology: For a Logic of Future Coexistence*. New York: Columbia University Press.

Olsen, B. and Pétursdóttir, Þ. 2016. "Unruly Heritage: Tracing Legacies in the Anthropocene." *Arkæologisk Forum* 35: 38–46.

Plueker, J. 2012. *Marina Zurkow 'Necrocracy' Exhibition Gallery Text*. Accessed 18 October 2019. www.o-matic.com/play/necrocracy/NecrocracyDiverseworks.pdf.

Sisavath, D. 2019. "The US Secret War in Laos: Constructing an Archive from Military Waste." *Radical History Review* 133: 103–116.

Vidal, F. and Dias, N. 2016. "Introduction: The Endangerment Sensibility." In *Endangerment, Biodiversity and Culture*, edited by F. Vidal and N. Dias, 1–38. Abingdon: Routledge.

Wilson, S., Carlson, A. and Szeman, I. 2017. *Petrocultures: Oil, Politics, Culture*. Montréal: McGill-Queen's University Press.

PART III
Affect

7

EMERGENT IMAGES

Matters of affect in heritage photography

Colin Sterling

FIGURE 7.1 John Thomson, 1866. *A Tower at Nakhon Thom [Angkor Thom], Cambodia*. Digital scan of modern contact print

Source: Wellcome Collection/CC BY 4.0. https://creativecommons.org/licenses/by/4.0.

Introduction

This chapter explores the ecological emergence of a highly specific heritage "thing" – in this case a single photographic image – across a variety of material conditions and modes of encounter. The photograph in question is one of the earliest images ever taken of Angkor in Cambodia. It was produced by John Thomson in 1866 and depicts a vast stone tower with serene carved faces in the middle of a dense forest (Figure 7.1). The tower forms part of a richly decorated temple complex built in the late twelfth or early thirteenth century by King Jayavarman II, now known as The Bayon. Over the past century and a half, the visual grammar we see deployed here – ruins engulfed in vegetation – has become an Angkorean cliché (Sterling 2016), a powerful symbol of the enigmatic "mysteries" of the site. This seductive reading continues to adhere to the image, which conjures impressions of an intrepid Western explorer uncovering the ruins of a long-lost Oriental civilisation (Barnett 1990). Significant work by heritage scholars has demonstrated the pernicious effects of such ideas in relation to Angkor (see Edwards 2007; Falser 2013; Winter 2007). Conscious of these critical agendas, my focus in this chapter is less on the visual-discursive construction of Angkor itself *as heritage* (see Di Giovine 2009) and more on the dispersed processes that constantly reconfigure the affective power of this particular photograph. Tracing the microhistorical trajectory of an individual image in this way is not simply an exercise in academic introspection. Nor do I seek complexity for complexity's sake. Instead, remaining alert to the socio-political work photographs perform in the world, this chapter focuses on the varied ecologies of production, circulation and use that underlie familiar engagements with "the photographic" across heritage practice. My turn to affect in this context seeks to emphasise the immanent, relational and multi-sensorial qualities of photography to better understand the agential capacities and power dynamics of the medium. Crucially, this model extends beyond human-centred "moods and sensibilities" to consider the ongoing influence of technologies, minerals, atmospheres and other non-human factors on public culture (Bennett 2010, 61), in this case seen through the lens of heritage and photography.

To help structure this investigation, three planes of emergence are highlighted: (1) an initial chemical-embodied moment of production in the forests of Cambodia, (2) a short period of visual circulation after Thomson had returned to Britain with his photographs in the late 1860s and (3) the contemporary deployment of this image in various heritage settings (including exhibitions, physical archives and auction sales). With each plane, I question a different reading of "ecology" as it relates to photography and heritage. My starting point here is the recent work of Joanna Zylinska (2017) and William Schaefer (2017), who stress in different ways the vibrant nonhuman agencies that contribute towards photographic production. I then move on to consider the early distribution of Thomson's image with reference to James Hevia's notion of the photography complex (2009) and Elizabeth Edwards' related definition of photographs as "objects of affect" (2012a). Finally, the ongoing digital and material (re)mediation of the photograph is interrogated via two worlds that are rarely considered together: algorithmic culture studies (Striphas 2015) and Caitlin DeSilvey's conceptualisation of "mnemonic ecologies" (2017). Clearly I cannot go into great detail on each of these theorisations in such a short space, but I feel it is important to acknowledge the range of "ecologies" (material, semiotic, atmospheric, chemical, computational) that may encircle and animate heritage and photography. This also helps to draw out the constantly shifting "politics of affect" (Massumi 2015) at work across these human and nonhuman assemblages.

Although the narrative I put forward here is temporally linear – moving broadly from photographic creation to visual reproduction and on to later contexts of care and consumption – it is worth highlighting from the outset that there are no strict boundaries between these planes of emergence. Indeed, some dimensions of the affective ecologies I question depend on reversal and feedback: "returning" an image to the space of production, for example, has become an important activity within heritage practice (this may involve physical display, memory workshops or even large-scale photographic projections onto the sites documented). The emergent image is just as prone to collapse and dormancy as it is constant re-awakening. Affect is typically brought to bear on such activities and experiences with reference to non-discursive sensations, intensities, atmospheres and feelings (see Stewart 2007). This builds on the foundational work of Spinoza, where affect designates "movements", "rests" and "charges": "individuating affective states of an anonymous force" (quoted in Thrift 2008, 13). As later writers have explored, such forces are not limited to the human and instead surface "in those resonances that circulate about, between, and sometimes stick to bodies and worlds" (Gregg and Seigworth 2010, 1). An ecology of affect then needs to address the uneven distribution of potentiality across human and nonhuman assemblages, asking where the force of change lies in any given situation but also how it might accrue in and around certain bodies and spaces. This, I would suggest, is where a politics of affect comes in, charting not simply endless states of emergence but also processes of closure and arrest. It is the tension between these two possibilities that underpins the current investigation.

Emergence one: chemical bodies

What do we *see* when looking at the above photo (Figure 7.1)? Initially, the eye is drawn to the two stone faces in the top third of the image, but it's impossible to rest here. Instead we follow the roots and branches of the trees, encroaching tendrils from the forest beyond that form a dense backdrop to the central monument. At the far right of the frame, a pillar and stone wall create a further layering effect: we are in the complex, not just gazing at it from a distance. Chunks of masonry litter the floor, intermingled with vegetation and broken trees. Detailed carvings, meanwhile, are overlain with moss and lichen, a symbiosis of human-fashioned rock and organic matter. And within this scene – easy to miss at first – a lone figure sits in the opening of the central tower. Although marginal to the composition as a whole, his presence upends our reading of the image, revealing the sheer scale of the edifice.

This sense of visual emergence – in which an embodied eye scours the two-dimensional surface of an image for compositional clues and aesthetic registers – is a familiar tactic within representational approaches to photography. In angling away from such questions, I do not want to suggest that ecologies of affect somehow transcend representation: what is in the image matters, as do the wider visualising practices in which such photographs are embedded. Indeed, the visual-discursive "ecology" of this particular image could be said to encompass ruin paintings of the Romantic period, French and British survey photography (Boyer 2003; Edwards 2012b), Ruskin's theories on architecture, and earlier sketches of Angkor by the French naturalist Henri Mouhot, which directly inspired Thomson's own expedition (Thomson 1867). At the same time, we must acknowledge that a focus on representation has occluded other agential forces involved in photographic production, which might be better served by a relational understanding of the medium.

To help introduce this alternative trajectory, it is worth viewing the photograph not as an image but as a material *thing*, all the while recognising that there is an "indissoluble, yet ambiguous, melding of image and form" in every instance of photographic representation (Edwards and Hart 2004, 2). The emergent ecology of the photograph *as object* gestures towards a network of more-than-human affects operating across a variety of scales. This is drawn out when viewing the glass plate negative from which the picture was created (Figure 7.2). Here the image itself fades from visual apprehension, obscured by scratches, blemishes and – at the top left of the frame – a hand-written caption with numbering details, suggesting a sequence or archive. This flawed surface speaks of an agitated movement through the world, one we might "read" in a variety of ways, informed by scientific investigation, archaeological methods or art historical knowledge. An ecological approach remains open to each of these perspectives, but for the sake of brevity I would like to focus on two main factors influencing the initial emergence of this image-object, namely its chemical formation and the conditions under which the photographer and his helpers operated.

When Thomson produced this image in March 1866, he used the most popular photographic technology of the time, known as the wet collodion process. First introduced by Frederick Scott-Archer in 1851, wet collodion photography was much faster than earlier daguerreotype or calotype processes and immediately free from patent restrictions. As a result, it swiftly became the predominant form of photographic invention and remained so until 1880, when more industrial methods became available. The wet collodion process was complex and required bulky equipment but resulted in images of incredibly high definition. As Thomson himself admitted much later in his career, the technology also had one other major benefit:

> The wet-collodion process, appropriately named, could not shed its ponderosity, and was hedged round with difficulties, as I had no reason to know and appreciate, and ill adapted for long journeys. It was the most chemically and mechanically exacting companion to be carried on any expedition, and its shortcomings were accentuated when my wanderings happened to be through forest and tropical jungle. One special virtue must be noted, and that is that the plate had to be exposed, developed and finished on the spot, so that one was enabled to judge of success or failure before striking camp.
>
> *(Thomson 1907, 16)*

The steps necessary to create the photograph can be broken down as follows. First, a glass plate was prepared by smoothing down the edges and polishing the surface with a solvent. The plate would then be carefully cleaned to remove any dust particles – a difficult task in the forests of Cambodia, as revealed in the various inconsistencies seen in the image below. Once ready, a sticky mixture of ether, alcohol, iodides, bromides and the collodion[1] itself would be poured onto the centre of the glass and distributed evenly across the surface using a method known as flowing the plate. The glass – now covered on one side in a syrupy

FIGURE 7.2 John Thomson, 1866. *A Tower at Nakhon Thom [Angkor Thom], Cambodia*. Digital photograph of glass plate negative

Source: Wellcome Collection/CC BY 4.0. https://creativecommons.org/licenses/by/4.0.

substance – would then be transferred to a portable darkroom and sensitised to light in a bath containing silver nitrate. After three to five minutes, the sensitised plate would be placed in a special holder, still damp, collodion side down. This holder would then be closed to block any light so that it could be removed from the darkroom and attached to the back of Thomson's camera. Once in position, the plate would be exposed for between 5 and 30 seconds, depending on available light. The photographer then returned to the darkroom with the holder, where the exposed plate could be removed. Using an even sweeping motion, a developing solution of either ammonio-sulphate of iron or pyrogallol would be poured onto the plate, followed by water to stop the development. The exposed and developed plate would then be placed in a bath of fixer to preserve the image. Finally, after drying, the plate would ideally be varnished to protect the fragile surface of the image from damage.

The expertise, arcane knowledge and bodily actions required to produce a photograph in this way open up the multisensory dimensions of the medium to critical scrutiny. Although focused on the production of a discrete visual artefact, the wet collodion process heightens a sense of embodied becoming in which the image is inescapably bound up with meteorological conditions, chemical compounds, malodorous scents, infinitesimal gestures and carefully rehearsed physical movements. The nonhuman agents at work here encompass the glass plates Thomson used, which were prone to deteriorate; the humid atmosphere and intense light encountered in the forests of Cambodia; the noxious and explosive chemicals required for the wet collodion process; and the hardwood cameras needed to withstand the tropical climate. The entanglement of human and nonhuman in this setting is brought vividly to the surface in one of Thomson's earliest pieces of writing, an article in *The British Journal of Photography* on "Practical Photography in Tropical Regions:" "I have felt, after a day's work in a tent, so thoroughly saturated with chemicals that I might almost be used for coating a plate or printing upon" (1866, 404).

In a recent essay titled simply "Photographic Ecologies", William Schaefer discusses the work of a number of contemporary Chinese artists who see photography as a "key site for staging and rethinking fundamental questions of the relations between culture and nature, landscape and ecology" (2017, 44). This is achieved – Schaefer argues – by recognising that both the materiality and the visual form of photographic images emerge from "the interaction of ecological processes", which in turn "allow the human to be seen as one among many contingent actors within such processes" (ibid). In making his argument, Schaefer singles out the work of artist Adou, who uses expired film to produce opaque and haunting images of mostly rural contexts. The already degraded chemical composition of such film means that – as with Thomson's glass plate – the "surface artifacts" of the emulsion are made visible: streaks and spots that "signify decay and age" (ibid, 49). Drawing on Jeff Wall's concept of "liquid intelligence", Schaefer uses Adou's work as a jumping-off point to consider film photography in general as a mode of representation that "explicitly stages the relationship between culture and nature, not as a relationship of Cartesian dualism or separation but rather as one of emergence" (ibid, 62). As he concludes:

> The medium of photography itself might be understood as ecological rather than as primarily representational. According to this ecological model, the pictures that take form through photography are emergent from but not equivalent to the self-organizing processes of the silver salts clumped together because of the action of light, of the development by liquid chemicals, of the layers of gelatin and cellulose that constitute it and gradually decay, and of the natural and cultural environment of which the film and photographer and the landscape depicted are all part.

> In photography, in short, the work of forming images is situated at once within and beyond human agency and culture.
>
> *(ibid)*

A similar line of thinking is articulated in Joanna Zylinska's notion of "nonhuman photography" (2017). This theorisation responds to the appearance of new digital imaging technologies and algorithmic processes of production, circulation and use, which invite us to "see the photographic image as first of all a node in the networked sequence of human-nonhuman processes of connection, identification, translation, and … invention" (ibid, 199). Working backwards from this twenty-first-century predicament (which I return to later), Zylinska suggests that photography has in fact always been subject to "nonhuman entanglement and kinship" (ibid, 199), with images generated through the interplay of light, chemicals and human creative intention. Taken to its extreme, this includes photographic practices from which the human may be entirely absent, either as "subject, agent, or addressee" (ibid, 5). In explaining this position, Zylinska draws on the work of William Henry Fox Talbot, whose publication *The Pencil of Nature* – printed in instalments between 1844 and 1846 – is widely considered one of the earliest examples of a photographically illustrated book. Here, as Zylinska records, Talbot emphasises the natural qualities of this remarkable new technology:

> The plates of this work have been obtained by the mere action of Light upon sensitive paper. They have been formed or depicted by chemical means alone, and without the aid of any one acquainted with the art of drawing. They are impressed by Nature's hand.
>
> *(quoted in Zylinska 2017, 109)*

By emphasising the "natural" emergence of such pictures, Zylinska does not advocate for a return to objectivity or "truthfulness" in the photographic image but rather underlines a sense of mutual becoming, which may support "an ecological model of perception as a more embodied, immersive, and entangled form of image and world formation" (ibid, 8). It is in this light that Zylinska sees photography as a "*formative practice of life*", shaping human and nonhuman existence through images, material impressions and "forms of perception" (ibid, 5–8, original emphasis).

The commingling of natural processes and cultural practices is immediately evident in Thomson's Bayon image. The unique material qualities of the collodion solution, glass plate, wooden camera and silver nitrate impress themselves upon the photograph. Thomson's aesthetic response to the monument depended upon an embodied engagement with these affective forces. Beyond this, however, we also need to account for the treatment of "nature" itself: the vibrant ecologies of plant and animal life in which Thomson operated. Here the words of H. G. Kennedy, Thomson's travelling companion on the expedition, are particularly useful, recording as they do a site "wholly overgrown with jungle and forest-trees" and "so thick with ruins of temples, palaces, and other buildings, that it would require at least two months to explore it satisfactorily" (1867, 308). The heat, humidity and sheer density of the Cambodian forest clearly took its toll on the travelling party (Thomson himself contracted "jungle fever" before reaching Angkor), but it also greatly influenced the expression of a certain visual language in relation to the site. This can perhaps best be understood as a kind of negative relationality, wherein the "forest trees, climbing vines, and thorny brushwood" necessitated "a hard day's cutting" to clear the tower before photography could be undertaken (Thomson 1867, 67).

At one level, we might read the chemical-atmospheric emergence of this image as a compelling example of the more-than-human entanglements through which photographs come into being. This understanding, however, risks eliding the power dynamics of photography if we do not also acknowledge the socio-cultural practices that motivated Thomson's aesthetic choices. This is felt most clearly in the positioning of a lone human figure as a scale against which to measure the monument and in the removal of vegetation from the site to allow the picture to be taken in the first place. Such practices speak to another emergent ecology: that of a European approach to preservation being imposed on indigenous lands and peoples. The technological possibilities of the camera, whilst rooted in "nature" and embedded in a broader ecosystem of matter, objects, bodies, spaces, surfaces, liquids and light (Schaefer 2017, 63), were central to the growth and spread of such ideas, as Thomson himself recognised in a later piece of writing on "Photography and Exploration": "We are now making history, and the sun picture supplies the means of passing down a record of what we are, and what we have achieved in this nineteenth century of our progress" (Thomson 1891, 673).

Outlining the emergent ecologies of photography can help us to understand in a more precise way the diverse material agencies involved in photographic production, but this should not be at the expense of a politicised engagement with the medium as a "formative practice of life" (Zylinska 2017). Moreover, as Schaefer asks, "what can this rethinking of photography as primarily ecological, not representational, tell us of what it means to be part of an ecosystem?" (2017, 66). I would like to suggest that one way of maintaining an ethical and political register when considering such ecologies is to highlight the affective, immanent and densely situated material-discursive environments in which the photograph as event unfolds. Here I am drawn to Karen Barad's insistence that "particular possibilities for acting exist at every moment, and these changing possibilities entail a responsibility to intervene in the world's becoming, to contest and rework what matters and what is excluded from mattering" (2003, 827). Seen from this perspective, rethinking photography and heritage as ecological is not about taking a step back to "map out" human and nonhuman ecosystems in their entirety (this would be impossible) but is instead part of a broader commitment to embedded relationality and responsibility. The emergent image in this sense evokes an always untapped potential for different ways of being and acting in the world.

Emergence two: distributed pictures

We can think about this another way by questioning the photograph's status as a permanent and singular visual record. As the previous quote from Thomson suggests, the "sun pictures" he produced in the forests of Cambodia were always intended to be a trustworthy record of his experiences: documentary artefacts which could be "passed down" to future generations. This positivist conception of photography was closely connected to an archival and mnemonic impulse, both of which drew on the specific material ecologies of the wet collodion process. The sense of movement and "vibrancy" (Bennett 2010) we find in the ecological emergence of the image within and across the surface of the glass plate negative needs to be balanced with the desire for permanence and stability which motivated Thomson's own use of the technology. Glass plate negatives and the wet collodion process afforded images of great precision, to be made with relative speed, and defined by a material obdurateness that other visualising techniques of the time lacked. They were created to last. This should not, however, be taken as the final word on "the archive" as envisaged by Thomson. As well as recording sites such as Angkor for posterity, Thomson – like most photographers – sought to promote his own work via publication and other distribution channels. This early diffusion of the image already marks out a second stage of emergence, one that encompasses ephemeral displays, printed matter and various text-based annotations, focused on a specific though constantly shifting visuality.

One such method of dissemination was Thomson's own illustrated volume, *The Antiquities of Cambodia* (1867). Here we find 16 albumen prints, with an accompanying narrative and plan drawing of Angkor Wat. The image I am particularly interested in here can be found on page 68 (Figure 7.3), preceded by an extended label that gives the photograph its enduring title: "Prea Sat Ling Poun" (Thomson tells us that this is a modern name given to the site by locals, meaning "the place where they play hide-and-seek"). Also worth noting is the fact that in this print, the human figure appears to be missing from the image. It is unclear whether this absence is the result of image manipulation (certainly the technology for such alterations existed at the time) or whether it points to the presence of a second, now lost, glass plate negative. It is through such visual modulations and discursive framing devices that the material ecology of the image-as-object starts to fade and the representational take precedence. This would depend, however, on the very materiality of the print itself, which is worth outlining briefly here.

The albumen print was introduced in 1850 and remained in general use until the end of the nineteenth century. Its popularity stemmed in no small part from the aesthetic advantages it offered, with precise definition, greater contrast and a wider breadth of shades than other printing techniques available at the time. To create prints using this technology, a high-quality sheet of paper would be coated with a thin layer of egg-white and salt, dried and then sensitised with a solution of silver nitrate. The paper would then be exposed under the negative in a printing frame as soon as possible after sensitisation. The visible image could take between a few minutes and one hour to emerge, with the print produced by exposure through the negative to sunlight. After this initial exposure, the image would usually be toned in a bath of gold chloride. This would bring a colder, reddish-brown tone to the print, depending on the composition of the gold bath. Crucially, this stage of printing also improved preservation by reducing chemical sensitivity. The toned proof would then be fixed in a bath of sodium thiosulphate. Such images, which lay on the surface on the paper, were able to capture a wealth of sharp detail and had a characteristically glossy sheen (see Falconer and Hide 2009, 172; Paris Photo 2018).

Although the aesthetic rewards of the albumen printed were considerable, it had two major drawbacks as a printed medium. These underline the affective work specific modes of dissemination might perform in the world, with each technological apparatus carrying its own material agency. Most immediately, pasting in albumen prints was a "cumbersome, expensive, labour-intensive and time-consuming process" (Ovenden 1997, 177). This greatly restricted the number of photographs that could be used in a given publication and meant print runs as a whole were fairly limited. The second and perhaps more worrying drawback concerned the tendency of albumen prints to fade, as can be seen in the image below. While this problem had been significantly reduced by the 1860s, the "specter of impermanence" (Szarkowski 1992, 177) continued to haunt the medium. I turn to issues of contemporary

FIGURE 7.3 John Thomson, 1866. *Prea Sat Ling Poun, Cambodia.* Digital scan of albumen print

Source: © The British Library Board.

care and conservation in the following, but here it is worth noting that albumen prints have long been considered to merit urgent attention in most collections, with approximately 85% suffering from yellowish-brown staining and overall fading (Reilly 1980).

While the albumen print captures the fragile and fluid materiality of this image in a form that is at once archival and communicative, this was not the sole means of encounter for Victorian audiences. In his biography of Thomson, Richard Ovenden outlines some of the many routes of circulation this image would have been caught up in in the late 1860s and early 1870s (1997, 11–12). As part of a "busy schedule of lecturing and writing", Thomson showed his photographs of Angkor to various special interest societies, including the Ethnological Society of London and the British Association for the Advancement of Science, and published a series of articles in the *British Journal of Photography*. In November, he exhibited his Cambodian images in a large exhibition organised by the Edinburgh Photographic Society held at the Museum of Science and Art and later that year lectured to the same society on his travels in Siam and Cambodia, illustrating his talk with lantern slides. Whilst we cannot be sure if the specific image I am discussing here featured in these more ephemeral activities, the Angkor photographs as a whole clearly garnered considerable interest, with further lectures and exhibitions for the Glasgow Photographic Association and the Architectural Institute of Scotland, and – in 1868 – a short article in the *Illustrated London News*.

Such differential spaces of encounter are typical of photographs, which are "specifically made to have social biographies. Their social efficacy is premised ... on their shifting roles and meanings as they are projected into different spaces to do different things" (Edwards 2012a, 222). This biography model is, however, limited. As Edwards continues, photographs are "divergent" and "nonlinear", spread over "multiple material originals and multiple, dispersed, and atomized performances" (ibid, 223). With each new performance a range of political, material and emotional factors are brought to bear on the image, from methods of display or printing to the narrative accompaniments that typically frame the photograph in written or spoken form. In this sense, photographs are to be seen as "tactile, sensory things that exist in time and space and are constituted by and through social relations" (ibid, 228). The visual, material and textual assemblages that constitute photographs as affective objects cannot be understood without this social element, but neither are they reducible to this human dimension. The image carries its own affective force and "acts back" in unexpected ways in a constant unfolding dialogue with wider photographic ecologies.

Writing in a similar mode – and drawing heavily on the work of Latour – James Hevia's notion of the photography complex describes "a network of actants made up of human and nonhuman parts" encompassing the photographer, the thing photographed, the camera itself (including its various parts), the negatives and chemicals used and the "staggering array of reproductive technologies through which images move and circulate" (2009, 81). This final gesture sheds light on the "transportation and communication networks along which all of these parts travel, and the production and distribution networks that link far-away places to end-users" (ibid). In a nod to the nonhuman foundations of photography, Hevia concludes this inventory with light itself, which he describes as an "author or actant" in the formation of the image (ibid). Photography assumes a very different ontological status through this reading: less a category or mode of representation than "a heading under which a range of agencies, animate and inanimate, visible and invisible, are clustered" (ibid).

The intersection of matter, technologies, processes, individual motivations and socio-political environments redirects questions of emergence away from the image itself and towards the constantly shifting relationalities in which photographs are always embedded. With respect to the initial circulation of Thomson's image, two modes of engagement are worth highlighting: the aforementioned lantern slide lectures and woodcut engravings in various publications.

These spaces of encounter encompass a range of materialities and affective capacities, linking text, embodiment and visuality in often divergent ways.

In a wide-ranging study of British survey photography in the late nineteenth century, Edwards describes the lantern slide as "a much underestimated player in the formation of historical consciousness and imagination" (2012b, 237). Thomson's use of the technology is well documented and aligns with a period in which the educative power of photographs was paramount. Lantern slides at this time depended on oxyhydrogen projection, a process in which oxygen was burnt with hydrogen, ether, acetylene or coal gas to create a flame, which was then directed on to a lime surface, which in turn created a more intense white light. This incandescent light was then focused onto the lantern slide using condenser lenses to project an image. Again, the photograph cannot be understood as a discrete mode of representation in this setting. The image as *a thing to be seen* only existed through a dense and volatile entanglement of various nonhuman agents, not least the gases themselves and the projection device. For Edwards, this "multisensory and all-absorbing experience" constituted a form of "affective knowing", with the viewer's entire body "wrapped in both darkness and light" (2012b, 237). Fully embodied, this sense of "material apprehension" (ibid) evokes a more open-ended understanding of photographic engagement than we are perhaps used to when considering nineteenth-century images. This should not, however, obscure the fact that such experiences helped to inform audiences of specific subjects (in this case the monumental remains of a distant civilisation) through a highly particular discursive frame.

The same may be said of the numerous books and journals that featured reproductions of Thomson's photographs as engravings in the late 1860s and early 1870s. Woodcut relief printing represented the cheapest and most efficient mode of circulating images at the time, but the resulting pictures often departed considerably from the original photograph. As Ovenden records (1997, 177), Thomson would have relinquished much of his artistic control for the production of an image such as we see in Figure 7.4, which appeared in the third volume of James Fergusson's *History of Architecture* (1876). Here the obvious alteration in scale (particularly of the central heads atop the tower) and the absence of the human figure suggest that the engraver has copied a photograph rather than the image being transferred directly onto the woodblock. This method of (re)transcription relied on a complex network of procedures and technologies, one that ultimately negates the visual stability of the image, even while it is founded on the established evidentiary value of the photograph. Emergence here implies a fundamental shift in the look and *feel* of the image but not necessarily its meaning.

Each of these spaces will have its own affective ecology: differential materialities and agential capacities that are connected with but not subservient to issues of visuality, representation and discourse. These dispersed ecologies are connected in a representational circuit through the content of the image but are not defined by this visual plane alone. Beyond this recognition of diffusion and multiplicity, what can an ecological perspective bring to our understanding of photography? One of the key points I want to make here is that despite this circulation, which sees the image enter into varied conditions of emergence and states of address (material, discursive, affective), this process does not necessarily lead to differences in signification. Instead, it is precisely through graphic deployments such as the book and multisensory experiences such as the lantern slide lecture that meanings begin to accrue and solidify. Emergence

FIGURE 7.4 Woodcut engraving based on John Thomson's photograph

Source: James Fergusson, 1876, History of Indian and Eastern Architecture, p. 680.

from this perspective does not automatically signify openness and may instead point towards a growing inertia bred through visual familiarity.

Emergence three: care and consumption

I mentioned at the start of this chapter that a linear narrative of ecological emergence would be misleading. Nowhere is this felt more clearly than in the affective economies of care and consumption encircling historic photographs in contemporary heritage practice. Here I follow Rodney Harrison's understanding of heritage as "an important language by which people globally express a sense of care for special objects, places, and practices", which in turn depends on a "series of contingent and emergent modes of caring for, valuing, and assuming an ethical stance towards the future" (2015, 39). Within this framework, the activities that surround "old photos", as Raphael Samuel described them (1994), are numerous. A partial list would include methods of storage, cataloguing and digitisation; exhibitions and other forms of display (promotional campaigns, event-based media, online narratives); domestic and vernacular engagements with the familial archive; reprints, reissues and private sales; local history publications documenting change in the built environment (e.g. "then and now" books and pamphlets); and the use of historic images on hoardings around building sites – an attempt to create visual continuity in the (usually) urban landscape (see Sterling 2020). Such practices are inherently nonlinear. Evoking a constant sense of return, they emerge from and respond to the specific material and visual demands of the photograph. At the same time, they are spatially and temporally distributed, with each new moment of encounter taking place within and against its own material-discursive ecology, expanding the photography complex in countless directions. Interestingly, Thomson's work has long been deployed in this way. As Samuel notes in *Theatres of Memory*, his photographs for the 1878 book *Street Life in London* provided "one of the sources of those atmospheric prints to be found of the walls of the 'period' pub" (1994, 316). While the visual record itself may appear fixed in such contexts (indeed, it gains value from this logic of stability), photographs are always buffeted by human and nonhuman flows of matter and meaning. To end this chapter, I want to briefly introduce two very different ways of thinking about such forces and their impact on the heritage field.

To open up this line of thinking, it is worth returning to the glass plate negative described previously, now held at The Wellcome Collection in London. As a unique material artefact, this image-object speaks to concepts of origin and singularity that may seem at odds with the various notions of ecology I have discussed so far in this chapter. While the nonhuman dimensions of wet collodion photography already undercut such a reading, this relational emergence did not end with the fixing of the image on the glass plate. The Library of Congress guidance on the "Care, Handling and Storage of Photographs" highlights the especially fragile status of glass plate negatives, which are prone to breaking when housed incorrectly. A variety of material interventions are recommended to preserve such collections, including high-quality pH 7–8.5 paper enclosures and specially made cabinets. Perhaps most importantly, climatic conditions must be stable, with glass plates ideally stored at 30–40% relative humidity to minimise decomposition and flaking. Such interventions are expensive and time consuming, and while The Wellcome Collection has been able to meet these conditions, not all photographic archives are so well cared for. This is also true for albumen prints, as a recent study in *Nature* highlights (Puškárová et al. 2016). Analysing two albumen photographs from the late nineteenth century using a mixture of spectroscopy and DNA extraction, the authors of this study found a range of microbial organisms, including bacterial microflora and various fungal species. Such findings are far from unusual and demonstrate the continued presence of nonhuman actants in relation to the materiality of historic photographs. These microscopic forces cause visible changes in the image – most notably foxing and an unevenness in tone. Whilst physically damaging, the patina of age that such processes engender is often valued for aesthetic reasons.

The Bayon image – whether as glass plate or albumen print – is managed in certain ways to protect against nonhuman forces of change and decay. As with most heritage artefacts, such processes are not viewed positively in practices of care and curation. Caitlin DeSilvey has recently put forward the notion of an "ecology of memory" to counteract this attitude (2017, 44). Rather than asking material artefacts to speak to a singular human past, this approach seeks to emphasise the more-than-human processes through which things "decay and disappear, reform and regenerate, shift back and forth between different states, always teetering on the edge of intelligibility" (ibid). This "entropic heritage practice" (ibid, 20) does not ignore human-based systems of value creation entirely but instead situates questions of care and responsibility in a broader ecology of matter, climate, organisms and other entities. As DeSilvey admits, this would require a fundamental shift in curatorial and interpretive practices, which "tend to come down heavily on the side of stasis and preservation" (ibid, 44). The fluid materialities of the Bayon photograph subvert notions of a static visual record, but they are still caught up in wider networks of valuing and attending to historic "things" in the world. This is a circular ecosystem – one in which processes of publicity, access and knowledge making (including the present chapter) emerge from and feed back into heterogeneous practices of care and attention.

To understand this feedback loop more clearly, we might consider the varied contexts in which Thomson's Bayon photograph currently circulates, whether in physical or digital form. Crucially, these two fields now bleed into one another in thought-provoking ways. The British Library, for example, included an albumen print version of the image from *The Antiquities of Cambodia* in its 2009 exhibition *Points of View: Capturing the 19th Century in Photographs* (see Falconer and Hide 2009, 69). In the process, a digital scan was made of the photograph, which is now available via the British Library online print shop in a variety of physical formats, including as a fine art poster, a framed picture or a greetings card (Figure 7.5). Here it is worth noting that the faded materiality of the image – a major concern for conservators – aligns perfectly with the nostalgic desires of the visual heritage economy, which tends to celebrate such signals of age value.

The role of algorithms in the consumption and circulation of photographs has received considerable attention from photography scholars in recent years. Zylinska, for example, uses the reconceptualisation of photography in "computational terms" as a jumping-off point to discuss broader nonhuman processes of photographic production and use (2017, 199). For Scott McQuire, the main significance of such socio-technical apparatuses is that they "foreground emergence as a general condition of the archive" (2013, 232). Writing from the perspective of memory studies, McQuire contrasts this sense of emergence with the foreclosure typical of traditional archival collections, where a "static, reproducible past" can be "seized objectively – which is to say, invested with the power of 'truth'"

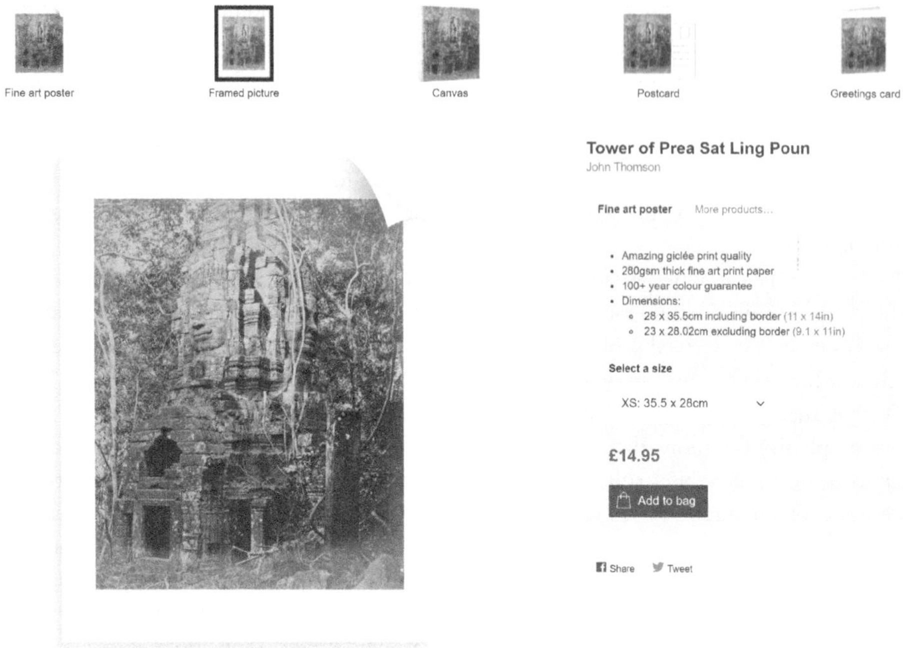

FIGURE 7.5 Screenshot from online British Library print shop: John Thomson, *Tower of Prea Sat Ling Poun*

(ibid). Others, however, are not so optimistic about the future of cultural production in the digital age. As Ted Striphas, one of the founders of "algorithmic culture studies", has argued, online algorithms have largely taken over the task of "reassembling the social" in the twenty-first century (2015, 406). This raises the question of who or what controls the flow of information and what degrees of openness truly exist in any moment of emergence. Ultimately for Striphas,

> what is at stake in algorithmic culture is the privatisation of process: that is, the forms of decision-making and contestation that comprise the ongoing struggle to determine the values, practices and artifacts – the culture, as it were – of specific social groups.
>
> *(ibid)*

We can see such processes at work in the various online platforms where The Bayon image may be found, from the Wellcome's own online collections to a scanned Xerox copy of *The Antiquities of Cambodia*, available through the *Gallica* digital library of the Bibliothèque Nationale de France (Figures 7.6 and 7.7).[2] The specific digital presence of the image in these discrete networks is brought together through the algorithmic capacities of Google and other search engines, but each iteration retains particular affordances, linked most clearly to resolution and the capacity for redistribution. One of the best-quality scans of the albumen print visible online, for example, is to be found on the website of Sotheby's auction house, where lot number 224 in a 2015 sale of "Travel, Atlases, Maps and Natural History" shows various images from *The Antiquities of Cambodia* (final sale price: £32,500),

FIGURE 7.6 Digital scan of photocopied page showing *Prea Sat Ling Poun* from John Thomson's *The Antiquities of Cambodia*, 1867

none of which can be saved or shared more widely. This differentiated digital emergence encompasses varied states of possibility and constraint, linked through the specific visual content of the photograph yet bound by distinct security protocols. The more-than-human ecologies of the algorithm are not necessarily more "open" than the material ecologies

FIGURE 7.7 Digital scan of *Prea Sat Ling Poun* from John Thomson's *The Antiquities of Cambodia*, 1867

Source: Bibliothèque Nationale de France.

found in practices of conservation. Qualities of stasis and emergence persist in both worlds, which are increasingly contingent on one other.

Building on the work of DeSilvey, the picture of emergence I have tried to put forward here sees material transformation as fundamentally generative rather than detrimental, even where wholesale destruction is concerned. We must acknowledge, however, that certain non-human processes are just as liable to close down possible meanings and routes of affect as are human interventions. Care in this sense becomes a constant process of negotiation between different forces of change – human, organic, atmospheric, algorithmic. It is not a question of "taking sides" but rather of accepting the relational and more-than-human politics within which heritage practices take place and unfold.

To understand the place of affect in these phenomena, we need to attend to a diverse range of scales, materials, agencies and forces within the contemporary photo complex. Intermedial and intersubjective, the emergent flows of heritage signification that coalesce around specific images and image collections are dependent upon and responsive to patterns of control and desire that transcend the human. This does not mean that human dimensions should be absent from our enquiries, only that the affective capacities of the image are not coterminous with social practices and personal mediations. The force of change often lies elsewhere and may prompt new forms of care, consumption and interpretation that upend familiar heritage ontologies.

Conclusion: what is heritage photography?

The intersection of heritage and photography cannot be reduced to the things photographed (historic sites, archaeological objects, ancient buildings, natural wonders) or the contexts in which "old photographs" are encountered (archives, books, museum displays etc.). Once created, any photographic image, whether digital or analogue, has an underlying heritage potentiality. Heritage photography, then, cannot be understood as a specific genre or aesthetic. Instead, it names a constantly moving ecosystem of image production, circulation and signification, made up of human and nonhuman parts, spatially distributed and inherently nonlinear. While some photographs may be created specifically to enter this ecosystem, heritage photography is not defined by these images. Furthermore, situated within a broader notion of heritage as a grammar of care, inheritance, value and ethics, heritage photography underlines issues of relationality and responsibility in image making and use. It is embodied and affective, capturing and conveying heterogeneous forces of change. Telling the story of photographic images and other heritage "things" in this way might help us to re-adjust our sense of attachment to a shared world. There is a tension here, however, between ecologies and narratives, with the former seemingly too open ended to allow for conventional articulations of the latter. Even the dispersed and contingent trajectories I have outlined in this chapter cannot take into account every more-than-human process and moment of emergence in the ecology of a single image. Something will always be left out. As Donna Haraway (via James Clifford) suggests, "we need stories (and theories) that are just big enough to gather up the complexities and keep the edges open and greedy for surprising new and old connections" (2015, 160).

What is at stake in the move from a representational view of photography to one that prioritises the ecological emergence of specific images and photographic assemblages? With each framing of "ecology", we find a different potential response to this question. For Schaefer, the key issue concerns the material formation of images beyond human agency and culture. Zylisnka meanwhile conceives of photography as a formative practice of life, one born of nonhuman entanglement and kinship. From the perspective of Edwards and Hevia, who are chiefly concerned with the broader networks in which photographs are put to work, the ecological points to nonlinear and divergent processes of photographic production, circulation and use. In the context of contemporary archives, this analysis must also attend to the algorithmic formations underpinning digital ecologies of visual consumption. Finally, viewing photographic images through the framework of DeSilvey's entropic heritage practice, we see how issues of material decay and preservation bring a multitude of nonhuman agencies to bear on questions of care and interpretation.

In mapping out just some of the many conditions and states of emergence in which we might encounter a specific photograph, this chapter has sought to determine the range of "ecologies" that may be of interest to heritage scholars moving forward. While more-than-human agencies have been stressed throughout, it is my contention that we also need to leave space for a continued engagement with the embodied and situated nature of heritage practice, understood here through the lens of affect. Issues of relationality and responsibility cut across all of the previous theorisations, as do questions of power and agency. Where does the force of change lie in any given situation, and who benefits from this affective potentiality? In addressing this line of enquiry, we need to acknowledge that processes of closure are just as likely to accrue around images as pathways of emergence, with meanings and possibilities solidifying just as quickly as they open up.

Heritage photography as I have described it here operates across all of these registers and thus represents a crucial area of investigation for the field as a whole. Building on this, I would like to suggest that photography is also a particularly fertile ground of research that may connect up critical heritage, creative arts practice and the environmental humanities. Felix Guattari's formative delineation of the "three ecologies" underlines the need to consider social relations, human subjectivity and the environment as a whole rather than in discrete silos (1989). In their more-than-human formation and nonlinear circulation, photographs intersect with each of these planes in ways that go far beyond representation. Heritage photography from this perspective may become a crucial component in fostering ecological principles that reflect the complex and always emergent relationality of things in the world.

Notes

1 Collodion is a solution of nitrocellulose (also known as guncotton) in alcohol and ether; when the solvents evaporate, a clear plastic-like film is formed (*Encyclopaedia Britannica*).

2 As the two images shown here indicate, the quality of an accessible online image will often depend on payment. Figure 7.6 is available for free, Figure 7.7 the fully reproduced photograph as available from the Bibliothèque Nationale imaging services. Financial barriers to access define the extent of many photographic ecologies.

References

Barad, K. 2003. "Posthumanist Performativity: Toward an Understanding of How Matter Comes to Matter." *Signs: Journal of Women in Culture and Society* 28(3): 801–831.

Barnett, A. 1990. "Cambodia Will Never Disappear." *New Left Review* 180: 101–125.

Bennett, J. 2010. *Vibrant Matter: A Political Ecology of Things*. Durham, NC: Duke University Press.

Boyer, M. C. 2003. "La Mission Héliographique: Architectural Photography, Collective Memory and the Patrimony of France, 1851." In *Picturing Place: Photography and the Geographical Imagination*, edited by J. M. Schwartz and J. Ryan, 21–54. London: I.B. Tauris.

DeSilvey, C. 2017. *Curated Decay: Heritage Beyond Saving*. Minneapolis, MN: University of Minnesota Press.

Di Giovine, M. 2009. *The Heritage-Scape: UNESCO, World Heritage, and Tourism*. Lanham, MD: Lexington Books.

Edwards, E. 2012a. "Objects of Affect: Photography Beyond the Image." *Annual Review of Anthropology* 41: 221–234.

Edwards, E. 2012b. *The Camera as Historian: Amateur Photographers and Historical Imagination, 1885–1918*. Durham, NC: Duke University Press.

Edwards, E. and Hart, J., eds. 2004. *Photographs Objects Histories: On the Materiality of Images*. London: Routledge.

Edwards, P. 2007. *Cambodge: The Cultivation of a Nation, 1860–1945*. Honolulu, HI: University of Hawai'i Press.

Falconer, J. and Hide, L. 2009. *Points of View: Capturing the 19th Century in Photographs*. London: The British Library.

Falser, M. 2013. "From Colonial Map to Visitor's Parcours: Tourist Guides and the Spatiotemporal Making of the Archaeological Park of Angkor." In *Archaeologizing Heritage? Transcultural Entanglements Between Local Social Practices and Global Virtual Realities*, edited by M. Falser and M. Juneja, 81–106. Berlin and Heidelberg: Springer.

Fergusson, J. 1876. *History of Indian and Eastern Architecture: Forming the Third Volume of the New Edition of the "History of Architecture"*. London: John Murray.

Gregg, M. and Seigworth, G. J. (eds) 2010. "Introduction: An Inventory of Shimmers." In *The Affect Theory Reader*, 1–25. Durham, NC: Duke University Press.

Guattari, F. 1989 [2014]. *The Three Ecologies*. Translated by I. Pindar and P. Sutton. London: Bloomsbury.

Haraway, D. 2015. "Anthropocene, Capitalocene, Plantationocene, Chthulucene: Making Kin." *Environmental Humanities* 6: 159–165.

Harrison, R. 2015. "Beyond 'Natural' and 'Cultural' Heritage: Toward an Ontological Politics of Heritage in the Age of the Anthropocene." *Heritage & Society* 8(1): 24–42.

Hevia, J. 2009. "The Photography Complex: Exposing Boxer-Era China (1900–1901), Making Civilisation." In *Photographies East: The Camera and Its Histories in East and Southeast Asia*, edited by R. C. Morris, 79–119. Durham, NC: Duke University Press.

Kennedy, H. G. 1867. "Report of an Expedition Made into Southern Laos and Cambodia in the Early Part of the Year 1866." *Journal of the Royal Geographical Society of London* 37: 298–328.

Massumi, B. 2015. *Politics of Affect*. Cambridge: Polity.

McQuire, S. 2013. "Photography's Afterlife: Documentary Images and the Operational Archive." *Journal of Material Culture* 18(3): 223–241.

Ovenden, R. 1997. *John Thomson (1837–1921): Photographer*. Foreword by Sir David Puttnam. Edinburgh: National Library of Scotland, The Stationery Office.

Paris Photo. 2018. *Albumen Print*. Accessed 12 August 2018. www.parisphoto.com/en/Glossary/Albumen-print/.

Puškárová, A., Bučková, M., Habalová, B., Kraková, L., Maková, A. and Pangallo, D. 2016. "Microbial Communities Affecting Albumen Photography Heritage: A Methodological Survey." *Nature Scientific Reports* 6: Article number: 20810.

Reilly, J. M. 1980. "The History, Technique and Structure of Albumen Prints." *AIC Preprints*, May: 93–98. Accessed 14 August 2018. http://albumen.conservation-us.org/library/c20/reilly1980.html.

Samuel, R. 1994. *Theatres of Memory: Volume 1: Past and Present in Contemporary Culture*. London: Verso.

Schaefer, W. 2017. "Photographic Ecologies." *October* 161(Summer): 42–68.

Sterling, C. 2016. "Mundane Myths: Heritage and the Politics of the Photographic Cliché." *Public Archaeology* 15(2–3): 87–112.

Sterling, C. 2020. *Heritage, Photography, and the Affective Past*. London: Routledge.

Stewart, K. 2007. *Ordinary Affects*. Durham, NC: Duke University Press.

Striphas, T. 2015. "Algorithmic Culture." *European Journal of Cultural Studies* 18(4–5). 395–412.

Szarkowski, J. 1992. *Photography Until Now*. New York: Museum of Modern Art.

Thomson, J. 1866. "Practical Photography in Tropical Regions." *British Journal of Photography*, 24 August: 404.

Thomson, J. 1867. *The Antiquities of Cambodia: A Series of Photographs Taken on the Spot with Letterpress Description*. Edinburgh: Edmonston & Douglas.
Thomson, J. 1891. "Photography and Exploration." *Proceedings of the Royal Geographical Society and Monthly Record of Geography* 13(11): 669–675.
Thomson, J. 1907. "Geographic Photography." *Scottish Geographical Journal* 23(1): 14–19.
Thrift, N. 2008. *Non-Representational Theory: Space, Politics, Affect*. London: Routledge.
Winter, T. 2007. *Post-conflict Heritage, Postcolonial Tourism: Culture, Politics and Development at Angkor*. London: Routledge.
Zylinska, J. 2017. *Nonhuman Photography*. Cambridge, MA: MIT Press.

8

AFFECTIVE ENCOUNTERS IN MUSEUMS

Marzia Varutti

Introduction

I had decided to visit the African Gallery at the British Museum. As I entered the main lobby, I headed towards the back, zig-zagging between tourists, eager to get through the crowds quickly. But in crossing the lobby, I instinctively turned my gaze to my right, where I was met with something that attracted my attention. Without rationalizing or questioning, I diverted my steps towards it.

The exhibition room was small and intimate, dimly lit and relatively quiet. As I entered, I could feel I was leaving behind me the noisy flow of tourists following their hurried trajectories straight into the heart of the museum; this smaller exhibition room on the side of the main entrance went unnoticed to many. There, alone, in a glass box at the centre of the room, sat the life-size bronze sculpture of a cat (Figure 8.1).

The four sides of the room around it were covered with text and images, a visual noise against which the dark, slender bronze figure cut a stark contrast. The sculpture conveyed poise and an effortless, relaxed alertness. Through hollow eyes, it seemed to gaze hieratically into timelessness, as if knowing it was destined to travel across thousands of years. Its profiles were neat, essential; it called for respect, intimated dignity, imposed its alterity. This was no pet. A God, rather – a Divine Cat revered in Ancient Egypt. It seemed to expect and to beckon admiration, and indeed, it felt appropriate to admire it. As much as the sculpture claimed my attention, it was straining to receive and hold such an intensity of purity of lines and elegance: its beauty was almost painful to look at. Too much to take in, at one single moment. I took respite in the surrounding texts. Reading some, browsing others, yet feeling slightly uncomfortable: turning my back to the sculpture in order to read the texts seemed not right. I felt compelled to return to it, forgo explanations, and just let myself become absorbed into its intense beauty. I knew nothing about the sculpture, but I didn't need to in order to fall under its spell. I eventually took the time to read the texts around it and became even more enthralled by this object, but in the process, it had indeed become *an object*. The texts did bring me to look back at it, now projecting on it my newly acquired knowledge, almost searching for traces of what I had read on the bronze surface accretions and patina. Now I was observing more than feeling.

FIGURE 8.1 The Divine Cat

Photo: Oxyman/CC BY-SA 3.0. https://creativecommons.org/licenses/by-sa/3.0/.

I have since forgotten most of the content of those texts. What I remember, from that visit on a day many years ago, is the cat, and how I felt in its presence.[1]

This chapter explores the concept of affect in the specific context of museums. I am interested in pondering some of the multiple affective dimensions of museums and in asking why it might be appropriate for museums to endeavor to integrate such affective dimensions in their work and mission. Could this open up new kinds of engagements with visitors? Might "taking affect seriously" recast what is considered "knowledge" in museums? More broadly, might a focus on affect lead to fresh perspectives on human and non-human relations?

I develop my arguments in several strands. First, I reflect on the methodological implications of working on and with affect. Second, I consider what affect entails in a museum and ponder how it relates to meaning-making and agency. I discuss some affective dimensions of museums, such as architecture, objects in the storage and on display, the exhibition environment and affective words. I then develop arguments in support of an active engagement with affect in museums and consider the implications and potential of such a renewed attention for affect – in museums and beyond. Last, I reflect on how allowing ourselves to be *affected* deeply denotes our humanity and at the same time it opens up for more equanimity and empathy in our relations with the non-human world, thus carrying significant ethical connotations.

Affect as method

As a concept, affect denotes an altered state of being, generated by a complex set of interacting factors (sensations, things, landscapes, atmospheres, memories, emotions etc.). Anthropologist Kathleen Stewart (2007, 1–2) talks of "ordinary affects" as "the varied, surging capacities to affect and to be affected that give everyday life the quality of a continual motion of relations, scenes, contingencies and emergences". She notes that "[*affects*] work not through 'meanings' *per se* but in the way that they pick up density and texture as they move through bodies, dreams, dramas and social worldings of all kinds" (Stewart 2007, 3). Emotions, sensory experiences, memories are then the stuff affect is made of, the channels through which affect makes itself manifest to the perceiver – and therefore they are the focus of a study of affect. It follows that the reflections presented in this chapter are largely based on autoethnography (e.g. Butz and Besio 2009; Ellis and Bochner 2000). I have drawn heavily on my own subjective responses to museum objects and environments. My methodological approach consisted in "letting myself become affected" and paying close attention to the process. My thoughts, emotions, sensory and embodied feelings are both my "object of study" and my method: they are what I "observe", "describe" and "analyze", but they are also instruments to delve deeper into my experience of being affected, pathways into developing a sharper awareness of myself as a human being relating to – *perceiving* and *being perceived by* – a part of the world, which happens to be a museum context. Rather than a limitation, I propose to think about subjectivity here as a research tool, whereby situatedness enables a form of "embodied objectivity", in line with Donna Haraway's conceptualization of situated knowledge as an antidote to "unlocatable, and so irresponsible, knowledge claims" (1988, 581, 583).

This methodological approach recognizes the need to challenge potential dichotomies between subject and object in studies of affect (in line with Marianne Liljeström 2016). One way to do this is to explore (also methodologically) the inherently dialogical and relational character of affect: we become affected by (and through) an encounter with someone or something else. Even though the site of *perception* of an affective state is necessarily personal (the individual), the site of *production* of that state is "transpersonal" in that it arises from the encounter with the world, mediated by embodiment and mutual permeability to affect. In Kathleen Stewart's words (2007, 128), ordinary affect is "transpersonal or prepersonal – not about one person's feelings becoming another's but about bodies literally affecting one another and generating intensities". My affective encounter with the Divine Cat illustrates this vividly: the bronze sculpture commanded my attention; it got under my skin and sent shivers down the spine. Those affective responses arose in and through the embodied encounter with the sculpture – a mutual "fringing" that challenges preconceptions and false certainties about subject/object and the directionality of exchanges.[2]

My methodological approach develops in three moments. First comes the "field observation" in museums, which sometimes involves taking notes and photos *in situ* and other times not, because being affected can neither be planned nor anticipated. Second, I revisit the experiences of being in the museum working with memory ex-post or, when available, by examining photos and notes. I have used memory and the practice of "trying to remember" as research methods: re-evoking memories enables me not only to reflect upon the content of the memory but also to interrogate the reasons those memories stayed with me over time. One cannot reflect about what one has forgotten, but it is still possible to reflect on how that which has been saved from oblivion relates to the overall experience – what makes some memories stick and others slip away? Could affect be part of the answer?

My third methodological step entails the transposition of my responses into a text. I think of this writing process as part of the methodology because in writing about affect, one cannot but write *with* affect, that is, write under the condition of *being affected*. Rather than aiming for "objectivity" and "verifiability", I endeavor to communicate my affective experiences and in that way also to possibly affect some readers. I open up for the possibility that the text may produce feelings of empathy in the reader so that my own personal reactions and reflections become, if not verifiable, at least plausible. In writing about affect, I am confronted with a "double estrangement": first, the challenge of verbalizing feelings, and second, of having to do so in a language that is not my mother tongue. This double estrangement develops around an inherent paradox: affect eschews meaning, it resists (or exceeds) interpretation and it is impervious to verbal or textual description. Yet words we must use, and stories we must tell, in trying to convey a situation that we experienced as affective. Despite its inherent challenges and paradoxes, the endeavor to write about affect, and with affect, seems worthy, as it opens for the possibility to engage in a slightly different way of thinking about and producing an academic text.

Affect and meaning

Museums are interesting contexts for discussions of affect because they subsume a tension between the museum's *plan to affect* visitors (through careful exhibition design and curation) and the fact that *affect cannot be planned for*. Museums and notably exhibitions are by definition designed, curated spaces where curatorial agendas, aims and intentions inform the organization of space and (at least in part) the experience of visitors. Texts and words have long been preferential tools for communication in museums. Exhibition titles, explanatory panels and labels have been the "traditional" tools for the production and dissemination of knowledge and meaning in museums. The centrality of texts in exhibitions is, however, being gradually displaced by non-textual elements such as memories, emotions and multisensory experiences, which contribute to create richer, more complex and more individualized – and therefore less predictable – interpretative frameworks (Varutti 2020). In one sense, affect, understood as an altered state, might be conducive to an enhanced attention for the textual messages and narratives in exhibitions. Here, however, I wish to take a slightly different analytical and argumentative angle in order to consider the relevance and potential of affect in its own right and for its own sake – irrespective of museum narratives.

Affect appears to be intertwined with a series of factors – sensory perception, memory, imagination, emotions – that mutually interact in establishing a relation between the perceiver (a museum visitor for instance) and the museum environment (or an element of it, such as an object, a sound or a smell). In this sense – and as illustrated by my own encounter with the Divine Cat – affect is (or can be) essentially the result of an encounter; its nature is fundamentally relational. This point emerges clearly in Navaro-Yashin's (2009, 14–15) words: "neither the ruin in my ethnography, nor the people who live around it are affective on their own or in their own right, but both produce and transmit affect relationally".

But if affect can be the result of an encounter, isn't that the same for meaning? How does affect then relate to meaning? For a museum visitor, usually the process of meaning-making occurs at the interface between the set of interpretative tools made available by the museum (museum texts; captions; and other contextualizing material such as audio-video material, digital platforms etc.) and another personal and mostly tacit set of interpretative tools which draw on the individual's inner sphere (personal memories, fantasies, moods etc.) (Hooper-Greenhill 2007, 2013). Often meaning emerges at the intersection of these two sets of interpretative tools. For instance, fragments of an object biography read in a label may engender connections with personal past experiences.

In other instances, a visitor might have a strong response to an object *irrespective of* or *prior to* accessing the museum's set of interpretative tools. An example of this is surprise. Unlike most human emotions, surprise is "neutral": it is neither positive (such as joy) nor negative (sadness); it can be both or neither. Either way, surprise is a powerful state of enhanced attention – and one that is often experienced in museums, as exemplified by children seeing a dinosaur skeleton in a natural history museum for the first time. Indeed, surprise and wonder have been at the heart of the very idea of the museum ever since the "cabinets of curiosities" of the Renaissance. Crucially, surprise often precedes, or stands independent from, explanations or interpretations. Surprise creates a "hook" to which meaning can be fruitfully attached.

The concept of affect is useful precisely in casting light on the experiences and encounters that precede interpretation, rationalization, explanation and meaning-making – though affect can influence the latter deeply. In studies of materiality and person-object relations, much attention has been rightly devoted to interpretation, symbolic and semantic meanings. However, as Kathleen Stewart points out (2007, 3) "affects are more directly compelling than ideologies, and more fractious, multiplicitous, and unpredictable than symbolic meanings". I suggest that the emphasis on meaning might have obfuscated the role that affect plays in shaping human and non-human relations. Anthropologist Paul Roscoe illustrates this point. Writing about art objects, Roscoe (1995, 9) cautions that "[a]uthors seem to allow that art has both affective impact and semanticity, but they collapse these properties under the polysemic term, 'meaning'". Rather, Roscoe continues, "art's affecting qualities are by no means secondary to whatever intellectual properties it may embody". In his ethnographic study of the Yangoru Boiken communities in Papua New Guinea, Roscoe focuses on *ka nimbia*, spirit houses considered powerful and dangerous. Roscoe (1995, 15, emphasis in the original) explains that

> it was anticipation of these affective repercussions that motivated the construction of a *ka nimbia*. Its creators' aim was not to communicate, to themselves or others, some semantic message along the lines of, "We are powerful and dangerous"; rather, it was to force themselves and others to *feel* this power and danger.

Here it appears that the experience of being affected is meant to precede, and possibly substitute for, the process of cognitive "making sense". In more general terms, affect and meaning appear to be deeply intertwined. Whilst affect can result from interpretation, it can also prescind it, and conversely whilst meaning can well be devoid of affect, philosophical and cognitive studies (among others) increasingly point at the connections between reason and affective factors such as emotions (see Nussbaum 2003; Damasio 2018; among others).[3]

The concept of agency, as elaborated upon by anthropologist Alfred Gell in his seminal *Art and Agency: An Anthropological Theory*, can provide further clues on the relation between affect and meaning. Affect and agency may appear to be interrelated and similar: both are useful concepts in exploring and explaining human-thing relations, and both predicate that things (or elements of the non-human world at large) can produce effects on the perceiver. Yet a distinction between the two can be made. In the case of agency, such effect entails an external manifestation, a "consequence" that translates into action: agency is about "doing something" (Gell 1998; Latour 2005). Conversely, in the case of affect, the effect produced on the perceiver is not necessarily expressed or made manifest to the external world. It may engender an intuition, a memory, an emotion, a physical sensation or other inner states that may or may not translate into action. Affect thus is not so much about doing as about *being*: as inhabiting a state of being whose origin lies in an experience or an encounter with an external stimulus. Whilst the perceiver might well be aware of such a changed internal state (that is, of being affected), it might be difficult to understand exactly in what ways one is affected, what exactly has induced this or the reasons we are affected in a certain way. Rationalization, explanations and possibly action come after (if at all) the realization and experience of being affected.

These considerations have potentially relevant implications for museums because museum texts and narratives are primarily about meaning and interpretation. Texts are often (consciously or not) carriers of ideologies, and museum pedagogy is often concerned with providing explanations and unfolding objects' symbolic meanings. The interpretations proposed by museum texts may or may not intersect with visitors' affective experiences – when they do, they can enhance, depress or in other ways alter each other. The point I wish to stress is the need to understand affect as something *independent* from explanations, symbolism, ideologies and meaning. Whilst affect may interact with the latter as part of meaning-making, it can also provide a different and self-contained experience. It follows that affect can hardly be leveraged by museums in their communication strategies: although, as mentioned (and as I will discuss subsequently), museum environments may be designed in order to engender affective responses, these can neither be fully anticipated nor controlled. And it is precisely because of this that affect retains its power.

But what, then, can museums do with affect? I suggest that a precondition to engage with affect is first and foremost to become aware of the multifaceted affective potential of museums. This awareness might not imply revolutionary changes in museum practice, yet its effect might be subtle, expressed in a different way of *thinking* and *feeling* museums, their collections and environments. It is then a matter of allowing for that affective potential to unfold, allowing for environments and objects to be experienced in unpredictable, non-normative ways and on levels other than (and independent from) the cognitive elaboration of meanings conveyed by texts. It's a matter of allowing for multifarious forms of encounters between objects, environments and visitors – encounters that are triggered not by interpretation but by intuition, creativity, individuality and that, as a result, might ultimately turn out to be more personal, democratic and ethical.

Affective museums

All museums, at their best, are affective. Whilst this might not always be explicit in their mission statements, museums have aimed at affecting visitors since their very origins: Renaissance cabinets of curiosity, as mentioned, were meant to affect viewers through the wondrous character of their contents. Later, in the nineteenth century, museums (especially folk and open-air museums) and exhibitionary complexes (such as world fairs and great exhibitions; see Bennett 1995) were loaded with emotional and sensory stimuli meant to instill, envision and materialize the ideologies of nationalism, romanticism and empiricism (DeGroff 2012; Stoklund 2003). The historical depth of affective experiences in museums reinstates that affect is a portal to connections between persons and things – and those same connections lie at the basis of everything that museums do and aim to achieve. Affect is therefore nothing radically new in museums, yet it is noticeable that affect has become, in the last decade, a prime area of exploration and experimentation in museum practice as well as theory (e.g. Baker 2013; Dudley 2012, 2013; Gregory and Witcomb 2007; Smith, Wetherell and Campbell 2018; Witcomb 2015). In part, this is due to historical and museological reasons linked to the rise and decline of the centrality of texts and narratives in museums (Varutti 2020), but it can also be seen as a refraction of a broader surge of interest for affect across the human and social sciences (Brennan 2004; Clough and Halley 2007; Frykman and Povrzanovic Frykman 2016; Stewart 2007; Thrift 2004).

Since affect is not a property held by objects or environments but the result of an encounter, one can think of museums as venues that *enable* encounters with and through objects, persons, environments; in turn, these encounters have the potential to engender affect. As mentioned, affective encounters may occur where one doesn't expect them, and each individual may respond (in different ways) to something that others consider insignificant. That said, it is possible to identify some museum aspects that particularly lend themselves to triggering affective encounters. I consider them in the following.

Affective architecture and museum environments

Since the origins of museums, the architecture of these institutions has been designed to be affective (similarly to other sites of knowledge and authority such as parliaments, law courts and national libraries). The earliest and most conventional museum architecture, reminiscent of Greek and Roman classic temples with colonnades, rising staircases and imposing gateways, was meant to affect the visitors approaching from the exterior, suggesting connections with acts of reverence and the presence of the sacred. Architectural styles might have changed over time, but their affective aim has perdured. Notably, since the 1990s, there has been a marked return of interest in museum architecture as part of larger processes of museums repositioning themselves as prominent social institutions in the urban fabric. The *Louvre* in Abu-Dhabi, the *Museum of Tomorrow* in Rio de Janeiro and the *M+* in Hong Kong are just a few examples of twenty-first-century museums that are defined by their stunning, purposefully affective architectural designs.

The affective potential of space and forms continues to unfold within the museum's walls; for instance, visitors may be affected by exhibition environments. This is confirmed by anthropologist Yael Navaro-Yashin. Elaborating on Deleuze's theory of affect, she writes (2009, 12) "[t]he reference-point for affect (which used to be, singularly, subjectivity) has been radically altered and multiplied in this approach, making it possible to read many other things, such as space and the environment, as affective". The affective potential of museum environments is probably most evident in the case of former residences of public figures subsequently turned into museums; here, the display environments may be imbued with the charismatic aura of former residents. For instance, architecture historian Nuala Hancock (2010, 2012) has written movingly about the former home of Virginia Woolf at Monk's House, leaving little doubt either on the affective potential of this house (now turned into a museum) nor on the affected condition thanks to which the author was able to compose such lyrical texts.

Affective museum objects

Whilst both museum architecture and interior environments may be designed with the specific purpose to affect visitors, this is not necessarily the case for museum objects. Indeed, most artifacts in museums were not even intended to be publicly displayed. In the words of Susan Vogel (1991, 1) "[a]lmost nothing displayed in museums was made to be seen in them". Yet (or possibly because of this) objects may be potently affective.

Objects do not necessarily need a specific curator-designed environment in order to affect us. This is not least conspicuous in museum storage. Have you ever tried to walk into a museum storage at night and felt somewhat uncomfortable? The discomfort might be due to a lack of visual control – as objects lie silent in the darkness of shelves, boxes and drawers – and the possibility of an unforeseen, perhaps unpleasant encounter. Because of this unique environment, the encounter with an object in the storage can be quite powerful: we can be taken by surprise or perceive the object's potency as silent presence in the storage, a potency that owes nothing to curatorial intervention but rather is inherent to the materiality of the object and the immediacy of the encounter. Conversely, for objects on display, the glass case (or rope, raised platform or cone of light) are all elements that provide a framework: they are saying "look at this" and at the same time "keep some distance". The display is normative: it attempts to direct the visitors' attention, and ultimately it may attempt to also engender affect. This may or may not work.

Affective words in museums

In his essay "The Storyteller", Walter Benjamin (2007 [1968]) cautions against the risk that an over-concern with information and meaning might bring us to lose "the ability to exchange experiences" (2007[1968], 83), notably in his case the experiences generated by storytelling. Benjamin notes (2007[1968], 89) that "no event any longer comes to us without already being shot through with explanation". But whilst explanations and information are tied to time – they are relevant here and now – a story is not. A story retains a dimension of performativity; it engenders an experience with a potential to generate memories, wisdom and imagination over time. This is what makes a story timeless:

> The value of information does not survive the moment in which it was new. It lives only at that moment; it has to surrender to it completely and explain itself to it without losing any time. A story is different. It does not expend itself. It preserves and concentrates its strength and is capable of releasing it even after a long time.
>
> *(Benjamin 2007[1968], 90)*

I take Benjamin's words as an invitation to explore the affective potential of words beyond information and explanations. Affect can indeed be generated by encounters with words – poetry provides examples of this. In museums, words are mostly used to carry interpretation and explanations – and sometimes also storytelling in Walter Benjamin's sense. Performative, experiential storytelling (as well as poetry) can find its place in museums, too. Hitherto, research on affect in museums has mostly focused on objects and materiality (e.g. Frykman and Povrzanovic Frykman 2016), and less attention has been paid to the affective potential of words. This may emerge when words are not used as auxiliary to information and explanation (as in the case of labels and captions, which can produce factual, dry categorizations and characterizations of objects, places, events and people). Rather, words can become affective when they are uttered as performance, as a form of communication and artistic expression in its own right: they are (and offer) an *experience*; they can conflate emotions, memories, imagination and sensory feelings – words can send shivers down the spine.

Could museums become more aware of (or just curious about) the affective potential of the environments, objects and words they use? And if so, what is there to gain? Much, I argue.

Affect as knowledge

In the opening paragraphs, I recounted my encounter with a special museum object – the Divine Cat from Ancient Egypt – and how I had been able to establish a connection with that unknown object, a connection informed by a mix of curiosity, wonder and aesthetic appeal, which triggered a sustained, focused, almost spellbound-like form of attention for an exhibit, something I have only occasionally experienced in a museum. What makes that connection noteworthy to me is that it was independent from the information provided by the exhibition. The display layout – a single glass box at the center of the room, bathed in strong and sharp light – certainly helped in drawing my attention to the object (would this artifact have caught my attention had it been placed at the bottom of a crowded wall case? I wonder). The display had been designed in order to facilitate an encounter: it presented the object, quite literally, in its best light at the same time as it directed visitors' attention towards it. In reflecting on my affected response, I cannot therefore discount the role of curatorial intervention. However, the display layout in itself (as opposed to the exhibition as a whole) did not do more than *invite* the encounter; it was up to me to read the panels and take in the other information presented in the exhibition – which would eventually alter my relation to the object. I want to suggest that the connection I established with the Divine Cat was special precisely because my instinctive, affected response had not yet been filtered through the exhibition's edited message (or, to use Walter Benjamin's words, it had not yet been "shot through with explanation"

(2007[1968], 89). Upon entering the room, I was under the spell of the Divine Cat's aura: the calm, silent power of its presence claimed my undivided attention. After reading the exhibition texts, my perception changed: the Divine Cat had become an illustration – of a civilization, a place, a time, a bronze casting technique, an aesthetic taste … it had become a "mere" sculpture, a collectible, a colonial trophy, a museum object, an artifact used by some humans in order to tell stories about other humans. I do not wish to diminish the importance of explaining, interpreting and contextualizing this object; after all, this only bears witness to its uniqueness. But I am interested here in what the exhibition *did* to my encounter with the object. The analytical texts segmented my attention and directed my gaze: I was invited to examine the bronze casting technique, to observe the x-rays revealing the internal structure, to zoom in on the accretions on the bronze surface. The Divine Cat had become a sum of its parts, an object of analysis, and the magic of its mysterious presence was lost.

Increasingly, museums aim to reach out to their visitors not only through cognitive approaches (by providing information and meaning) but also through complementary, parallel approaches. Sensory approaches are one example (eg. Edwards, Gosden and Phillips 2006; Levent and Pascual-Leone 2014), and a renewed engagement with emotions is another (e.g. Smith, Wetherell and Campbell 2018; Watson 2013). Would it be possible to create opportunities for meetings with objects (or environments) where time, space and legitimacy are given to affective encounters on the same footing as contextualization?

Placing the importance of emotional and personal experience on an equal level with contextualization and interpretation opens the door to an understanding of knowledge not only as being *located* in the body but also as *emerging* from the interaction of the human body with other bodies as a result of the porousness, or leakiness, of the human body to the world. In anthropologist Tim Ingold's (2012, 438) words

> Things can exist and persist only because they leak: that is, because of the interchange of materials across the ever-emergent surfaces by which they differentiate themselves from the surrounding medium. The bodies of organisms and other things leak continually; indeed, their lives depend on it.[4]

The inter-relations between the human body and the environment have been the object of extensive study in the fields of philosophy (ever since Merleau-Ponty's *Phemomenology of Perception*, 2013 [1945]), neuroscience (e.g. Damasio 1999) and anthropology (e.g. Ingold 2000), among others. Yet, museum studies have only started to explore this: whilst much attention has been devoted to representations of the body in museums (e.g. Hjemdahl and Opdahl Mathisen 2017; Qureshi 2011; Sandberg 2002), studies of the body as a site and instrument of knowledge are rarer (exceptions being e.g. Helen Rees Leahy's study of how museums have shaped bodily practices over time; 2012).

As mentioned, opportunities to engage the senses and emotions in museums are more and more widespread. But are such experiences recognized as knowledge on the same footing with the contextualization provided through curatorial intervention? Should they be? Is "knowledge" the most suitable term to capture the essence of these experiences? Is it possible to integrate affective potential in curatorial thinking so that visitors' personal experiences and curatorial intervention are set into dialogue and can thus build on each other? Once the opportunity for a sensory, emotional or otherwise affective experience is produced, how can both museums and visitors further engage with it and build on it? As argued previously, affective responses cannot be directly induced. Curators have no real control over affect – what will affect visitors and in what ways. But is it possible to actively enable affect by becoming more aware of the affective potential of museum environments and objects? This could involve, for instance, display techniques and layouts that enable a personal encounter with the museum environment or objects, allowing for the possibility (the space and the time) to become aware of one's responses (memories, emotions, feelings) *before* the attention gets caught by something else happening in the exhibition, such as contextualization and meaning.

Allowing for affective encounters in museums entails a set of specific curatorial skills, as curators need to be able to conceive a museum space that is enabling such encounters. In order to do this, curators most likely need to let themselves become affected. I develop this point in the next section.

Affected curators, affected (museum) objects

If, as suggested, one approaches affect as the outcome of an encounter, and if we postulate that we experience the world as a series of encounters, we realize that as much as we (humans) encounter the Other (human or non-human), so the Other encounters us. It is then possible to take a non-anthropocentric position in order to open up for an understanding of the world that includes relations between humans and non-human as co-creators of reality. Stretching such understanding further through our imagination, we might ponder (or just conceive the possibility of) non-human perspectives: what might the world look (feel, smell, taste, sound or vibrate) like when experienced from a non-human perspective?

Let us consider a museum context. Admiring a Greek sculpture in an exhibition entails exposing that sculpture to human presence. This has mundane, concrete implications (in terms of exposure to light, air, temperature and bacteria, among others) as well as more subtle implications: what does our attention do to museum objects? What about the affects created by the new relations that objects forge between themselves, for instance, through the material and chemical processes of decay taking place in the storage or through displacement, whereby "ruptures can be transgressed and connections forged" (Dudley 2017, 40). And what if we dared taking the object's perspective (as Dudley does, among others, 2017) and begin to ask: what does it feel like to have been closed in a box for a hundred years? What literally does it feel like to be touched by inquisitive human hands (many of them, covered by plastic) or not touched at all? Can the act of displaying or the act of seeing and (not) paying attention affect the object? These questions gain saliency when we consider objects that were not made to be displayed in a museum or were expressively meant *not* to be publicly seen. What does the display do to the object in those cases? What did it do to the Shuar and Achuar shrunken heads on public display at the Pitt Rivers Museum to be closely examined and handled by generations of anthropologists, curators and conservators and to be pointed at, or be laughed at by, hordes of school children every day?

Let us consider the *malanggan*, wooden ritual sculptures from Papua New Guinea believed to carry the life force of a deceased person. *Malanggan* are meant to be seen only for a few hours or days before being destroyed or thrown away (Strathern 2001, 259–260). What does it do to these ritual objects to be preserved in museums "for eternity", publicly displayed in museums and galleries as art objects? What does all this do to their sacred character and ritual power? These questions do have museological relevance. By raising them, can we begin to imagine the world from a different, non-human perspective? And even so, why should we bother? Curators who let themselves become affected by their encounters with objects in museums will be in a privileged position to begin to engage with these questions. Affect can then become part of curatorial practice, a tool to enhance understanding of the collections. Again, affect can be a method and being affected a vantage position from which to experience the world, as I suggested in the opening of this chapter. More broadly, affect can operate a potentially powerful shift in the way we, museum scholars, professionals and visitors, can come to think about, relate to and feel not only museum collections and museum environments but also the non-human world at large. I elaborate on this in the next section, where I expand the frame of my reflection beyond museums and suggest some considerations that apply more generally to the opportunities that open to us, humans, when we let ourselves become affected by encounters with the non-human world, specifically objects.

Affect and what it means to be human

The previous line of questioning might seem outlandish, but I suggest that these questions are good food for thought. It takes attention, empathy and imagination to begin to answer those questions. When, instead of experiencing affect strictly on a self-centered level, we begin to ask questions about the world that surrounds us and how we might affect it, affect takes up an ethical dimension. Asking the previous questions, I suggest, is a worthwhile exercise because the capacity to stretch our understanding of the world through attention, empathy and imagination is a measure of our own humanity.[5] This should not be read as yet another rephrasing of the dichotomy between the human and non-human. Rather, I want to suggest that affect – as a condition, an opportunity, an experience – might not be exclusive to human beings, and we can gain this insight precisely because we can thrust our imagination beyond our humanity-bound experiences and perspectives.

Affect, then, becomes a way to engage with the world, to recognize that we are not insensitive to objects and indeed to anything in the worlds we inhabit and share. Letting ourselves become affected entails a specific kind of attention as well as capacity for empathy and imagination. We pay attention to how our own modes of being (thoughts, feelings and bodily sensations) are responses to, and impact, the outer world. There is reciprocity in affective encounters; as mentioned, affect is essentially relational. Attention to our affective responses can then be extended to humans and non-humans through empathy. Singer and Klimecki (2014, 875) define empathy as "the capacity to share the feelings of others. … Empathy makes it possible to resonate with others' positive and negative feelings alike". Empathy – the ability to shift from a self-centered perspective to adopt (however superficially, partially and for a fleeting moment) the perspective of the Other – is considered a distinctive trait of humanity, indeed, a human virtue. But is it possible to be empathic with the non-human world? We do it more or less effortlessly with animals, but what about objects?

Pondering how the Other (the animal, the plant, the object) may experience the world – that is, practicing empathy through affect – is a humbling and illuminating exercise. But it is also a skill, something one learns and hones, a muscle that needs to be exercised. This is not without challenges: when we let ourselves become affected and empathic and try to make this shift away from a human-centered perspective, the point is *not* to project on the Other our own human experience. Rather, it is about distancing ourselves for a moment from what we know and experience and taking another direction; it's about turning our back on what is familiar and jumping into the unknown. As Karen Barad (2012, 164) points out

> living compassionately requires recognizing and facing our responsibility to the infinitude of the other, welcoming the stranger whose very existence is the possibility of touching and being touched, who gifts us with both the ability to respond and the longing for justice-to-come.

We shouldn't be lured into thinking that we can ever get to know what it means to be Other. We should pause in front of Otherness and recognize it as we recognize a solid materiality in front of us: it is precisely the impossibility to move through it that brings us to know it as solid. Accepting the Otherness of the non-human is a portal into a form of understanding that begins with respect – respect of its alterity, of its difference. That means not trying to force it into our epistemological boxes, to make it useful for us humans, to project our values on it, to predicate its future – but allow it to be, with equanimity. Allowing, for instance, the alterity of the Divine Cat to radiate rather than disregarding it or coercing it into classificatory schemes and narrow perspectives that project and impose human interpretation. We need to preserve alterity (together with our ability to respect it) if we want to live in a world that is rich in diversity, where things can be whatever they are and not be reduced to tools serving human needs. Equanimity helps us to accept the limitations of our thoughts.

But we can imagine. That step of distancing from what we know, from our humanity, is essential. It is this attempt to shift perspective that counts. This attempt to imagine the Other is a crucial step in order to recognize that there may be other-than-human experiences and other points of view. Empathy with the non-human worlds demands that we acknowledge and respect radical alterity, full-blown Otherness and that we try to *imagine* what that might be like. Imagination is our key ally in this venture. We must draw on imagination about non-humanity if we want to cast our own humanity, and the condition of being affected is what triggers and enables empathy and imagination.

Affective encounters in and beyond museums

Affect and the related ability to expand our awareness of being variously connected with other human and non-human worlds can help us forge an ethics of human and non-human relationships that is not anthropocentric. Affect entails that we open a window of attention towards our entanglements with the world. Affect, in this sense, is a mode of life, a way of being *in, of* and *with* this world, as well as a way of making sense of it from a human perspective that is not exclusive or excluding – a human perspective that gains saliency and nuances precisely because it is engendered by a connection with the non-human. Here the inner state dissolves into a multi-faceted, multi-perspectival "worlding", a being with the world that may be confrontational or co-operative in relation to other beings but is always already *engaged.*

In this chapter, I have focused on perspectives that conceptualize affect as an encounter between the human and non-human worlds, and I have chosen to reflect on how these encounters might play out in the context of museums, considering specifically encounters between persons (visitors) and things (museum objects). Museums offer a unique context for this reflection since they are highly revelatory frames: they give away much about what we (humans) value and why. Museum objects have long been considered symbols – of the past;

of cultural diversity; of human ingenuity and creativity; of power relations, colonial greed and social discrimination. Museum objects hold up mirrors to ourselves, our virtues and limitations, our anguish and dreams.

But if our reflection had to stop here, we would remain trapped in an anthropocentric perspective. I have suggested trying (the intention is key) to shift away from this understanding of objects as symbols, as this problematically disregards and overrides the alterity of objects. Recognizing the capacity of objects to affect us affects this shift in our interpretations away from symbolism and iconography, for it allows us to encounter objects in their objecthood, peeling away the multiple layers of meanings and references that so often narrowly reflect human values and concerns.

I set my affected imagination free, then, and envision a museum that considers affect not an accidental and inconsequential side-product of pedagogic or leisure activities but a form of knowledge in its own right, as well as a tool to expand that knowledge through empathy and imagination. This is a museum that acknowledges both the human and the non-human as co-producers of experiences, a museum that recognizes, enables and encourages the patient work of listening, of staring back, of just being in the presence of and being perceived by. A museum that slows down the pace of the visit enough for visitors to experience affective encounters and to become aware of it; to ponder what or whom lies at the other end of these encounters and to ask questions; to use empathy to "welcome the stranger", make outlandish guesses and open imaginative horizons about what lies beyond our humanity and then, perhaps, accept the outcomes of this exercise and its limitations with equanimity.

Notes

1 *Divine Cat: Speaking to the Gods in Ancient Egypt*, exhibition at the British Museum, November 2007–January 2008. See https://www.britishmuseum.org/collection/animals/gayer-anderson-cat
2 I am grateful to the editors for bringing this point to my attention.
3 I am grateful to the editors for bringing this point to my attention.
4 I am grateful to the editors for this reference to Tim Ingold's work.
5 A rapidly increasing number of works in the fields of ecology and philosophy, among others, are exploring this line of enquiry, producing fascinating accounts and opening ground-breaking perspectives on human and non-human relationships (see, for instance, Abram 1996, or more recently, Mueller 2017).

References

Abram, D. 1996. *The Spell of the Sensuous*. New York: Random House.

Baker, J. 2013. "Anarchical Artifacts: Museums as Sites for Radical Otherness." In *The International Handbooks of Museum Studies*, edited by A. Witcomb and K. Message, 63–77. Hoboken, NJ: John Wiley & Sons.

Barad, K. 2012. "On Touching – The Inhuman That Therefore I Am." In *The Politics of Materiality*, edited by S. Witzgall and K. Stakemeier, 153–164. Zurich: Diaphanes.

Benjamin, W. 2007 [1968]. "The Storyteller. Reflections on the Works of Nikolai Leskov." In *Illuminations*, 83–109. New York: Schocken Books.

Bennett, T. 1995. *The Birth of the Museum: History, Theory, Politics*. London: Routledge.

Brennan, T. 2004. *The Transmission of Affect*. Ithaca, NY: Cornell University Press.

Butz, D. and Besio, K. 2009. "Autoethnography." *Geography Compass* 3(5): 1660–1674.

Clough, P. T. and Halley, J., eds. 2007. *The Affective Turn: Theorizing the Social*. Durham, NC: Duke University Press.

Damasio, A. 1999. *The Feeling of What Happens: Body and Emotion in the Making of Consciousness.* New York: Harcourt.

Damasio, A. 2018. *The Strange Order of Things: Life, Feeling, and the Making of Cultures.* New York: Pantheon Books.

DeGroff, D. A. 2012. "Artur Hazelius and the Ethnographic Display of the Scandinavian Peasantry: A Study in Context and Appropriation." *European Review of History: Revue Européenne d'histoire* 19(2): 229–248.

Dudley, S. H., ed. 2012. *Museum Objects: Experiencing the Properties of Things.* London: Routledge.

Dudley, S. H. 2013. "What, or Where, Is the (Museum) Object? Colonial Encounters in Displayed Worlds of Things." In *The International Handbooks of Museum Studies*, edited by A. Witcomb and K. Message, 41–62. Hoboken, NJ: John Wiley & Sons.

Dudley, S. H. 2017. "The Buzz of Displacement: Liminality among Burmese Court Objects in Oxford, London and Yangon." In *The Inbetweenness of Things: Materializing Mediation and Movement between Worlds*, edited by P. Basu, 39–56. London: Bloomsbury.

Edwards, E., Gosden, C. and Phillips, R. B., eds. 2006. *Sensible Objects: Colonialism, Museums and Material Culture.* Oxford: Berg.

Ellis, C. and Bochner, A. P. 2000. "Autoethnography, Personal Narrative, Reflexivity: Researcher as Subject." In *Handbook of Qualitative Research* (2nd edn), edited by N. K. Denzin and Y. S. Lincoln, 733–768. Thousand Oaks, CA: Sage Publications.

Frykman, J. and Povrzanovic Frykman, M., eds. 2016. *Sensitive Objects: Affect and Material Culture.* Lund: Nordic Academic Press.

Gell, A. 1998. *Art and Agency: An Anthropological Theory.* Oxford: Clarendon Press.

Gregory, K. and Witcomb, A. 2007. "Beyond Nostalgia: The Role of Affect in Generating Historical Understanding at Heritage Sites." In *Museum Revolutions: How Museums Change and Are Changed*, edited by S. Watson, S. MacLeod and S. Knell, 263–275. London: Routledge.

Hancock, N. 2010. "Virginia Woolf's Glasses: Material Encounters in the Literary Artistic House Museum." In *Museum Materialities: Objects, Engagements, Interpretations*, edited by S. Dudley, 114–127. London: Routledge.

Hancock, N. 2012. *Charleston and Monk's House: The Intimate House Museums of Virginia Woolf and Vanessa Bell.* Edinburgh: Edinburgh University Press.

Haraway, D. 1988. "Situated Knowledges: The Science Question in Feminism and the Privilege of Partial Perspective." *Feminist Studies* 14(3): 575–599.

Hjemdahl, A.-S. and Opdahl Mathisen, S., eds. 2017. "Theme Section: Bodyworks." *Nordisk Museologi* 1: 3–72.

Hooper-Greenhill, E. 2007. "Interpretive Communities, Strategies and Repertoires." In *Museums and Their Communities*, edited by S. Watson, 76–94. London: Routledge.

Hooper-Greenhill, E. 2013. *Museum, Media, Message.* London: Routledge.

Ingold, T. 2000. *The Perception of the Environment.* London: Routledge.

Ingold, T. 2012. "Toward an Ecology of Materials." *Annual Review of Anthropology* 41: 427–442.

Latour, B. 2005. *Reassembling the Social: An Introduction to Actor-Network-Theory.* Oxford: Oxford University Press.

Leahy, H. R. 2012. *Museum Bodies: The Politics and Practices of Visiting and Viewing.* Farnham: Ashgate.

Levent, N. and Pascual-Leone, A., eds. 2014. *The Multisensory Museum: Cross-Disciplinary Perspectives on Touch, Sound, Smell, Memory, and Space.* Lanham, MD: Rowman & Littlefield Publishers.

Liljeström, M. 2016. "Affect." In *The Oxford Handbook of Feminist Theory*, edited by L. Disch and M. Hawkesworth, 16–38. New York: Oxford University Press.

Merleau-Ponty, M. 2013 [1945]. *Phenomenology of Perception.* New York: Routledge.

Mueller, M. L. 2017. *Being Salmon, Being Human: Encountering the Wild in Us and Us in the Wild.* White River Junction, VT: Chelsea Green Publishing.

Navaro-Yashin, Y. 2009. "Affective Spaces, Melancholic Objects: Ruination and the Production of Anthropological Knowledge." *Journal of the Royal Anthropological Institute* 15(1): 1–18.

Nussbaum, M. C. 2003. *Upheavals of Thought: The Intelligence of Emotions.* Cambridge: Cambridge University Press.

Qureshi, S. 2011. *Peoples on Parade: Exhibitions, Empire, and Anthropology in Nineteenth-Century Britain*. Chicago, IL: The University of Chicago Press.

Roscoe, P. B. 1995. "Of Power and Menace: Sepik Art as an Affecting Presence." *The Journal of the Royal Anthropological Institute* 1(1): 1–22.

Sandberg, M. B. 2002. *Living Pictures, Missing Persons: Mannequins, Museums, and Modernity*. Princeton, NJ: Princeton University Press.

Singer, T. and Klimecki, O. M. 2014. "Empathy and Compassion." *Current Biology* 24(18): 875–878.

Smith, L., Wetherell, M. and Campbell, G., eds. 2018. *Emotion, Affective Practices, and the Past in the Present*. London: Routledge.

Stewart, K. 2007. *Ordinary Affects*. Durham, NC: Duke University Press.

Stoklund, B. 2003. "Between Scenography and Science." *Ethnologia Europaea* 33(1): 21–36.

Strathern, M. 2001. "The Patent and the Malanggan." In *Beyond Aesthetics: Art and the Technologies of Enchantment*, edited by C. Pinney and N. Thomas, 259–286. Oxford: Berg.

Thrift, N. 2004. "Intensities of Feeling: Towards a Spatial Politics of Affect." *Geografiska Annaler: Series B, Human Geography* 86(1): 57–78.

Varutti, M. 2020. "Vers une Muséologie des Émotions." *Cultures et Musées* 36: 171–177.

Vogel, S. 1991. "Always True to the Object, in Our Fashion." In *Exhibiting Cultures: The Poetics and Politics of Museum Display*, edited by I. Karp and S. Lavine, 191–204. Washington, DC: Smithsonian Institution Press.

Watson, S. 2013. "Emotions in the History Museum." In *The International Handbooks of Museum Studies*, edited by A. Witcomb and K. Message, 283–301. Hoboken, NJ: John Wiley & Sons.

Witcomb, A. 2015. "Toward a Pedagogy of Feeling: Understanding How Museums Create a Space for Cross-Cultural Encounters." In *The International Handbooks of Museum Studies*, edited by A. Witcomb and K. Message, 321–344. Hoboken, NJ: John Wiley & Sons.

9

A GENTLE SHOCK OF MILD SURPRISE

Surface ecologies and the archaeological encounter

Tim Flohr Sørensen

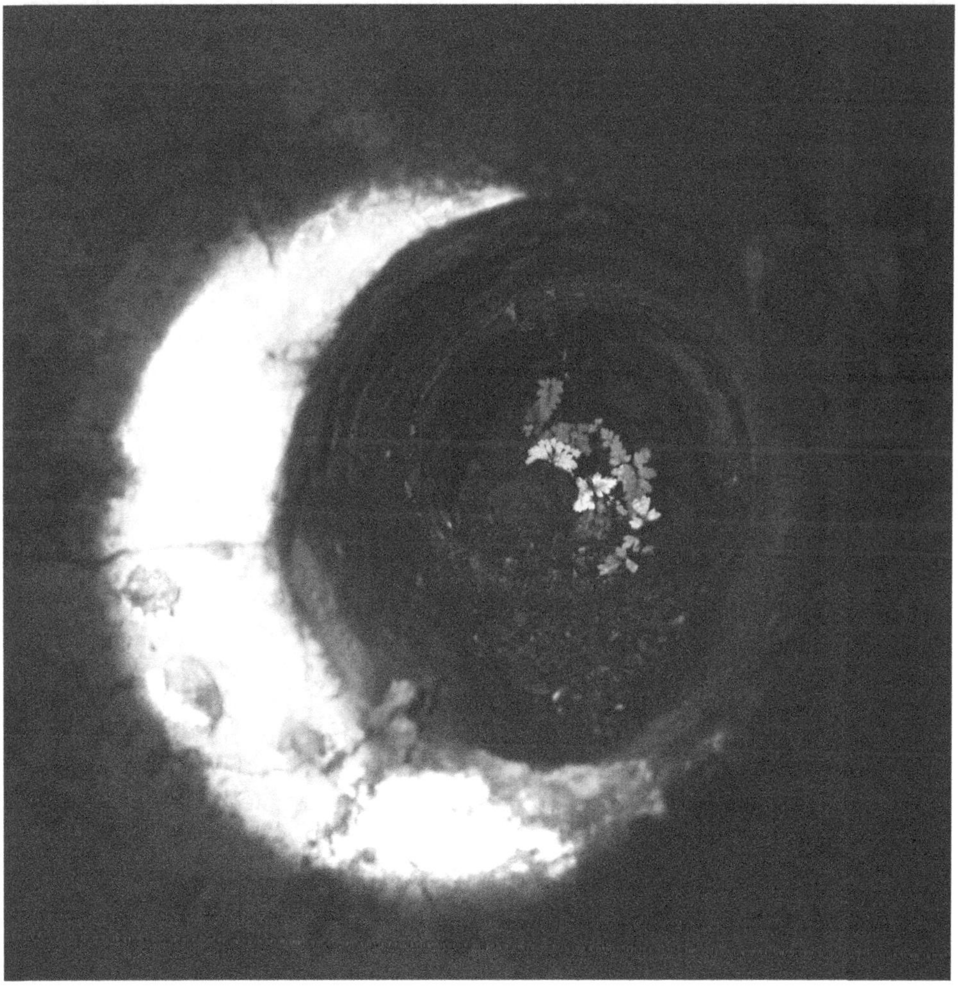

FIGURE 9.1 Life in the dark. Prøvestensfortet, Denmark, 12 January 2018

Photo: T. F. Sørensen.

The traction towards things

In January 2018, I had the opportunity to visit an otherwise inaccessible, discontinued naval fortress on the waterfront of Copenhagen, Denmark. Walking through the somewhat ruinous and rather empty fortress, called Prøvestensfortet, I entered a small, dark room with no windows. In the back of the room, the openings of four ventilation shafts emerged in the concrete wall, each of them about a hand's breadth in diameter. The openings disappeared into pitch-black wells that once allowed air to circulate through the thick concrete walls of the fortification. Today, various forms of matter have clotted up in them. A strange white substance fills one of the shafts with what looks like a mixture of salt and chalk, which almost seems to ooze out from the depths at an unperceivably slow pace.

I turned to one of the other openings in the wall, where, at first sight, dark soil appeared to block this shaft. Then, when I pointed my flashlight into the dark hole, I found to my surprise that a plant occupied this space, emerging quietly from a damp black soil looking like rich, coarse sphagnum (Figure 9.1). The plant was tiny with leaves no larger than a fingernail, sitting shyly, half an arm's length into the shaft behind a chunk of soil. The leaves appeared healthily green and delicate. I thought to myself that this must be a modest creature, residing in utter darkness, where it flourishes on the moisture of the soil, congealed in the shaft. Without much deliberation, I grabbed my camera and quickly took a couple of photographs of the plant in the mouth of the shaft, then turned away to continue my tour of the fortress.

The tiny plant only started to stir a curiosity in me some time after my visit to Prøvestensfortet. I did not visit the naval fortress for the sake of botanising, nor in search for secrets hidden in dark ventilation shafts. I had arranged with a student of mine to visit the fortress, because he was writing a Master's dissertation on historic ruins in Copenhagen, and the port authorities had granted him permission to visit the site. Given the opportunity, I decided to tag along, partly as a form of on-site supervision but also because Prøvestensfortet is a place you cannot visit ordinarily, so I was curious to see this discontinued and abandoned site.

The naval fortress seemed to match the focus of my current research, revolving around active archaeological formation processes. I had been visiting numerous decaying and abandoned places before, encountering the material life that unfolds once human maintenance ceases and the autonomous reordering of buildings, plants, roads, trails, vehicles and animals takes over. Most of the places I had been visiting previously were former homes, farms, cemeteries and industry, and I was keen to complement such sites with a military fortification. My interest in the different sites was not so much driven by their various cultural and functional categorisations. What attracted me was more the various material qualities offered by, for instance, a domestic house built in 1927 and abandoned in 2005, a farm built in 1760 and abandoned in 2010, a cemetery consecrated in 1915 and discontinued in 1968, a glassworks built in 1825 and abandoned in 2012 and now a coastal fortress built in 1863 and abandoned in 1963.

In fact, it is imprecise to say that these sites were abandoned, because numerous people have been revisiting them after formal use was terminated: archaeologists, urban explorers, vandals, museum staff, curious teenagers, role players, bored teenagers, politicians, bankers, redevelopers, journalists and former inhabitants or workers. If 'abandoned' means 'deserted' and 'unvisited', the places I have worked at hardly qualify. Even though the sites I have been working at were in many cases off limits, sometimes even condemned, fenced off, locked up and monitored by video surveillance, some individuals were nevertheless granted access, while others made their way without being invited. All of us noticed something at these sites which made us pause, focus and reflect, perhaps intervene, or perhaps eventually compelled us to leave the site.

In this chapter, I wish to explore the lure of objects and object constellations and how even the most unlikely and withdrawn of them somehow draw my attention. Elsewhere, I have described this as the *traction* towards things (Sørensen 2017a, 141). Here, I want to take the opportunity to wonder what it is about particular objects and the ways in which they hold together and break apart that makes me linger and return. In this chapter, it is this latency of things – their delayed traction – which I frame as 'heritage', and I want to explore the relation between my initial encounter with things and their affective persistence. The things I dwell on in this chapter are not culture-historically significant, they are not speaking to contemporary social or environmental issues and perhaps they are even irrelevant as objects of archaeological attention, falling way off the heritage grid. Yet nevertheless, they seem to stick with me, and, regardless of their irrelevance, I cannot quite seem to shake them off me. They lead me on somehow: they are redundant yet also redolent objects.

Specifically, I am interested in probing into how I direct my archaeological gaze and use photographs and photography – both my own and that of another – as a means of encountering, composing and remembering what I term 'surface ecologies'. I want to explore the subtleties of archaeological traces and the slownesses of transformation processes that seem to *seize* my gaze in various ways: how encounters with the unexpected emerge in the present or, perhaps rather, how I become vulnerable to their affecting presence (Armstrong 1971). In the cases I draw on, photography and the exploration of existing photographs have constituted my encounter with surface ecologies rather than excavation and collection. In fact, the photographs introduced in this chapter emerged during ephemeral and condensed fieldwork, and I will focus my attention on the chance encounters with unlikely and unimportant phenomena that nevertheless stuck with me.

The notion that the surface is a boundary against a deeper substance, and that it is superficial to remain on the surface, rests on the contention that the surface covers up more important things, otherwise unknown, forgotten, repressed or subconscious; that is to say, the surface can be distinguished from sub-surface. However, in my experience, the stratigraphic principles of the surface ecology do not allow for the isolation of distinct surfaces; surfaces are always in the plural, and they are interlaced. Such notions of the 'blending' or 'mixtures' of multiple, emergent, and heterogeneous, flat ontologies, in addition to the decentring of the human, have of course been rehearsed abundantly with the recent onset of new materialisms, object-oriented philosophies, assemblage theory and other post-humanist orientations. With this chapter, I attempt to contribute to such studies by probing into the methodical implications of a post-humanist focus, that is, how to do archaeological fieldwork from a post-humanist perspective. I believe one consequence of the post-humanist orientation is to 'decentre' the ordinary discursive mode of inquiry, revolving around meaning, purpose, linear narrative, plot and explanation, and instead allow questions, curiosity and confusion to emerge in the delayed, purposeless encounter with surfaces. This may allow for a heritage ecology that attends to and cares for objects that otherwise go unnoticed as 'irrelevant' to humans.

Surface ecologies

A while after visiting Prøvestensfortet, I started wondering what it was about the tiny plant in the ventilation shaft that preoccupied me. It puzzled me, how long the plant had lived there and how it arrived there in the first place. I kept speculating what kind of life this is. Not a bad one, I presume, at least for humble and patient creatures that do not crave attention, finding, perhaps, dignity in solitude. Returning to my two almost identical photographs of the plant, I was continuously unable to decipher why I was drawn to it. Considering I visited the fortress in an archaeological errand rather than to do botanical studies, I had to ask myself why I was so preoccupied by my photographs of a humble and unnoticeable herb resulting from the chance encounter instead of dwelling with careful and systematic observations of more historically significant parts of the fortress.

Prøvestensfortet is a nineteenth-century coastal fortress, located 5 kilometres from central Copenhagen. Currently, it is an abandoned and rather ruinous site, despite being scheduled for protection by the Danish heritage authorities as a historical landmark. Originally built between 1859 and 1863, the coastal fortress is an artificial island, which was part of a string of naval defensive works on the seaside of Copenhagen. Prøvestensfortet was decommissioned as part of the coastal defence in 1922, and it was turned into an oil and petrol storage facility for the Danish Navy. In 1934, Copenhagen Port took over the facility, continuing to store fuel, and expanded the artificial island significantly, while the actual fortress was used for storage. Today, the area, managed by the port operator Copenhagen Malmö Port, also harbours large amounts of gravel for construction work, in addition to receiving and embarking shipment of other kinds. This means security around the area is strict and that all visits to Prøvestensfortet must be arranged with the port operator for a member of security staff to chaperone the visit. My visit therefore only lasted for about an hour and was a one-off visit.

So, why does the fleeting happenchance encounter with a shy lifeform capture me, overshadowing other places in the fortress, where I would be able to reconstruct past uses, modifications, sequences of events and historical trajectories? Why dwell on an object that escapes intentionality, purpose, meaning and discourse in a space otherwise heavily charged historically and politically? And why focus on my photographs of the herb and not the plant itself, which I did not spend much time on at the site? There may seem to be something counterintuitive and perhaps unproductive about attending to a withdrawn plant, sitting in dark seclusion, being even 'irrelevant' to the history of the coastal fortress. There is perhaps even something narcissistic about the preoccupation with the photographs, marking my subjective and largely aesthetic engagement with the archaeological.

In an exploration of ruin photography as an "engagement with things", Þóra Pétursdóttir and Bjørnar Olsen (2014, 8) argue for how photography *relates* to "the aesthetic" rather than simply *being* aesthetic or a form of aestheticisation, focusing on the "instant response to an immediate material encounter" (Pétursdóttir and Olsen 2014, 21). They discuss photography and its role in archaeological fieldwork, research and dissemination, turning some of the alleged shortcomings of ruin photography on its head. They contend the purported deficiencies are actually strengths, because they allow for a different archaeological engagement with things and ruins. Hence, the superficial, the subjective and the aesthetic are here reclaimed as tenets of a methodological approach that are not second to the notion of "analytical depth" in academic writing (Pétursdóttir and Olsen 2014, 9).

Elsewhere (Sørensen 2014), I have argued for a strategically 'superficial' archaeology. The aim of this is to challenge the almost unquestionable need in archaeology to 'unveil' or 'unravel' what is 'behind' or 'underneath' the surface in the pursuit of what is oftentimes perceived as the more genuine 'real' residing within the depth. This is of course not simply a syndrome of archaeological thinking (Harrison 2011; Pétursdóttir 2016; Thomas 2004) but also relates to wider psychoanalytical, Marxist, existential and semiotic perspectives, defined by Frederic Jameson as various "depth models" (Jameson 1993, 12). Like much archaeology, these models tend to mistrust the surface as the token of something superficial and shallow, that is, not only lacking depth but also substance. However, surfaces do not preclude depth, and there need not be any opposition between surface and depth (Harrison 2011, 156–157), as argued, for instance, by Nadia Bartolini, advancing the concept of 'brecciation' from geology (Bartolini 2013a, 2013b). She uses the concept as a figure to describe objects that are coherent but nonetheless consist of elements with different origins and no shared connections. In breccia, elements from the past are "*cutting through* the present and introducing a *new* surface of time, giving the present *depth*" (Bartolini 2013a, 524). As a result, surfaces constitute ecologies in their own right, not as a distinct and separate domain but simultaneously as interfaces or interstices of objects.

Rehabilitating the surface as "the restoration of social life itself", Tim Ingold (2017, 105), promotes an approach to the surface as 'complexion', contending that the bifurcation of "interior and exterior are not set apart by the surface" (104; compare with Levinas 1978, 53):

> The veil covers, but does not cover up; it dresses but does not dress up. It is not a disguise but a revelation. Encrusted with its own material, the veil is not just a self-shadowing but a self-adornment. Far from hiding the depths *behind* the surface, it allows us to feel the depths *in* the surface. Why else do we use the verb 'to wear' to mean, in the same breath, both adornment and erosion? The worn face, wrinkled and ruddied by long exposure to sun and wind, *wears* its expression; worn hands *wear* the folds and calluses that come from years of working with materials, ancient stone steps *wear* the patina of thousands of passing feet, worn ground *wears* the texture of its wearing away.

I want to suggest this surface ecology is observable only by *trusting* the immediate encounter, about not yielding too readily to the need to efface the inexplicable confrontation with the surface in the search for explanation or meaning. Instead, trusting the surface, remaining open to its allure, might allow the encounter with the object to endure for just a little while longer, sustaining an important, if not decisive, moment of one's interest in the surface in the first place.

This trust may engender a different kind of knowledge akin to qualities highlighted by Pétursdóttir and Olsen (2014, 20), referring to Roland Barthes' famous distinction between *studium* and *punctum* in photography. Where *studium* denotes the analytical, discursive, culturally informed and consciously reflected interpretation of the photograph, and "to encounter the photographer's intention" (Barthes 1989, 27), *punctum* signifies the personal experience of the photograph that "rises from the scene, shoots out of it like an arrow, and pierces me" (Barthes 1989, 26). Unlike *studium*, "it is not I who seek it out" (Barthes 1989, 26), and it is rather the experience of the element of the photograph that issues forth as an unexpected "accident that pricks me (but also bruises me, is poignant to me)" (Barthes 1989, 27). In this sense, the *punctum* is 'aberrant' and a disruption that confronts the expected encounter with an image or scene,

FIGURE 9.2 Cave-in. Truelshøj, Denmark, 14 July 2017

Photo: T. F. Sørensen.

which means it is difficult to capture the scene with words: "what I can name cannot really prick me. The inability to name is a good symptom of disturbance" (Barthes 1989, 51).

Chance encounters

The experience of having to remain open to unexpected surface encounters resounds with many places where I have conducted fieldwork, oftentimes more or less deserted or discontinued locales. Importantly, however, all these sites are different, and the way they hold together and break apart lends itself differently to trust. In some cases, the traction of things has happened with a slight delay upon my initial encounter, as in the case of the herb at Prøvestensfortet. In other cases, the patience of my trust has been tested, leaving me for a shorter or longer while with a feeling that there was 'nothing to see'.

At a smallholding in northwestern Zealand, for instance, I encountered a heavily decaying property, last occupied some ten years earlier. Unlike the tiny secluded plant at Prøvestensfortet, I did not stumble across a little curiosity right away in this case. Rather, having passed by the place on numerous occasions, I was puzzled by how much the site changed from time to time. Sometimes, the vegetation completely engulfed the ruin, while at other times, it faded or withered, exposing rather than embracing the ruin. When visiting the site, I found that the buildings were far more dilapidated than I had expected when planning the visit, and they seemed to be so fragile that I considered them too dangerous to enter (Figure 9.2), and I chose only to survey them from the outside. This was of course not an ideal condition for doing fieldwork, especially in the fertile summertime, when bushes and trees encroached the entire property (Figure 9.3) so I could only see through windows and doors in a few places (Figure 9.4). Except for taking exterior photographs and pondering the surface of the buildings and the

FIGURE 9.3 Overgrowth. Truelshøj, 14 July 2017

Photo: T. F. Sørensen.

FIGURE 9.4 Window. Truelshøj, 14 July 2017

Photo: T. F. Sørensen.

entangled vegetation, there was only so much I could do at the site. I left, somewhat dissatisfied, after a rather brief encounter with the place of about half an hour, looking mainly at a dilapidated building through its embrace of thicket and vines (Figure 9.5).

Being unable to study the actual house and its interior, I began searching for historical documents relating to a dilapidated house, finding a copy of an old photograph of the property in the local history archives (Figure 9.6). Information in the dossier states that the photograph dates to the 1920s, that the name of the property is Truelshøj and that the woman in the photograph was named Anna Rasmussen. Anna stands on the front stairs at the main entrance to the house with a dog sitting by her side, while they pose patiently for the unnamed photographer. According to the national Buildings and Housing Register, the house was built in 1927. If dating to the 1920s, I imagine that the photograph might have been taken with the intention of producing a keepsake of the new home: a new beginning. At my visit to the house in

FIGURE 9.5 Main entrance. Truelshøj, 14 July 2017

Photo: T. F. Sørensen.

July 2017, I found myself standing on the same spot as Anna did some 90 years earlier, yet I was standing at the other end of the biography of the house; it was heavily deteriorating and advertised for sale with the declared aim of demolition rather than restoration.

Looking at my photographs from my visit, I cannot help but compare them to the photograph from the late 1920s. There is a conspicuous difference in the conditions of the buildings; a radically diverging distance between the newly built house on the old photograph and the crumbling mess I encountered in 2017. Then, I notice how the copy reveals that the original photograph from the 1920s is in fact also marked by ageing. It is scratched and slightly torn, and its wrinkles create an uneven surface. Looking carefully at the copy of the photograph, I can trace tiny folds, fissures, dints and bulges, some of which distort or blur the motif. The photograph is itself a ruin, deteriorating as an "anterior future" in the words of Barthes (1989, 89), offering us "a glimpse into the deep past and the deep present of physical things"

FIGURE 9.6 Anna Rasmussen and Truelshøj, late 1920s

Source: Photographer unknown. Reprinted by kind permission from Odsherred Lokalarkiv.

(Mitchell 2005, 182). In this sense, the folds, dints and bulges are not simply signs of dissolution but contribute to the photograph by *adding* distortion and blur as a topographic build-up.

The photograph does not offer full disclosure. Looking at the photograph in detail, I have a hard time discerning the facial features of Anna, leaving me with no impression of her mood or sentiment. Looking even closer, I am unsure whether it is in fact a dog next to her or if it might be a child sitting on the stairs or perhaps some other form of object. I find myself on a trajectory resembling the one Barthes lays down in the pursuit of details in a faded image of his deceased mother: "I live in the illusion that it suffices to clean the surface of the image in order to accede to *what is behind*. … Alas, however hard I look, I discover nothing" (Barthes 1989, 100).

As time has passed, I realise I have become more concerned with the photograph than the ruined building, and my attention now revolves just as much, if not more, around a small, fragile photograph instead of the volume that is the modern ruin, which was at first the intended object of my research. The old photograph was somehow out of place in terms of what I planned to focus on, and much like the tiny plant at Prøvestensfortet, the photograph was the product of brief chance encounter rather than a careful and purposeful methodical study. Furthermore, juxtaposing the two photographs also seem to add to their traction for me, because they both seem to endure in ways that surprise me.

Openness and wonder

Returning to the plant at Prøvestensfortet, I want to turn from the photograph itself to exploring, why I was compelled to take the photograph in the first place. While Barthes discusses the relationship between the photograph and the perception of the photograph as two modes of imaging a scene, I want to include the moment of taking the photograph as the very necessary first instance of imaging. This is the initial encounter, where something catches the photographer's attention, leading her or him to frame an image and press a shutter release button. Are there not in fact two photographs at play at once, or at least two moments of imaging? Is there not first a moment of encountering the tiny plant (the moment in which I took the photograph) and then a second moment of looking at the photograph on my computer screen, in print or in whatever form I revisit the photograph? This implies that the two photographs are conditioned on different grounds. At Prøvestensfortet, I took a photograph of a plant in my own encounter with the object, whereas in the case of Truelshøj, I encountered a photograph taken by someone else. Looking at the photograph of Anna, I am not revisiting an initial encounter in the field, and there is no delayed traction but perhaps more a renewed attention to the dilapidated house, another layer in its surface ecology.

This is slightly different in the case of the plant at the naval fortress, where the way in which the image emerged was, at least in part, a question of how I directed or did not direct my gaze but also about letting constellations emerge autonomously. What I encounter at the naval fortress and in the photograph from Truelshøj are thus an *excess* of my intentionality. Yet the images emerge through two different gazes: in one case through my own delayed response to a chance encounter in the field and in the other through my encounter with another person's gaze. What the photographs have in common is that I did not plan for them; they 'happened' to show themselves to me.

In Pétursdóttir's study of drift matter (Pétursdóttir 2017, 2018; also Pétursdóttir and Olsen 2018), objects emerge at their own bidding on beaches in the North Atlantic, whether anyone goes looking for them or not. They wash ashore, compiling in massive constellations of heterogeneous assemblages with no sense of chronology, purpose or identity (known to us at least) apart from forming a new ecology: drift matter. In a sense, my photographs are another form of 'drift matter', constituted by *my* drift towards things and not the drift of the objects. In this surface ecology, I claim it would not make sense to go searching for surprising objects, because then the encounter with the objects would not be surprising. Rather, surprise and wonder cannot be planned but have to be realised after they take effect, as "an experience in which we find ourselves already moved, already affected, already opened up to what is before us" (Malpas 2012, 256).

The state of wonder, however delayed, surpassed the observations I might have been aiming for when visiting the sites. This state was only possible because I became sensitive to surface impressions and started attending to the occurrences that are at the same time mundane and unexpected. Kathleen Stewart describes how such "poesis of the ordinary draws attention and becomes habitual because things don't just *add up*. *Some*thing throws itself together and then floats past or *sticks* for some reason" (Stewart 2008, 74). Accordingly, the surface ecology is not a boundary covering up an essence I have to 'get right'; the surface ecology is itself a 'something' that for some reason may endure, whether I notice it or not.

For this receptiveness to be possible, for something to 'stick', a methodical disposition is necessary, allowing for an "opening onto *something*" (Stewart 2008, 72). Then, the question is, of course, what is nested in this elusive 'something'. Obviously, the 'something' is difficult to identify and categorise; otherwise, it would be a particular thing and not a 'something'. In turn, it is difficult to name it, which brings us back to Barthes' notion of the *punctum* as escaping labelling. Altogether, the 'something' seems to revolve around that which denies definition, explanation, naming, causality; it is breaking free of our expectations and instead issues forth as something autonomous, solitary and mysterious.

For Jeff Malpas, such wonder "requires no effort on our part – no decision or act" (Malpas 2012, 256). However, I question if this is entirely true, or at least to what degree the encounter with the wondrous is effortless. According to Martin Heidegger, the opening "toward things and the openness to the mystery never happen of themselves. They do not befall us accidentally" (Heidegger 1966, 56). For this reason, the opening or allure is not simply a matter of leaving things alone – about doing nothing and waiting for things to happen. The surface ecologies I refer to in this chapter – from Prøvestensfortet and Truelshøj – were all encountered by me and allowed to linger with me, because I put myself in situations that made them possible. I made an effort to go somewhere, making the confrontation with a surface ecology possible, and I made decisions that impressed in me details for which I was unprepared. I actively pointed my flashlight into the ventilation shaft at Prøvestensfortet, and I looked into the local archive in the case of Truelshøj – looking for *something* without knowing what exactly I was looking for or necessarily why I was looking.

Similarly, it seems to me, when Pétursdóttir describes the wondrous, mysterious and strange in ruins and drift matter (Pétursdóttir 2014, 2016, 2018), she refers to encounters occurring in the context of archaeological fieldwork at desolate sites in remote locations. She decided to put herself in situations and places, where this responsiveness became possible. In neither of the cases was the wonder planned for, but nor is the openness to the wonder simply about being unprepared and inactive. It is, perhaps, instead about being susceptible to what we are not prepared for, about putting ourselves in situations and contexts that allow unlikely surface ecologies to press themselves onto us. Importantly, for the openness to be possible, we cannot be conditioned by a discursive need to explain, use or valorise things. It has to be given the privilege of entering into "the ethos of letting be" (Introna 2009, 37). In this sense, it does not make sense to 'reveal' objects, conserve them, explain them or possess them (also Witmore 2014, 223). Rather, it is necessary to 'wait' for them; a waiting that should not be equated with doing nothing.

This, of course, raises the question whether the instant of taking the actual photograph corresponds in any way to the openness or the *punctum*. Is the capturing of a photograph a way of taking possession of the object, betraying the very openness? Can the *punctum* only issue forth through photographs and only in photographs taken by someone else? If I reflect on the first moment of imaging, the moment of taking the photograph, is it, then, not more about intention and reflexivity, that is, the very opposite of the *punctum*? From noticing the object to pressing the shutter-release button might not take a very long time, but it still amounts to a procedure that transforms the immediacy of the gaze to the production of an image. It stretches from being drawn to the object and deciding more or less consciously to photograph it, over aligning the camera with one's gaze, composing and cropping the motif, checking aperture and shutter, depth-of-field, being conscious that the camera should not shake too much. Do these steps imply that I have started to compromise my initial lure towards the tiny herb, turning it into something betraying the instantaneous superficial encounter? A manufactured object of reflection, a meditated and staged product of my gazing?

Moreover, the photographs included in this chapter are obviously not simply reproductions of the immediacy of the initial encounter with objects and object constellations. I have, unashamedly, processed and manipulated them in a number of steps in an image-editing program. I have cropped my photographs to a square format, I have converted them from colour to monochrome, I have processed most of them with an infrared filter and adjusted light/shade in all of them. Accordingly, the steps that the photographs go through in the editing process are just as formative of the image as the initial act of photographing and therefore also need to be considered in a discussion of photography as a method for being open to surface ecologies.

For this opening to lend itself to the indeterminate *something*, the *something* cannot be eclipsed by the predefinition of the surface as hiding a deeper array of analytically accessible objects beneath. The methodical disposition therefore needs to frame an attitude to the unexpected surface encounter without immediately leading from wonder to explanation. It must sustain the mystery in order to do justice to the unexpected. And this also pertains to the objects constituting the surface. Stewart calls us "to wonder where they might go and what potential modes of knowing, relating, and attending to things are already somehow present in them as a potential or resonance" (Stewart 2008, 73). In this, one might celebrate the unedited photograph as a more authentic representation of the potential mode of knowing. However, this implies that the untouched is somehow closer to the truth, as if my human touch (in the form of first photography and then image editing) creates a distance between the 'authentic' object and its perception. Conversely, we might see the act of photography and image editing as a way of wondering where objects might be heading, or a way of "feeling forward" (Simonetti 2015).

Looking away

Although I believe there is every reason to wonder whether the act of photographing sustains the immediacy of the initial encounter with the object, I do not regard the viewfinder as an instrument for unravelling a deeper meaningfulness or explanation. Rather, in the moment of pressing the shutter-release button, gazing through the viewfinder, I believe to be framing the lure or the traction towards the thing itself, because the act of photographing implies I was somehow not done with the object upon seeing it. Taking the photograph was a point of departure for my engagement with the object and not its conclusion. I did not seek out the tiny herb; I responded to its allure, its 'accidence' or 'disturbance'; I was drawn into a surface ecology, whose elements I am not necessarily able to name, being quite incapable of explicating the reason for my traction towards things.

While taking the photograph of the tiny plant at the naval fortress was a moment of attending to detail, revisiting my photograph of the plant has largely been conducted without looking at the actual photograph. Writing these lines about the plant, I do not have the photograph in front of me, yet it stands clearly to me as a real object. This, I suppose, makes it even more important to consider the moment of capturing the image, because the question is how the image endures? Consider Barthes' line of reasoning:

> the *punctum* should be revealed only after the fact, when the photograph is no longer in front of me and I think back on it. I may know better a photograph I remember than a photograph I am looking at, as if direct vision oriented its language wrongly, engaging it in an effort of description which will always miss its point of effect, the *punctum*. … I had just realized that however immediate and incisive it was, the *punctum* could accommodate a certain latency. Ultimately – or at the limit – in order to see a photograph well, it is best to look away or close your eye.
>
> *(Barthes 1989, 53)*

Contending that an object is seen better by looking away, by giving way to wonder, and to dwell with the affective response to the encounter can seem iconoclastic in the field of archaeology, where exhaustive accounts and meticulous analyses of objects are normally celebrated as a cardinal point of the disciplinary methods. Much archaeology even treats the immediate, sensory and affective traction of things with scepticism or outright disdain for "signifying a naïve and fetishistic inclination incompatible with rational and analytical thought" (Pétursdóttir and Olsen 2014, 12). In a sense, there is something paradoxical about this affective allure of the object of photography. Pétursdóttir and Olsen contend that this is what leads ruin photography to being described as 'ruin porn', because the instant and sensory aspect is associated with a pornographic and purely self-centred gaze (Pétursdóttir and Olsen 2014, 12).

Yet if this self-centred, narcissistic gaze is marked by a 'certain latency' and an inability to name its object – and if the object is indeed approached best by looking away – how can I know that there is anything there at all? Hans Ulrich Gumbrecht (2014) argues that while we may know that something 'latent' is present, we cannot say where our certainty about its presence comes from. Moreover, since our appreciation of the latent is delayed, we cannot know where the latent object is located now, whether it has changed or if it is still there at all. He argues that what has "passed into latency" cannot be accessed by any 'method' or 'interpretation': "In order to be accessible to interpretation – that is, to permit us to identify a meaning that seems to lie 'under' a surface – the latent would have to bring forth … 'propositional content'" (Gumbrecht 2014, 23). This leads him to ask: "how can we be so certain that something latent is 'really there' if it eludes our very perception?" (ibid). Still, "we sense that something (or somebody) is there that we cannot grasp or touch – and that this 'something' (or somebody) has a material articulation, which means that it (or he, or she) occupies space" (ibid).

In this sense, the surface ecology may trigger a powerful affective agency in shaping resonance between perceiver and perceived, revolving around something utterly different from the "deeper meaning" of things, revealed through critical analysis (Pétursdóttir 2016, 383). Like in Ingold's discussion of the surface, its fissures and bruising – its wear – is instead a testimony to the inseparability of surface and depth, fragment and whole, the superficial and the profound. This is an invitation to attend to the vibrancy of fragments and of the minute in how 'constellations' come together in moments of surprise, or "in a flash with the now" (Benjamin 1999, 462). By extension, I ask why the 'minimal' – that is, details, textures, broken surfaces – is so important for my intuitive and affective response to the objects I encounter in my fieldwork. Traditionally, archaeology has emphasised unique discoveries, representative assemblages and the large-scale and the spectacular in its search for explanation and interpretation. However, my intuitive attention to details, the 'little' and the 'few' (Benjamin 1999, 160), is more in line with Walter Benjamin's emphasis on the "rags, the refuse" (1999, 460) in his attempt to define montage as method, emphasising how the otherwise humble, withdrawn and insignificant can turn out to trigger resonance, curiosity and wonder (see also Greenblatt 1990). Peter de Bolla argues that this is a way to understand why we are sometimes moved so powerfully by objects, in his case artworks, contending that their "state of dignity chimes so clearly with the vulnerability, the evanescence of art's wonder" (de Bolla 2001, 145).

Drawing closer to exploring how a tiny herb and a wrinkled photograph were able to stir an affective connectivity in me, triggering a longer-lasting resonance, I note again that Barthes emphasises the latency of the *punctum* (Barthes 1989, 53). Similarly, de Bolla (2012) describes such experiences as offering "a gentle shock of mild surprise" (referring to Wordsworth; see

Wordsworth 1977, 87), arguing that the gentle wonder may in fact have a longer-lasting effect than the more conspicuous appreciation of the art object. In this sense, the more powerful and sudden the shock, the quicker the surprise "runs out of steam, loses its appeal, fades into the familiarity of being shocked" (de Bolla 2012, 159). He contends that "the milder the surprise, generally speaking, the more enduring the shock" (de Bolla 2012, 158). And this capacity of the surprising, unexpected encounter to engender wonder is in turn the reason why it is a heresy in conventional 'scientific' terms, because, according to de Bolla:

> wonder is the recognition of ignorance. Wonder requires us to acknowledge what we do not know or may never know, to acknowledge the limits of knowledge. It is, then, a different species of knowledge, a way of knowing that does not lead to certainties or truths about the world or the way things are.
>
> *(de Bolla 2001, 143)*

The consequence of this observation is that a space must be made available in academic practice for encounters that cannot be exhausted by the forms of knowledge ordinarily circumscribing the "conventional modern understanding of scholarly reasoning" (Pétursdóttir 2016, 372). Based in intuitive and sensory experiences that have affective bearing, surface ecologies open up a path towards a different scholarly engagement, even "a different science language" that may "care for personal experiences and convey moments of wonder and affect" (ibid).

Continuing from this, it is also possible to return to Pétursdóttir and Olsen's discussion of the alleged detriments of ruin photography, where it is sometimes claimed that this genre excludes people and the social, moral and political dimensions of decaying buildings (Pétursdóttir and Olsen 2014, 9). However, the experience of people having withdrawn from ruins is hardly the surprising element, for instance in contemporary Denmark, where dilapidated architecture is stigmatised socially, politically and morally (Sørensen 2017b). So even though I argued earlier that these sites are rarely entirely unvisited, oftentimes they do indeed offer a sense of spending most of their time as abject matter in solitude. Hence, they can feel far removed from everyday spaces of bustle and conviviality. Yet, is the solitary not also an aspect of the social, another mode of being? And does the aesthetic inquiry into the uninhabited ruin through photography and surface engagements necessarily exclude social and political concerns? In the aesthetic arises a different form of attention to the intensity of surface ecologies that are, inevitably, assemblages of phenomena with a host of inseparable dimensions, including aesthetics, affect, the social, the political, the ethical, the archaeological and so on.

For example, when I found myself in the former living room of a severely decayed farmhouse in Northeastern Zealand, Denmark, having to make a decision about whether it was safe to move further into the space under the partially collapsed ceiling (Figure 9.7), I cannot say I was oriented towards one particular quality of the farmhouse – archaeological, political, aesthetic and so on. A curiosity drove me to learn more about how plants, herbs and fungus became part of brick, veneer, paint, tiles, flagstones, window sills, thatched roofs and concrete and how these elements were also breaking apart (Figure 9.8). I was interested in the material traces of past and possibly present lives within the place, wondering how I might understand its transformation. At the same time, I felt a nauseating weight of the ceiling right above me. I wanted to move a little closer to a particular vantage point, and I considered crawling under the caved-in ceiling (Figure 9.9). Yet *if* the ceiling should collapse on me, it would trap me in a place where nobody would come to my rescue. This realisation suddenly triggered a sense

FIGURE 9.7 Collapse. Ullerød, Denmark, 30 August 2016

Photo: T. F. Sørensen.

of being in the loneliest place on the planet. I got the overwhelming impression that the entire property was not just abandoned but depressingly neglected and unloved. Someone had failed to care for the place, and it was now left on its own. This experience clearly stood out of the ordinary, even though I was witnessing an utterly mundane archaeological formation process. The two aspects do not exclude one another, and they may indeed issue forth as a "coupling of the remarkable and the *ordinary*, of the strange and the *familiar*, of that which is outside of any explanation and that which is *explicable*" (Malpas 2012, 260).

The outermost boundary

The fulcrum of this chapter is centred upon short-lived, abbreviated fieldwork with my attention directed towards chance encounters with unlikely and insignificant occurrences

FIGURE 9.8 Storage. Ullerød, 30 August 2016

Photo: T. F. Sørensen.

encountered as photographic images. Altogether, this constitutes what I have termed a surface ecology, cutting across otherwise separate domains, such as the superficial and the profound, the mundane and the surprising, becoming and dissolution, and these confused and complex ecologies harbour both the inexplicable and inexhaustible. Thus, the surface ecology is "an outermost boundary" (according to Oxford English Dictionary); a borderland or frontier inhabited by fleeting traces of the past and of ephemeral monuments to the relentless metamorphosis of objects. These traces testify that even the most superficial surface is an assemblage of both old and new surfaces that are folded, intermingled, juxtaposed and blended (Harrison 2011, 154). The surface – to the extent that there is ever any surface in the singular – is anything but zero-dimensional, one-dimensional or two-dimensional; it is "deep", "multi-layered", "thick" (González-Ruibal 2011, 165), and the surface is where things emerge, transform and dissolve.

FIGURE 9.9 Gravity. Ullerød, 30 August 2016

Photo: T. F. Sørensen.

This is not exclusive to the surfaces we might connect with our own present day but equally those produced in the context of archaeological excavation. When carefully scraping off thin layers of soil, archaeologists generate surfaces that had never existed in the same way before, only to reconfigure them shortly after, before removing them entirely. Archaeological excavation can thus be considered "as a process of creating and exposing a series of archaeological surfaces" (Harrison 2011, 156). However, it might be helpful to add an approach to such surfaces that does not automatically construct them as distinct strata but as a cohesive 'mess'.

Surface ecologies emerge through surprise encounters, drift and wonder, and they 'stick' if we somehow attend obliquely or indirectly to them, because surfaces constantly emerge, morph or disappear. The duration of surface ecologies cannot be predicted, and there is no fixed chronology for them. They have their own rhythm, their own pace, their own life, their own death, their own pasts, their own futures, their own presence – holding together and breaking apart. This

FIGURE 9.10 Skylight. Ullerød, 30 August 2016

Photo: T. F. Sørensen.

existence at the outermost boundary is liberated of the trivial sequences of past-present-future, because surfaces are permeable, branded by frictions that go both ways (Figure 9.10). In the surface ecology, we readily see how elements that do not have any common origin whatsoever fuse together and form a whole, which does not preclude them from pulling apart subsequently. In this sense, surfaces are far from lacking substance. They may instead be the mark of intensity.

Acknowledgments

I am grateful that Þóra Pétursdóttir and Torgeir Rinke Bangstad invited me to contribute to this volume, offering me an opportunity to rethink and formulate my methods of making archaeological observations. Also, their comments on an earlier version of this chapter have been invaluable. Finally, I owe my thanks to Amalie Donsbjerg for the reference to Peter de Bolla's work.

References

Armstrong, R. P. 1971. *The Affecting Presence: An Essay in Humanistic Anthropology*. Urbana, IL: University of Illinois Press.

Barthes, R. 1989 [1980]. *Camera Lucida*. Translated by Howard, R. New York: The Noonday Press.

Bartolini, N. 2013a. "Critical Urban Heritage: From Palimpsest to Brecciation." *International Journal of Heritage Studies* 20(5): 519–533.

Bartolini, N. 2013b. "Rome's Pasts and the Creation of New Urban Spaces: Brecciation, Matter, and the Play of Surfaces and Depths." *Environment and Planning D: Society and Space* 31(6): 1041–1061.

Benjamin, W. 1999. *The Arcades Project*. Translated by Eiland, H. and McLaughlin, K. Cambridge, MA: Belknap.

de Bolla, P. 2001. *Art Matters*. Cambridge, MA: Harvard University Press.

de Bolla, P. 2012. "The State of Wonder." In *Museum Objects: Experiencing the Properties of Things*, edited by S. Dudley, 158–165. London: Routledge.

González-Ruibal, A. 2011. "In Praise of Depth." *Archaeological Dialogues* 18(2): 164–168.

Greenblatt, S. 1990. "Resonance and Wonder." *Bulletin of the American Academy of Arts and Sciences* 43(4): 11–34.

Gumbrecht, H. U. 2014. *After 1945: Latency as Origin of the Present*. Stanford, CA: Stanford University Press.

Harrison, R. 2011. "Surface Assemblages: Towards an Archaeology in and of the Present." *Archaeological Dialogues* 18(2): 141–161.

Heidegger, M. 1966. *Discourse on Thinking*. Translated by Anderson, J. M. and Freund, E. H. New York: Harper & Row.

Ingold, T. 2017. "Surface Visions." *Theory, Culture & Society* 34(7–8): 99–108.

Introna, L. D. 2009. "Ethics and the Speaking of Things." *Theory, Culture & Society* 26(4): 25–46.

Jameson, F. 1993. *Postmodernism, or, the Cultural Logic of Late Capitalism*. London: Verso.

Lévinas, E. 1978. *Existence and Existents*. Translated by Lingis, A. The Hageu: Martinus Nijhoff.

Malpas, J. E. 2012. *Heidegger and the Thinking of Place*. Cambridge, MA: MIT Press.

Mitchell, W. J. T. 2005. *What Do Pictures Want? The Lives and Loves of Images*. Chicago, IL: The University of Chicago Press.

Pétursdóttir, Þ. 2014. "Things Out-of-Hand: The Aesthetics of Abandonment." In *Ruin Memories: Materialities, Aesthetics and the Archaeology of the Recent Past*, edited by B. Olsen and Þ. Pétursdóttir, 335–363. London: Routledge.

Pétursdóttir, Þ. 2016. "For Love of Ruins." In *Elements of Architecture: Assembling Archaeology, Atmosphere and the Performance of Building Space*, edited by M. Bille and T. F. Sørensen, 365–386. London: Routledge.

Pétursdóttir, Þ. 2017. "Climate Change? Archaeology and Anthropocene." *Archaeological Dialogues* 24(2): 175–205.

Pétursdóttir, Þ. 2018. "Drift." In *Multispecies Archaeology*, edited by S. E. Pilaar Birch, 85–101. London: Routledge.

Pétursdóttir, Þ. and Olsen, B. 2014. "Imaging Modern Decay: The Aesthetics of Ruin Photography." *Journal of Contemporary Archaeology* 1(1): 7–23.

Pétursdóttir, Þ. and Olsen, B. 2018. "Theory Adrift: The Matter of Archaeological Theorizing." *Journal of Social Archaeology* 18(1): 97–117.

Simonetti, C. 2015. "Feeling Forward into the Past: Depths and Surfaces in Archaeology." *Time and Mind* 8(1): 69–89.

Sørensen, T. F. 2014. "The Outermost Boundary." *A … Issue* 1(1): 12–13.

Sørensen, T. F. 2017a. "Art and Archaeology: Looking at Holmegaard Glassworks." *Journal of Contemporary Archaeology* 4(1): 141–147.

Sørensen, T. F. 2017b. "I tidens fylde – samtidslevn og arkæologiske de/formationsprocesser." *Arkæologisk Forum* 35: 23–30.

Stewart, K. 2008. "Weak Theory in an Unfinished World." *Journal of Folklore Research* 45(1): 71–82.

Thomas, J. 2004. *Archaeology and Modernity*. London: Routledge.

Witmore, C. L. 2014. "Archaeology and the New Materialisms." *Journal of Contemporary Archaeology* 1(2): 203–224.

Wordsworth, W. 1977. *The Prelude, 1798–1799*. Ithaca, NY: Cornell University Press.

10

FROM-THE-HIP

Rocks and critical heritage ecology in the Western Australian Pilbara

Janice Baker

The affective agency of things and nonhumans is discernible in encounters with rocks in the Pilbara, one of Australia's largest mining zones. By reimagining the formative power of rocks and minerals in the West Australian Pilbara, ores acquire agency beyond their exchange value in industrial production or, not unrelatedly, as signifiers of a deep past. Either way, as exchange or temporal markers, elemental materials are miscast in humanist systems that commodify nonhuman encounters through discourses of othering. In the mining context, processes of othering are tethered to a subjectivism where ores and minerals are substantiated to service humans. In distinction to such "orephosis", we can forward an ecological thinking more in alignment with the aesthetic and affective entanglements of an earth-bone-rock ontology. For the big mining companies operating in the Pilbara, this fundamental temporal disruption is a threat. If rocks are transversal existents with affects and agency, they are no longer timeless or delegated to deep time, a necessary distancing to justify their extraction on a massive scale. Instead, in their present-ness, rocks affect the past in the future. The Pilbara, and particularly the Burrup Peninsula, is difficult heritage for big mining: "It threatens to break though into the present in disruptive ways" (Macdonald 2009, 1).

Aboriginal rock petroglyphs at the Burrup Peninsula are also distanced in the temporal framing of rocks that supports the extraction industries. Bringing Aboriginal culture too close is precarious terrain for the big miners. It is through intimate entanglements that the affect of temporal slippage is formative. Affects, in this sense, are not linear or reasonable; rather, they circle and slide, they "pass body to body (human, nonhuman, part-body, and otherwise), in those resonances that circulate about, between, and sometimes stick to bodies and worlds" (Seigworth and Gregg 2010, 1). Affects are powerful in their hiddenness; they cannot easily be interpolated into the temporal language that supports the mining industrial complex in north Western Australia.

The discourse associated with othering the earth defines what is human by separating human from nonhuman existence. New or "post" humanist ecological thinking exposes the limits of this discursive divide by acknowledging the symbiotic attachment of all entities. This ecological thinking informs a significant direction for critical heritage studies. Sharon Macdonald coined the term "difficult heritage" to address "a past that is recognized as meaningful in the present that is also contested and awkward for public reconciliation with a positive, self-affirming contemporary identity" (2009, 1). We can attend to this contestation in relation to things and nonhumans by considering an ontological intimacy with the earth that gives rocks a volcanic presence out of accord with their diminished status as deep time. These wayward stones then resurface in our thinking in defiance of the othering that is the result of their utilitarian representation as a distant, unimaginable past. Affectively, this breaks the linear thread of instrumental logic to bring rocks into the present as felt and embodied. Approaching rocks thus, through the entangled temporality of ecological thinking, provides a critical response to ideologically driven industrial futures where ores are no more or less than commodities.

In Australia, development ideology sustains the status of multinational mining companies (Rio Tinto, Woodside, BHP Billiton etc.) and is connected to the way the nation's past is conceived through colonial Anglo-settler narratives. This past overwrites the impact of Aboriginal dispossession with a frontier myth of progress based on industrial and land development that attributes Earth and Indigeneity as without time, as timeless. Affectively, however, and as I explore ahead, the past/future is carved in stone at a site in the Pilbara known as Murujuga. Such a claim realizes that a significant aspect of 'heritage' is the temporal alterity that is inherent in post-humanist ontology. This turn toward alterity is vital to allow for hybrid and entangled existences. The need of mining companies, however, is to deflate a growing sense of nuanced ecological thinking by manifesting to an increasingly concerned public that mining the earth is the only rational and reasonable way to progress.

The challenge is taken up in approaches and methodologies that acknowledge the critical valence of affect, ways forward that are not threatened by the strange and discordant connections of alterity. The philosophical integrity of affective approaches is pragmatically inflected in recent heritage studies on abandonment and ruination, such as Þóra Pétursdóttir's engagement with the aesthetic of detritus at a herring station storage house in Iceland and also in political ecologies that work with aesthetics such as Lucy Lippard's project *Undermining*. Inspired by gravel pits in the landscape, Lippard's book conveys the often ruthless politics of industrial extraction around her home in New Mexico and the ways that Indigenous people and artists respond locally and imaginatively to the behemoth of corporate mining (2014). Such resistance is profoundly driven by awareness of affects, of the way the earth enters the bones of people living in a particular place. Lippard shows how art significantly nuances ecological thinking, as in projects supported by the Centre for Land Use Interpretation (CLUI), where artists have a disingenuous dedication "to the increase and diffusion of information about how the world's lands are appropriated, utilized and perceived" (2014, 28).

The Pilbara in northern Western Australia is not unlike the arid desert landscape of Lippard's home – which perhaps bears similarity to an Icelandic landscape – and is also rich in mineral deposits. Murujuga, or the Burrup Peninsula (the Burrup), is a strip of land that juts into the Indian Ocean – a 118-km island linked to the mainland by a road and rail causeway. Here, massive industrial infrastructure overlaps with areas of extensive Aboriginal petroglyphs. The fact that uses of rocks in the region represent incommensurable differences in land use is clearly apparent (see FARA.com.au). The region's official history glosses over such differences and largely avoids that this is difficult heritage. I reformulate this neglect by imagining a future when mining fossil fuels and locating heavy industry at the Burrup has stopped. With this cessation, mining infrastructure is in a generative condition of ruination. This ruse or speculative realism perhaps aligns with what Fredric Jameson describes as a negative utopianism, "transforming our own present into the determinate past of something yet to come" (Demos 2016, 255). For T.J. Demos, this suggests "the future preordained in the conditions of our present" (2016, 255).

My focus in this speculation is not the built environment as such – the infrastructure of ports, towns, railways, mines – but the status of rocks encountered differently, in their

FIGURE 10.1 Murujuga rock petroglyphs, Pilbara, Australia

Photo: Colin Baker.

present-ness. This is how we physically experience rocks – walking on stony ground, crouching in a cave, climbing rocks, and so on. Rocks are felt earth; they are tactile, solid and present, holding us in place. They have gravitas. Yet we feel enabled to extract and crush the earth; rocks, pebbles and gravel exist at the very end of a spectrum of the non-sentient, nonhuman. I want to consider how we might approach a critical intimacy with rocks, where rocks from an industrial, fossil-fuelled age are "existents" rather than raw material.

In the ecological thinking this complex continuity of entities admits, the key actors – ores, oil, natural gas, steel, lithium and so on – are not victims. As Timothy LeCain observes, the earth is not passive: "Coal shaped the humans who used it far more than humans shaped Coal" (2015, 21). The affect is destabilizing, forcing us to elide the western philosophical separation of nature from culture: that "notion of Nature as an autonomous ontological domain, a field of inquiry and scientific experimentation, an object to be exploited and improved" (Descola 2014, 69). As we move away from discursively taming the earth and start to render alterity into our encounters with nonhuman others, it follows to acknowledge that any environmental degradation enacts our own geo-bio destruction.

To impart a critical role to affect in this way might seem a contradiction; isn't affect precognitive and autonomous (Massumi 2002)? It is, but in the sense that such autonomy leads to new thought. A shock to thought. Affect theory engages with sidelined realms of feeling, with embodied beings-in-the-world; it treats affects as a form of intelligence (Sedgwick 2003; Thrift 2008; Gregg and Seigworth 2010). In affective mode, "things do not have to be draped with our conceptual apparatus in order to be acknowledged" (Anderson 2014, 39).

Mining the past: the blinkered effect of orephosis

The Western Australia Pilbara is about the size of Spain, a hot and arid landmass stretching from the Indian Ocean across the Great Sandy Desert toward the Northern Territory border. One of nine Western Australian governmental regions, the Pilbara spans

507,896 square km and, given its size, is sparsely populated, with an estimated 65,859 people resident in the area in 2015 (ABS 2015). It is difficult to accurately gauge population size, as this fluctuates with mining activity, most of which is performed by fly-in fly-out (FIFO) workers or carried out remotely from the distant capital city, Perth. The region has vast deposits of iron ore, petroleum and off-shore natural gas, the extraction of which largely informs the national identity of the region; "most of the (very few) books written about the Pilbara have focused on its [mining] companies, founding figures and the leading entrepreneurs" (Ellam 2017, 4).

A discourse of isolation, remoteness, distance, maverick characters and tenacious going-it-alone is attached to the Pilbara (Layman 1982) and continues to support the autonomy of the big miners. In a study of the de-unionizing of labour in the Pilbara mining industry, Bradon Ellam suggests, however, there is a mischaracterization in the perception that the region is isolated:

> If the Pilbara is physically isolated, then in economic terms it is anything but that. It lies at the heart of global production networks, linked to the Asian steel mills buying its ores and to mining companies listed on stock exchanges in Britain and funded from all over the world. Its wealth puts it at the heart of political networks criss-crossing Australia.
>
> *(2017, 6)*

The significance of industrial extraction to the Australian economy is effectively attached to national identity. Malcolm Knox argues it is little understood how integral mining is to "Australians' perception of themselves", nor is the industry's role in maintaining this perception recognized. The industry harnesses a deeply held belief in mining as the "backbone of the country" (Knox 2013, xiii). Guy Pearse, a close observer of Australia's "quarry vision", notes wryly that "Faith in mineral and energy commodities is such an important part of the national psyche that TV newsreaders routinely assume the nation will not sleep without knowing the closing BHP and Rio Tinto share prices" (2009, 8).

Elsewhere, I describe a condition I call orephosis – the subjectivism that supports the extractive industries through the blinkered logic of anthropocentrism (Baker 2017). Orephotic thinking is the "common sense assumption that the mining of natural resources and the progressive evolution of the human species are inexplicably connected" (ibid). Here we find an assumed common sense that through economic growth temporality *is* progress. The orephotic instrumentalism of contemporary mining, where rocks are determined by their exchange value, is significantly exposed through the relations and entangled existences of an earth-bone-rock ontology. In this, in a bleak optimism for those of us interested in facilitating new subject positions, the earth is on our side. As Deborah Bird Rose observes: "The Anthropocene is now doing what decades of insightful critique never quite succeeded in doing: it is forcing the truth upon us" (2013, 209). In this vein, anticipating the future of a thriving industrial zone as an abandoned ruin might seem sci-fi, yet it resonates with the felt truth of our entangled existences. A methodology of abandonment as a pragmatic reconciliation with rocks situates a nuanced heritage ecology in the midst of the problematic attached to the production of the new. This problem requires the critical and creative energy to stay abreast of mining companies perpetuating their control over public discourse through orephotic rhetoric in an example "where the human capacity for innovative thought has become a new realm of exploitation" (O'Sullivan and Zepke 2008, 2). Félix Guattari suggests not only that ethical aesthetics is the way to affectively penetrate the control of capital but optimistically that it's fundamentally immanent:

> Perhaps I'm delirious, but I think we're in a period of absolutely fantastic production, creation, and revolution with regard to the emergence of a people. This is what I mean by molecular revolution: it's not a slogan, it's not a program, it's something I feel, something I live though in encounters, in institutions, in affect.
>
> *(cited in O'Sullivan and Zepke, 10)*

In the Pilbara, large-scale mining followed the lift in 1966 of a federal embargo on the export of iron ore. Mining and development has since grown across "boom and bust" cycles driven largely by fluctuations in Asian commodity markets. The engine of this industrial activity is on the Burrup Peninsula. The region's Indigenous Yaburara people aptly named this peninsula "Murujuga", meaning hip-bone sticking out. The area is sacred to Indigenous people, with archaeological evidence documenting human habitation in the area for at least 30,000 years (Donaldson 2009, 17). The Murujuga Aboriginal Corporation's website informs us that since European occupation of the region and with the massacre of Yaburara people, the area is considered orphan land by neighbouring Indigenous groups, who took over custodial duties for the area (Murujuga n.d.).

Geologically, the area is significant not only for vast mineral deposits, but outcrops of granite rocks at the Burrup are over three billion years old and "among the oldest remnants of the early Earth's crust" (Donaldson 2009, 15). These rocky outcrops cover the Peninsula and feature countless diverse Aboriginal petroglyphs. While, as geologist Mike Donaldson notes, there is no detailed inventory of this vast body of art, "some researchers estimate that there are more than a million engravings on the peninsula and surrounding islands" (2009, 16). This astonishing number of cultural artefacts is difficult to comprehend, not least as this indication of activity contradicts the colonial expediency that prior to Europeans, the area was unoccupied.

Following the commencement of mining in the late 1960s, rock sites at the Peninsula were registered under the Aboriginal Heritage Act of 1972 (FARA n.d.). This offered limited protection, with mining companies securing approval to either move or destroy sites, "no one knows how many sites were destroyed prior to 1972, it is estimated that many thousands of petroglyphs have been destroyed since then" (Witcomb and Gregory 2010, 312). Archaeologist Ken Mulvaney observes that since the first iron ore shipping facilities were constructed at the Peninsula the impact "is not just the footprint of commerce ... the visual and audio amenity of the place is also altered by industry" (2015, 766). Mulvaney is a passionate advocate of the rock art's global importance, pointing out these artefacts are "older than the pyramids or Stonehenge" (see ABC Open Pilbara).

Areas at the Burrup have been "put aside" for heritage protection of rock art, but Mulvaney observes that, while some consider a balance has been achieved between heritage protection and economic prosperity, "in reality the management of heritage is framed in such a way that ensures that planned developments, the capacity of existing industries to expand and provisions of utilities are not constrained" (2015, 770). José González Zarandona puts it bluntly: "Indigenous sites like Murujuga are destroyed because the mining companies consider them to be obstacles to fulfilling their operation" (2015, 465). He proposes that what is happening at Murujuga can be understood as "landscape iconoclasm" and that we should aim "to create a *hericlash*: the destruction of a definition of heritage in which invisible meanings and unseen values are not recognized" (González Zarandona 2015, 468). Given this contention, it is complicated to situate the 2018 process for UNESCO World Heritage nomination for the Burrup Peninsula (Government of Western Australia 2018).

An irony attached to the protection of the Burrup's rock art is that experts involved with protecting its heritage value are often either directly or indirectly remunerated by the mining multinationals. This tension is one of many challenges in ecologizing the region. Mining companies proceed with strategic awareness that managing cultural heritage ensures their social license to operate is not compromised. The state of play underpinning this form of expediency aligns with what Elizabeth Povinelli calls a "cunning recognition", which disarmingly connotes what is at stake:

> As the nation stretches out its hands to an ancient aboriginal law in order to embrace its own ideal body, indigenous subjects are called on to perform a complex set of semiotic maneuvers in exchange for the good feelings of the nation and the reparative legislation of the state.
>
> *(2002, 55)*

Fuelling the situation at the Burrup is a contention that no one really "owns" the land: "In 2003 the federal government denied outright native title to local groups on the basis that no one group could demonstrate a continuous association with the area" (Witcomb and Gregory 2010, 311). And as González Zarandona notes: "Destroying Indigenous sites that are considered to be cultural signifiers continually enhances the political goals of diminishing the Indigenous past and weakening Indigenous claims to the land" (2015, 465).

Responding to these inequities through a kind of affective destabilization sits outside the dominant view that heritage is solely related to efforts to memorialize the past by fixing a site to its official meaning. In destabilizing the Burrup industrial zone as a site that is exemplary of progress, the context shifts to that of large-scale extraction as a destructive force, as a material manifestation of damage to entangled human and nonhuman relations. We interpret ideological harm through its material ruination. This opens subjective experience to the valence given to the excesses of capital through the condition of decay left in its wake. An existential question seems formative to the measure of such an interpretation. This acknowledges existence as a matter of impermanence and transition and might be seen as a productive trajectory of the perspective suggested by Caitlin DeSilvey's evocative term "palliative curation" (2017).

In the context of recent industrial heritage and experimental preservation of "curated decay", DeSilvey qualifies that this does not assume traditional heritage is not important but that there are also occasions for "an alternative to the stability and stasis we are accustomed to experiencing in heritage spaces" (2017, 115). Relatedly, a speculative approach to the Burrup industrial zone also extends the insight of Pétursdóttir and Olsen to "how modern ruins and thing-oriented perspectives may help rethink heritage in relation to aesthetics and ethics" (2014, 5). In this regard, the action of ruination becomes the significant affect of a heritage of abandon at the Burrup. "The ruins of the recent past … display themselves in the ongoing process of ruin*ing* – where ruination itself, the active process of withering and decay, becomes conspicuous and draws our attention" (Pétursdóttir and Olsen 2014, 7).

In a vision of the Burrup without industry, a visitor to the area might encounter abandoned earth in various states of ruination – crushed ores, steel girders, gravel beneath railway tracks, all left to weather according to the agency of their own duration. Rust never sleeps. There is an impact here of abandon that provides alternative existence to the anthropogenic othering of rocks, a kind of passing the reins to matter, a matter-dom. Such materialism respects the duration of rocks not as signifiers of deep time but by an intersectionality that is akin to Guattari's transversal approach to ethico-aesthetic practices (2006). His anticipatory subjectivism does not separate the environment from a subject but is a way of moving through multiple worlds *as* the present rather than thinking through the neat division of being as time fixed to a single past and future.

Pétursdóttir's study of material abandonment gets to the gritty sense of matter as decaying and weathering stuff that touches our own condition of mortality (2014). Likewise, what happens at the Burrup when we let go of the metal and steel structures that form our extractive mentality? Such decay would of course happen over millennia, outside human duration. The site would normally, probably, be turned over, stripped and then left. But we are letting it alone in its abandon, to decay over countless generations, out "mattering" everyone alive today. Pétursdóttir has studied the aesthetic affect when things are left to themselves: "It was not something to be made sense of but rather something that had to be sensed in its strangeness at that very moment – here and now" (2014, 357). We have here a sense of the production of something new. In this otherness "that is not of anyone's making, 'things' strangeness was genuinely theirs" (ibid). Such affect, she observes, defies our rigorous ideas about what things are, and how they should be.

Transversal existents: toward a post–fossil fuel future

The Western Australian gas industry plans massive expansion at the Burrup, dismissing local and global protests to cease mining fossil fuels, and far removed from deep reflection on the politics of fossil fuel extraction such as philosopher Reza Negarestani's conception of oil. In his exploration of the effect of oil on the Middle East, Negarestani's theory-fiction novel (*Cyclonopedia* 2008) treats oil as a form of sentience. In his novel, oil's agency imparts the power that oil holds over the politics and theology of the Middle East. The relevance of such a political ecology is apparent, as eco-activist Bill McKibben determines: "The richest industry in human history (oil) wants to keep on their current path for a few more years, even if it means dragging the whole planet over a cliff" (cited in Demos 2016, 26). Stimulated by the disturbance oil's sentience enacts in Negarestani's writing – where oil lubricates the Middle East and the so-called War on Terror – we can frame a disturbing political ecology in the Pilbara. This would have rocks as seductive players in cognitive capitalism: constructs of desire made manifest in their transition from underground minerals to crushed ore to giant steel bodies piped gas, and liquid lithium to the banks of mega mining multinationals.

Negarestani invents a spectral ecological text that brings affect and oil into politics. He reveals Earth as a politically historicized feint through concepts that construct a darkly thought oil heritage. This oily thinking perhaps sits alongside Timothy Morton's dark ecology with its manifestation of hyperobjects (2013). The speculations enabled by such conceptual inventions circulate in my thinking to suggest an alternative vision affected through conceiving industrial abandonment.

From a post-humanist vantage point, the demise in the exploitation of the extractive industries requires attending to what are currently deemed uncanny engagements with the earth,

FIGURE 10.2 Murujuga landscape. Rock petroglyphs and Karratha Gas Plant

Photo: Colin Baker.

such as listening to the earth speak. Attending to the uncanny is not, as Morton observes, a usual human thing to do:

> What is most uncanny about human being is its attempt to rid the world of the uncanny. Or, and this is putting it in its most ecological register: human being disturbs Earth and its lifeforms in its desperate and disturbing attempt to rid itself of disturbance.
>
> *(2016, 64)*

A way to approach the uncanny is to "let it be", which is where new awareness of temporality and duration are both effecting and affective. In the Pilbara, such an engagement with rocks as transversal existents is imminent in Aboriginal culture. The ontological incommensurability between western and indigenous time comes into play here.

Murujuga: aligning to earth-bone-rock ontology

At Murujuga – the hip-bone–shaped island extending into the stunning turquoise Indian Ocean – images of fish, animals, people and other things belong to the rocks that acquired these markings by removal of the surface through techniques of "abrading, carving, engraving, scratching, incising, pecking, pounding or rubbing" (Donaldson 2009, 16). Created and re-patinated across tens of thousands of years, five sequential changes in the rock art are thought to reflect shifting responses by Indigenous people to flora and fauna across the region during the Pleistocene and into the Holocene.

Features in the rock art appear to parallel cultural events still taking place in the region. In more recent rock images, for example, "the figures appear to hold thin linear objects, similar to the dance sticks in use today" and Ken Mulvaney observes: "These sticks are made by shaving the bark of the Pilbara Wild Cotton (*Gossypium robinsonii*), leaving it attached to the thin stem, which dries into a tight curly ball" (2015, 764). A photo of men and women of

Roebourne community show them performing public ceremony holding these dance sticks, "suggesting such ritual practice has many millennia of use" (Mulvaney 2015, 763). The poignancy of this enactment is intense when we open to the involvement of time. For Aboriginal people engaging in this public ritual, the past is the future. Deborah Bird Rose discerns that in the disorientation of western time inherent in Indigenous experience, such as the dance stick ceremony, there is the opportunity to engage "quite differently with Aboriginal people" (2013, 11). She cites Derrida's concept of hauntology to suggest the awareness of facing the past as this comes to us from the future. The future is already with us. Bird Rose notes that in the Aboriginal world, the future is those who come behind us, "the behind mob". "Those behind us walk in our footsteps, as we walk in the steps of our old people. It is easy to experience a dizzying sense of the past jumping ahead, or of time running backwards" (Bird Rose 2013, 14).

As we attune to being present with rocks, as we walk on, up and around the rocky outcrops at Murujuga, the certainty represented by the adjacent vast industrial infrastructure is temporally destabilized. There is a transmission of affect that, as Teresa Brennnan observes, "if only for an instant, alters the biochemistry of the subject. The 'atmosphere' or the environment literally gets into the individual" (2004, 1). There is something here of Lucy Lippard's point that "rock art quietly absorbs us into its place, even when we understand very little about the messages we are getting" (2014, 91).

On this remarkable piece of the planet, there is a transversal aesthetic, equivalence among things actual and virtual that can be somehow felt through the earth's crust, through the metaphor of a hip bone, and through the rock imaginary of countless generations. This earth-bone-rock ontology is incompatible with the land interpreted in Anglo-settler discourse where the first inhabitants are conceived as placeless nomadic hunter-gatherers. The notion of Aboriginal nomads was a "fundamental component of the doctrine of terra nullius [empty land], an idea pivotal to the British assertion of the right to colonise" (Coleman 2017). Claire Coleman writes that her Aboriginal forbears "were not simple hunter-gatherers before white people came, rather we were farmers using methods and technologies unrecognisable to the British invaders". She observes it was the British who were the actual nomads "by the true definition of the word, travelling to find pasture for their herds" (Coleman 2017).

The land speaks so differently to the colonial record that we surely have to feel ashamed of human acts made through the sensible auspices of western reason. Perhaps the shame of having compromised with a way of thinking is what enables us to acknowledge the importance of new thinking. Perhaps the shameful nature of humanity becomes a form of empathy. The strong affect of shame, Elspeth Probyn suggests, "radically disturbs different relations of proximity: to our selves, bodies, pasts" (2010, 86).

At Murujuga: "Vast galleries of images occur in association with other archeological features, artifact scatters, shell middens and stone arrangements" (Mulvaney 2015, 760). Standing stones, quarries, stone tool-manufacturing sites, short-stay meal-time sites and longer-term campsites present "a cultural landscape, where the mark of countless millennia of human activity is imprinted onto the natural landscape" (Mulvaney 2015, 766). I look at a sun motif impressed into a rock face at the top of a basaltic outcrop rise in the Burrup, known today as the Valley of the Sun. Across tens of thousands of years, I envisage family groups in rock shelters, their activity adapted to the country where, during summer months, the maximum daily temperatures exceed 32 degrees C (90°F) and not uncommonly 45 (113°F) (Pilbara Weather n.d.). Mirage forms shimmer in the heat. The sun of red-brown rock has a dark yellow core from the imprint of pounding the surface.

I'm conscious of objections to the pursuit of the aesthetics of rock art as misleading and limiting but see with Thomas Heyd that art historians are not as careless as they once were, and it is recognized that nineteenth-century conceptions of art are anachronistic (2012, 287). The ethics of knowing rocks interacts with the surfaces and their markings, all tugging on the way I think and walk with attentiveness to the stony ground. This is not some kind of sublime encounter that manifests in the western tradition as an existential lack, nor a terrible beauty to be countered by finding my way back to reason. There is simply too much bodily stumbling over rocky terrain in hot dry conditions, a pragmatic clinging to what is underfoot. A journal entry and later comment on 1860s explorer A. E. Anderson note something of the same:

> The country … is difficult to travel over, the hills covered with stones, a horse could do not more than 25 miles a day and must be well shod. [Explorer A.E. Anderson] is referring to the iron ore stone country … between the Ashburton River and Hamersley Range. This reference to stone, the fact that one's boots wore out quickly and that they had to blast to sink a well through blue stone rock or solid opaline rock are the only hints which indicate that they were in fact on what we now know to be vast mineral deposits.
>
> *(Edgar 1989, 78)*

The ethical-aesthetic of stony Murujuga is not about absence of reason but being present. A pragmatic condition of land underfoot recorded in pictures of its fish, creatures, food and people while nearby, the iron and steel infrastructure supporting massive industrial extraction makes a slow rusty transformation to dust – this at least is the speculation – the duration of petro-capitalism's orephotic mark in deep time.

This reverse aesthetic – where petro monuments are in palliative care and the carved rocks voice a future of clean air – attunes the body to country. In thinking more generally why certain rocks within specific landscape settings are chosen, and what this does to the body, Chris Tilley and Wayne Bennett bring affect to the task of archaeology of identifying objects:

> Going beyond vision what do the rocks and carvings feel like? Might changes in tactile sensations be important? What about the relationship between carved rocks and auditory dimensions of experience that can still be recorded in the surrounding landscape, such as the sound of the near or distant sea?
>
> *(2004, 223)*

This phenomenological approach brings to mind David Mowaljarlai's ritual knowledge of his country in north Western Australia. He attuned to the country "because of all the cultural work he performed", which included "lodging painted figures in caves, determining sight-lines to other sacred zones, bouncing songs of cliff faces" (Gibson 2006, 23.4). Rocks matter as markers of the health of country, and Mowaljarlai's audio-visual awareness of subtle shifts imparts through rocks in his country swinging around his body.

We can add to our stumbling-over-rock intimacy that everything is constituted of the geological and biological matter that eventually becomes matter for reformulation. Artists know this. "We are like volcanoes, producing new landmasses on a micro scale" says Ilina Halperin (2015, 82). Her art practice queries where the boundary blurs between the biological and geological. We form geology: "What is the relationship between a body stone and a volcanic

FIGURE 10.3 Rock petroglyphs with Karratha Gas Plant in the background

Photo: Marius Fenger. Murujuga National Park. Burrup Peninsula, Pilbara Western Australia.

rock?" (Halperin 2015, 82). So it is that, given we are transversal matter, increasing acidity from toxic industrial pollutants transform both us and the environment, changing the rocks and transforming our physiology at a micro scale.

Close attention to Indigenous painting imbues the patina of such a corporeal geology. As I write, I turn to a recent painting of the Hamersley Ranges in the Pilbara by Yindjibarndi woman Celia Sandy. The canvas inscribed with her marks is part of her geology, a vivid activity of red-purple, white and ochre geometric forms that manifest the undulating rhythm of the Pilbara hills. The colours, shapes and vibrancy de-centre any desire to stabilize the picture into a three-dimensional perspective. There is no horizon line to give the perspective of depth or distance; rather there is paint *as* the terrain of ancestral Yindjibarndi country (Yinjaabarni n.d.). This privilege of seeing through the ease of painting country as affect does not accord with a gaze that attributes nature as the other of human sentience. In breaking from this logic of difference in the western philosophical schema, aesthetic affect in such Aboriginal art does not impart the narrative of land use encountered at places of settler heritage.

Where are the rocks? Settler heritage and the present earthhuman

The rocky outcrops in the arid Pilbara were never a fit with the European vision of colonizing a picturesque landscape. Not that this prevented the illusion of taming and developing the country in the European way of cultivation: "the advancement of pastoral and mining interests in Australia has always depended on drawing a picturesque veil over reality" (Carter 2004, 22).

The material impact of human intimacy with the region's rock sites goes largely unrecorded in representations of Anglo-settler history at official heritage sites in the Pilbara. These sites interpret material remains of homesteads and buildings through the lens of hardy pioneers taming and developing unoccupied arid land. This narrative is iterated through objects and trails to impart a tradition "in which the pioneers revealed moral courage and physical strength in the face of hardship and opportunity" (Witcomb and Gregory 2010, 10). Discourse around mining and economic growth in the Pilbara extends the pastoralist story as one of European settlers overcoming difficulty to develop the land for the betterment of all. A 1980s history published by the Pilbara-based mining company Hamersley Iron is indicative: "The dust, flies and excessive heat should have been enough to dampen the spirits of even the hardiest but it seems the challenge of finding unknown riches was sufficient to keep many out in the scorching wilderness" (Hamersley Iron 1986, 17). Heritage sites in the Pilbara at Karratha, Roebourne and other towns signify this tenacity and reiterate the ideological impetus we see promoted as a measure of value of today's industrial extraction. Additionally, such rhetoric is attached to the useful qualities of iron ore, which provides a tacit distance from the ancient rock art. Rock art is ontologically dangerous ground for the big miners, as it conveys an alternative mode of existence, that of earthhuman assemblages, a symbiotic entanglement to threaten the economic ground for mining by attributing a meaning for geology not in step with the exchange value of geological formations as trading commodities.

Before mining, and prior to the 1860s advent of pearling and pastoralism in the Pilbara, between 20 and 30 groups of Indigenous people thrived in the region, interacting to share knowledge and manage the land. This fundamental attachment to country was dismissed in the colonists' view that Aboriginal people were nomadic and not attached to place. Yet we have seen the rocks inscribe Indigenous attachment to particular sites over countless generations. The dissembling activity of history finds the status given to European occupation dismissing Aboriginal knowledge. With the arrival of British pastoralists, Aboriginal people worked as shepherds on the large stations (Edgar 1989, 75) so they could remain on country. Paid in rations, they worked in the pastoral industry until 1968, when, following the declaration of an equal pay award, many were evicted from the stations (Bolton 2008). Assimilationist policy took children away from families, placing them in missions and rupturing continuities with country and culture (Haebich 2000).

Settler stories in the Pilbara filter out or pay scant attention to the dispossession of Aboriginal people (Paterson and Gregory 2013). Funded by a 1988 bicentennial grant to celebrate the arrival of Europeans on the east coast of the continent 200 years earlier, Pilbara heritage trails continue to commemorate European settlement rather than exposing its dislocations, such as, for example, the Emma Withnell Heritage Trail named after the first white woman in the Pilbara. Known as the "mother of the northwest", Emma Withnell and her husband took up a lease at Mallina Station in the Pilbara in 1863. A "treasured local legend" (Bolton 2008, 57) tells of the discovery of gold on the Mallina Homestead in January 1888 by their young son shying a stone at a crow. The story goes that by 1890, there were up to 150 miners working around the station; from this a township grew, and the region was named 'Pilbara' after a local creek (WA Heritage Council).

Today the Withnell Trail is a 52-km self-guided tour of historic sites at the towns of Cossack, Roebourne, Wickham and Point Samson. An implication of non-Indigenous heritage as a representation of development, and Indigenous culture as the other of the colonial settler narrative, is to ignore the powerful meaning of this settler history for Indigenous people (Paterson and Gregory 2013). The reality is that the pastoral industry was backed by what was essentially a "paramilitary police force" in northern Western Australia, and, given this, "Aboriginal imprisonment rates soared to levels far higher than in any other colony from the 1880s" (Haebich 2000, 210). The trails, statues and monuments erected to pioneer settlers stand as a refutation of this other difficult history contributing to an authorizing discourse that simplifies historical messages, leaving "little conceptual room for alternative uses of heritage" (Smith 2006, 41).

The Yaburara Heritage Trail is a 3.5-km walking trail of the Pilbara's Karratha district natural history and cultural heritage and "in particular the heritage of the traditional owners of this land, the Ngarluma people, and their neighbors the Yaburara people" (City of Karratha n.d.). However, whilst the trail map acknowledges dreaming sites of Indigenous people who used to live in the region, there is a general vagueness in this and other heritage interpretation about the 1868 Flying Foam Massacre at Murujuga. This was an "incident" involving a series of killings of Aboriginal people by police following what appears to have been the theft of flour (Gara 1983). The Trail map avoids the region's difficult heritage and conflicting land use by separating information about Aboriginal culture from industrial development. There is a substantial section titled "Karratha's Development" with several long paragraphs extolling industry, such as this excerpt:

> While Hamersley Iron was developing its iron ore project, on the nearby tidal flats Dampier Salt Company was constructing the largest solar salt fields in the southern hemisphere. The discovery of major natural gas reserves on the North West Shelf in the early 1970s heralded another major phase in the consolidation of the Pilbara's development. Woodside Offshore Petroleum has invested $11 billion in the exploitation of these reserves in the 1980s with the construction of the North Rankin 'A' gas platform, a 13km submarine pipeline, a domestic gas plant, a 1600km pipeline to supply natural gas to Perth and the South West and a liquefied natural gas plant at Withnell Bay to produce Liquefied Natural Gas (LNG) for export to Japan.
>
> *(City of Karratha n.p.)*

Where are the rocks? The lack of ecological context given to industrial development and the reality of significant and ongoing contested histories has been explored by Alistair Paterson and Kate Gregory who describe, for example, the Old Roebourne Gaol as an incongruous heritage site; it is both a business tourist centre and a place of memory in the present of brutality toward Indigenous people (2013).

That the Flying Foam Massacre at Murujuga is overlooked is reflective of a national approach to frontier conflict in Australia. The notion of massacre is deeply confronting for the Australian psyche, and the nation is only gradually approaching this shame. Mark McKenna notes: "Australians have yet to accept that they live in a country with two ways of knowing the past" (2018, 43). With new scholarship providing the evidence that there are probably "more than 500 massacre sites nationwide" (ibid), there is a sense that this situation has begun to change. The 1868 massacre at Flying Foam at the Burrup is neglected in official heritage; however, a group of 96 erect standing stones are believed to be Indigenous markers of the slain Yaburura (Mulvaney 2015, 766).

Ecological thinking: critical heritage and imperative civility

Ecological heritage draws attention to the way the industrial era is remembered materially and how we might manifest this materiality. For artists, this is not new. Andrej Zdravič has long manifested cinematic environments that present nonhuman durations, such as his film *Riverglass* (1997) which, filmed over four years beneath the surface of a river, slows the passage of time to enable an experience of the nonhuman (Brown 2014, 46). More recently, the artists HEHE produce images of strangely coloured skies reflecting the meaningless of scientific codes in picturing global warming and air pollution (2014, 34). What alternative ecological perspectives productively highlight is the careless defaulting to human privilege over and above the reality of our entangled existence with the nonhuman world.

A speculative cessation of large-scale mining in the Pilbara is an ontological shift in sympathy with what Bruno Latour calls "suspension of the modernization front" (2013, 10). The north of Western Australia is currently a thriving industrial front, and my critique is fictional. However, my musing may be anticipatory, given that hovering over alternative contexts is the prediction that industrialized civilization may collapse within the next hundred years (Latour 2013, 57). As such, Latour's call to ecologize as an alternative to modernize is not as crazy as supporting the status quo, which gives a critical imperative to what it might entail to return affect to earth's systems. An ecological heritage of rocks as an imperative civility.

Abandoning steel and iron structures to rust and dismantle in the desert – Ozymandias of the Anthropocene – responds to an awareness that is actually underway even as the old economic mantra of growth and prosperity continues. To ship ore from Pilbara's Dampier Port to China to return it to the Pilbara as steel girders to construct structures to extract more ore is only rational in the economic logic of commodity trading. As ecological heritage, the abandoned infrastructure of this bizarre trade provides a millennial watching platform as earth reclaims its iron. For a while, perhaps decades, the site will remain under guard, then made safe for visitors, then a site from which to ponder what went wrong with economic misuse of the earth. Given that in 2018, the Port at Dampier is the world's second-largest bulk export port, this futurism will be quite a spectacle. Yet, as Torgeir Rinke Bangstad observes, loss of context in an obsolete site "makes the industrial relics carry a whole range of unauthorized and undetermined associations offering considerable leeway for new interpretations" (2014, 98). It is through oscillation between thriving industry and devastation of this modern feat that we might speculate on the affective possibilities for an industrial heritage landscape of radical ontological undoing.

An ecological heritage of abandoned metal and steel structures as the future legacy of the Burrup exposes processes of inclusion and exclusion that diminish affective voices indigenous to the region, human and mineral. While economic logic always makes it hard to encounter other agencies (Latour 2013), there is such clear wrongness in the presence of industrial complexes laying bare and polluting the earth that exploring the condition for ethical exits to a post-human future is increasingly imperative (Kroker 2014).

Our affective entanglement with things is increasingly activated in our thinking. Curiously, it is actually more difficult to know minerals and rocks as "energy" than it is to appreciate them as the earth underfoot. I felt a keen sense of this in the extensive coverage given to the 2018 rescue of boys and their football coach trapped underground in Thailand's Tham Luang Nang Non cave system. The caves were not represented in this coverage as monstrous or a force to be overcome – nor was the monsoon rain that trapped the group in the cave – rather there was respect ceded to the weather and vast underground network of spaces. Respect for

FIGURE 10.4 Rock petroglyphs with Karratha Ammonia Plant in the background

Photo: Marius Fenger. Murujuga National Park, Burrup Peninsula, Pilbara, Western Australia.

earth's agential capacities and formative power is increasingly discerned as awareness grows of the violence imparted in fuelling human life by extracting and burning fossilized earth. Rocks can no longer effortlessly be diminished to either riches or wonders, which is how they signify as natural resources or sublime objects in the western tradition. This shift in thinking suggests elemental entwining as an affecting experience, an ontological alterity that is good to reframe the rhetoric of economic growth underpinning human relations with things as commodities.

The turn to ecological awareness is underpinned by affects that put us down a rabbit hole to encounter the state of the earth as our own. The idea that things and nonhumans are continually affected and themselves affect rejects the notion of earth as dead inert matter and places the onus on us to find ourselves in zones of alterity. The earth is the expert on the earth, so let's pass it back the mantle. Of course, humans may not be around to witness the earthhuman future of Murujuga's abandoned industrial infrastructure. Life may have moved into an extinction scenario, in which case the petroglyphs remain witness to the excesses of the humanist project. We must stumble more carefully over the stony earth.

References

ABC. 2014. *Open Pilbara*, 11 November. Accessed 3 May 2018. https://open.abc.net.au/explore/83052.

ABS. 2015. Accessed 17 May 2017. http://stat.abs.gov.au.

Anderson, D. T. 2014. "Trusted Vagueness: The Language of Things and the Order of Incompleteness." In *Ruin Memories: Materialities, Aesthetics and the Archeology of the Recent Past*, edited by B. Olsen and Þ. Pétursdóttir, 33–40. London: Routledge.

Baker, J. 2017. "Metal Fictions." Theorizing the 21st Century. *CTheory.net*. Accessed 7 May 2018. https://journals.uvic.ca/index.php/ctheory/article/view/16858/7179.

Bangstad, T. R. 2014. "Industrial Heritage and the Ideal of Presence." In *Ruin Memories: Materialities, Aesthetics and the Archeology of the Recent Past*, edited by B. Olsen and Þ. Pétursdóttir, 92–105. London: Routledge.

Bolton, G. 2008. *Land of Vision and Mirage: Western Australia Since 1826*. Perth: University of Western Australia Press.

Brennan, T. 2004. *The Transmission of Affect*. Ithaca, NY: Cornell University Press.

Brown, A. 2014. *Art and Ecology Now*. London: Thames & Hudson.

Carter, P. 2004. *Material Thinking: The Theory and Practice of Creative Research*. Melbourne: Melbourne University Press.

City of Karratha. n.d. *Yaburara Heritage Trail Map*. Accessed 23 July 2018. https://karratha.wa.gov.au/yaburara-heritage-trail.

Coleman, C. G. 2017. "When We Encountered the Nomads." *Meanjin*, Summer. Accessed 5 May 2018. https://meanjin.com.au/essays/when-we-encountered-the-nomads-2.

Demos, T. J. 2016. *Decolonizing Nature: Contemporary Art and the Politics of Ecology*. Berlin: Sternberg Press.

Descola, P. 2014. *Beyond Nature and Culture*. Translated by J. Lloyd. Chicago, IL: The University of Chicago Press.

DeSilvey, C. 2017. *Curated Decay: Heritage Beyond Saving*. Minneapolis, MN: University of Minnesota Press.

Donaldson, M. 2009. *Burrup Rock Art: Ancient Aboriginal Rock Art of Burrup Peninsula and Dampier Archipelago*. Mt Lawley, WA: Wildrocks Publications.

Edgar, I. 1989. "History of the De Grey Station 1863–1913." In *North of the 26th*, edited by H. Weller, 75–80. Perth: Reeve Books.

Ellam, B. 2017. *The Pilbara: From the Deserts Profits Come*. Perth: UWA Publishing.

FARA. n.d. *Friends of Australian Rock Art/FARA Activities/Research*. Accessed 23 July 2018. www.fara.com.au/category/research/.

Gara, T. J. 1983. "The Flying Foam Massacre: An Incident on North West Frontier, Western Australia." In *Archaeology at ANZAAS*, edited by M. Smith, 86–94. Perth: Western Australian Museum.

Gibson, R. 2006. "Spirit House." In *South Pacific Museums: Experiments in Culture*, edited by C. Healy and A. Witcomb, 23.1–23.6. Clayton: Monash University ePress.

González Zarandona, J. A. 2015. "Toward a Theory of Landscape Iconoclasm." *Cambridge Archeological Journal* 25(2): 461–475.

Government of Western Australia. 2018. *McGowan Government Progresses Burrup World Heritage Listing*. Accessed 17 May 2018. www.mediastatements.wa.gov.au/Pages/McGowan/2018/08/McGowan-Government-progresses-Burrup-World-Heritage-listing.aspx.

Gregg, M. and Seigworth, G. J., eds. 2010. *The Affect Theory Reader*. Durham, NC: Duke University Press.

Guattari, F. 2006. *Chaosmosis: An Ethico-Aesthetic Paradigm*. Translated by P. Bains and J. Pefanis. Sydney: Power Publications.

Haebich, A. 2000. *Broken Circles: Fragmenting Indigenous Families 1800–2000*. Fremantle: Fremantle Arts Centre Press.

Halperin, I. 2015. "Physical Geology/The Library." In *Art in the Anthropocene: Encounters Among Aesthetics, Politics, Environments and Epistemologies*, edited by H. Davis and E. Turpin, 79–84. London: Open Humanities Press.

Hamersley Iron. 1986. *Early Mining in the Pilbara. Hamersley Iron 1986: The 20th Year of Hamersley Iron*, edited by K. Wearne. Perth: Hamersley Iron.

Heyd, T. 2012. "Rock 'Art' and Art: Why Aesthetics Should Matter." In *A Companion to Rock Art*, edited by J. McDonald and P. Veth, 276–293. London: Blackwell Publishing.

Knox, M. 2013. *Boom: The Underground History of Australia, from Gold Rush to GFC*. London: Viking/Penguin.

Kroker, A. 2014. *Exits to the Posthuman Future*. Cambridge: Polity.

Latour, B. 2013. *An Inquiry into Modes of Existence*. Translated by C. Porter. Cambridge, MA: Harvard University Press.

Layman, L. 1982. "Development Ideology in Western Australia 1933–1965." *Historical Studies* 20(79): 234–260.

LeCain, T. J. 2015. "Against the Anthropocene. A Neo-Materialist Perspective." *International Journal for History, Culture and Modernity* 3(1): 1–28.

Lippard, L. R. 2014. *Undermining: A Wild Ride Through Land Use, Politics, and Art in the Changing West*. New York: The New Press.

Macdonald, S. 2009. *Difficult Heritage: Negotiating the Nazi Past in Nuremberg and Beyond*. London: Routledge.

Massumi, B. 2002. *Parables for the Virtual: Movement, Affect, Sensation*. Durham, NC: Duke University Press.

McKenna, M. 2018. "Moment of Truth: History and Australia's Future." *Quarterly Essay*, Issue 69. Melbourne: Black Inc.

Morton, T. 2013. *Hyperobjects Philosophy and Ecology after the End of the World*. Minneapolis, MN: University of Minnesota Press.

Morton, T. 2016. *Dark Ecology: For a Logic of Future Coexistence*. New York: Columbia University Press.

Mulvaney, K. 2015. "Burrup Peninsula: Cultural Landscapes and Industrial Hub, a 21st Century Conundrum." *Landscape Research* 40(6): 759–772.

Murujuga Aboriginal Corporation. n.d. *Stories from the Stones*. Accessed 12 July 2018. www.murujuga.org.au/history/the-burrup.

Negarestani, R. 2008. *Cyclonopedia: Complicity with Anonymous Materials*. Melbourne: Re-Press.

O'Sullivan, S. and Zepke, S. 2008. *Deleuze, Guattari and the Production of the New*. London: Continuum.

Paterson, A. and Gregory, K. 2013. "Commemorating the Colonial Pilbara: Beyond Memorials into Difficult History." Paper. *Investigating Catastrophe: Commemoration, Accountability and the Public Record of Disaster. International Workshop*. Perth: UWA, 5–7 June.

Pearse, G. 2009. "Quarry Vision: Coal, Climate Change and the End of the Resources Boom." *Quarterly*, Issue 33. Melbourne: Black Inc.

Pétursdóttir, Þ. 2014. "Things Out-of-Hand: The Aesthetics of Abandonment." In *Ruin Memories: Materialities, Aesthetics and the Archeology of the Recent Past*, edited by B. Olsen and Þ. Pétursdóttir, 335–364. London: Routledge.

Pétursdóttir, Þ. and Olsen, B. 2014. "An Archeology of Ruins." In *Ruin Memories: Materialities, Aesthetics and the Archeology of the Recent Past*, edited by B. Olsen and Þ. Pétursdóttir, 3–29. London: Routledge.
Pilbara Weather. Accessed 20 June 2018. www.weatherzone.com.au/wa/pilbara.
Povinelli, E. A. 2002. *The Cunning of Recognition: Indigenous Alterities and the Making of Australian Multiculturalism*. Durham, NC: Duke University Press.
Probyn, E. 2010. "Writing Shame." In *The Affect Theory Reader*, edited by M. Gregg and G. J. Seigworth, 71–90. Durham, NC: Duke University Press.
Rose, D. B. 2013. "Anthropocene Noir." *Arena Journal* 41(42): 206–219.
Sedgwick, E. K. 2003. *Touching Feeling: Affect, Pedagogy, Performativity*. Durham, NC: Duke University Press.
Seigworth, G. J. and Gregg, M. 2010. "An Inventory of Shimmers." In *The Affect Theory Reader*, edited by M. Gregg and G. J Seigworth, 1–25. Durham, NC: Duke University Press.
Smith, L. 2006. *Uses of Heritage*. Abingdon: Routledge.
Thrift, N. 2008. *Non-Representational Theory: Space, Politics, Affect*. London: Routledge.
Tilley, C. and Bennett, W. 2004. *Materiality of Stone: Explorations in Landscape Phenomenology*. Oxford: Berg.
WA Heritage Council. Accessed 14 March 2018. http://inherit.stateheritage.wa.gov.au/Public/Inventory/Details.
Witcomb, A. and Gregory, K. 2010. *From the Barracks to the Burrup: The National Trust in Western Australia*. Sydney: UNSW Press.
Yinjaabarni. Accessed 15 January 2018. https://yinjaa-barni.com.au/artist/celia-sandy.
Zdravič, A. 1997. *Riverglass: A River Ballet in Four Seasons*. Antara Soundfilm and Video, Ljubljana.

PART IV
Memory

11

MENDING SHATTERED TIME

22 July in Norwegian collective memory

Helge Jordheim

Ghosts from the past returning to haunt the present may cause time to come out of joint, well known from the famous scene in Shakespeare's *Hamlet*.[1] A present-day, real-life terrorist, on the other hand, may shatter time completely. This is what happened when Anders Behring Breivik, a 32-year-old white supremacist, carried out an attack on the Norwegian Social Democratic Party – *Arbeiderpartiet* – on 22 July 2011. The first part of the attack took place in the centre of Oslo, where he blew up the most prominent high-rise of the Government Quarter, housing several ministries, killing eight people. After planting the bomb, he drove out to the summer camp of the Social Democratic Youth Organization – AUF – at the island Utøya in the lake Tyrifjorden, just northwest of Oslo, and began a killing spree that did not stop until he had killed 69 people, mostly children and youngsters, when he was apprehended by Norwegian police. By brutally ending 77 human lives, most of them still in their beginnings, Breivik changed Norwegian history forever.

In this chapter, I want to suggest that one way of describing the impact of the 22 July terrorist attacks is that it shattered the temporal fabric of Norwegian society by effectively breaking up the social and historical web that connects past, present, and future. At least since the works of Benedict Anderson, Reinhart Koselleck, and Eviatar Zerubavel, we recognize that societies rest on a shared set of temporal experiences, memories, and expectations organized according to certain structures or rhythms that regulate how members think about their lives and the life of their communities (Anderson 2006; Koselleck 1985; Zerubavel 2003). When something entirely unexpected happens that disrupts the connections between prior experiences and expectations for the future, and with them the feeling of belonging to the same social time, we can talk about a *time-shattering event*.[2] This chapter is a study of the attempts made by various official actors in Norway, in the government, the police, the health sector, and the art world, as well as the press, to put this shattered temporal fabric back together again – to "reassemble" (Latour 2007, 1) or to "mend" it, if you like – in order to project new optimistic futures for the many collectives that make up Norwegian society.

What I will argue in the following is that the attempts to reintegrate the 22 July terrorist attacks in Oslo and at Utøya in Norwegian history and not least in Norwegian futures, deal with time in two different ways, or rather that they target two different forms of time: on the one hand, universal, quantitative clock- and calendar-time, for which one dominant representation is the timeline; on the other hand, experienced or phenomenological time, manifest in individual and collective memory as well as concepts of heritage, and given material shape in memorials and memorial sites. To reintegrate the terrorist attacks into clock- and calendar-time, a large number of timelines have been produced, in which the 22 July event is divided up into discrete incidents and time intervals, preceding and succeeding, causing and following from each other. For the reconstitution of experiential time another 'time tool' has come into use: the construction of a site of memory, including a concrete, physical memorial, by means of which the event is given a permanent place in Norwegian collective memory and thus turned into heritage.

This distinction between absolute, homogenous, quantitative clock-time, on the one hand, and relative, heterogeneous, qualitative phenomenological time, on the other, corresponds to the distinction made in most time research (Hoy 2012). What this chapter will show, however, is that faced with time-shattering events and in the process of reassembling time in the aftermath of catastrophe, these distinctions are blurred or even obliterated. Instead time emerges as a multifarious and heterogeneous mess of durations, speeds, scales, and rhythms. In the attempts to reassemble social time, specific temporal forms are targeted, inherent in particular institutions, practices, or materials. However, these attempts also pull at other threads in the temporal fabric of society, causing it to unravel even more. These examples demonstrate that social times are never clearly and permanently demarcated but become entangled with other forms of time, at work in and between the kinds of life they set out to organize and control, other "lifetimes".[3] Before I come to these entanglements, I will discuss some of the analytical approaches and theoretical concepts that are accessible to us in analyzing *time-shattering events*, most prominently the concepts of trauma and heritage.

Recasting heritage in multiple lifetimes

Along with many other similar publications, this book attests to a shift in the way we talk about the past. Not the pastness of the past, like in the case of historicism, but its continuous presence is emphasized, mobilized by concepts like "cultural memory" and "heritage". The rather recent drift from the first to the second, to "heritage" as *Grundbegriff* in the Koselleckian sense (Koselleck 2011), reflects an increasing focus on materiality and practice, which have been less prominent in more cognitively geared concepts like "cultural" or "collective memory" (Wilson 2009). In recent discussions about the end of "the modernist regime of temporality", heritage concepts and practices are furthermore perceived to be symptoms of a collapse in the modern past-present-future nexus. The assessments whether this collapse represents something positive or something negative varies, depending on the observer. Whereas the French historian and theorist of history François Hartog understands the turn to heritage as a symptom of an ever-expanding presentism (2015, 107–114), his German counterpart Aleida Assmann recognizes the emergence of a new "temporal regime" in which the past is no longer the residual element of modernist future-orientation but the source of endless knowledge that can be used to decode presents and futures (2013, 131–208). To Hartog, heritage emerges as an example of what he calls the "double indebtedness" of the present toward the past and the future, which he recognizes as a "hallmark of our contemporary experience" (2015, 201). After the collapse of the modern "regime of historicity", Western societies are mired in "presentism" (ibid). The present extends into the past, he argues, by means of "concepts such as responsibility and the duty to remember, the drive to make everything into heritage … , and last but not least, the notion of 'debt'" (ibid).

Regardless of the larger normative frameworks, the concept of "heritage" offers an effective tool for controlling and curating the past in the name of the present and the future. In work by Lynn Meskell, Sharon MacDonald, and others, the concept is expanded also to include "negative heritage", which Meskell discusses in relation to World Trade Center and the Bamyian Buddhas (2002), and "difficult heritage", which MacDonald exemplifies with the remnants of Nazi architecture in the German city of Nürnberg (2008). The list of examples goes on indefinitely. Deploying "heritage" to deal with, or, in Meskell's words, to "master", the haunting presence of the terrorist attacks on 22 July in Norwegian public culture would belong in the same category as I will discuss toward the end of the chapter looking at plans for memorials and memorial sites.

As tools for handing down past experiences to future generations, as well as to deal with the material presence of difficult pasts, the concepts and practices of heritage are still indebted to ideas about "history" and "historical time" in terms of a temporal framework tailored to contain human lives, activities, and experiences. What I want to suggest here, however, is that "historical time" – as a specific form of time defined by an idea of the human – is part of the problem rather than the solution (Smail and Shyrock 2011, 5). To understand what heritage is in the twenty-first century, we need to radically expand our notions of time. "History", "historical time", and, indeed, "heritage" need to be opened up to include other "lifetimes", in the sense of other temporal durations, rhythms, and speeds embodied in different forms of life. The concept of "lifetimes" bridges the nature-culture distinction and replaces the dichotomy of natural time, on the one hand, and social and historical time, on the other, by a continuity of scales. Life – both human and non-human – unfolds at various time scales, from seconds to millions of years, which converge in events and their aftermaths. To account for all these different times in our ways of thinking about the past is central to *Heritage Ecologies* which explicitly wants to "reflect critically on the anthropocentric assumption that heritage is an insular human property detached from its messy worldly relations" (Bangstad and Pétursdóttir, chapter 1). Even in the case of the terrorist attacks in Oslo and at Utøya, more-than-human times, technological, geological, and institutional, come into view, which cannot be ignored. In the attempts to "mend" the social and historical time shattered by the mass murder, other lifetimes are mobilized, which might be "out of sync" with the times that the governmental and non-governmental actors are addressing. A time-shattering event gives rise to new temporal arrangements, more or less able to deal with the feelings of grief, responsibility, anger, fear, or hate produced by the event itself. This is where the concept of heritage comes up against the concept of trauma.

Trauma and the shattering of time

In a recent paper, the psychologist Gilead Hirschberger defines "collective trauma" as "a cataclysmic event that shatters the basic fabric of society" – and he adds: "Aside from the horrific loss of life, collective trauma is also a crisis of meaning" (2018, 1). The terrorist attacks in Oslo and at Utøya on 22 July 2011 fit this description to a T. To what extent the "basic fabric" of Norwegian post-war social democratic society was shattered can be gleamed from the stream of reactions, official or personal, first to the event itself, then to each of the administrative steps taken to understand and process what had happened and to make sure something similar cannot happen in the future. The immediate and strong public resistance to the first forensic evaluation of the perpetrator Anders Behring Breivik, which concluded that he had a psychotic disorder and thus was legally unaccountable, also testified to the need of the Norwegian public to find meaning in this heinous crime and establish a political and historical context around it (Melle 2013).[4]

In discussions of trauma as a cultural phenomenon, in works by Shoshana Felman, Dori Laub, Mieke Bal, Jeffrey Alexander, and others (Alexander 2012; Bal 1999; Felman and Laub 1992; Laub 2003), the dominant temporal form, from which traumatic events and memories set themselves off and into which they need to be reintegrated in order to help the traumatized subject, is the narrative. In the introduction to the path-breaking *Acts of Memory* from 1999, Mieke Bal operates with three kinds of memory: habitual memory, narrative memory, and traumatic recall, which she defines as "the painful resurfacing of events of a traumatic nature". Characteristic of traumatic memories is that they "resist integration"; they "cannot become narratives" but will always remain fragments or splinters (Bal 1999, viii). But apart from resisting integration into narrative structure, and entering into endless cycles of repetitions and indeed "hauntings", these scholars are less precise about what traumatic events and memories do with temporal experience, even with the timeline itself. In his book *The Evil Hours*, David J. Morris, a former Marine and war-correspondent suffering from post-traumatic stress disorder (PTSD), writes:

> Trauma destroys the fabric of time. In normal time you move from one moment to the next, sunrise to sunset, birth to death. After trauma, you may move in circles, find yourself being sucked backwards into an eddy or bouncing like a rubber ball from now to then to back again. August is June, June is December. What time is it? Guess again. ... Another feature of traumatic time is that it doesn't just destroy the flow of the present into the future, it corrodes everything that came before, eating at moments and people from your previous life, until you can't remember why any of them mattered.
>
> (2015, xii)

Morris describes his personal experiences of shattered time, of a shattered timeline. Linearity and regularity are lost, replaced by circular and bouncing movements; the same goes for the continuity between past experiences and future expectations. Both clock-time and calendar-time, usually the most solid defences against temporal confusion, are equally affected.

Personal experiences should not be taken as proof for collective conditions, also not in thinking about an event like the 22 July terrorist attacks. However, the almost compulsive use of timelines to make sense of the events offers us a material to study whether Morris' description of a collapsing personal timeline might also have a broader cultural significance. In his "theory of cultural trauma", Jeffrey Alexander expands the concept of trauma to include not just individuals but collectives. "Cultural trauma", he writes, "occurs when members of a collectivity feel they have been subjected to a horrendous event that leaves indelible marks upon their group consciousness, marking their memories forever and changing their future identity in fundamental and irrevocable ways" (Alexander 2004, 1). Again, time appears as central to traumatic as well as post-traumatic experiences. Memories of traumatic events in the past come to influence how groups think and act in the future. As Gerard Delanty points out, trauma is an integral of European heritage and has been at least since 1918. Trauma, he states, "can be normalized by repetition and enters into the dominant discourses, in which a collectivity interprets themselves around notions such as guilt, revenge, or responsibility" (Delanty 2017, 10). One way to describe this process of normalization, is in terms of a "work of synchronization", by which different lifetimes are aligned with each other, as well as with political events and processes, in order to form temporal arrangements which enable and constrain future plans and actions (Jordheim 2014). Timelines and memorials, I will argue, should both be understood as tools of synchronization, collecting, assembling, and aligning different lifetimes into new temporal fabrics within which new imaginaries for the future of Norway can form and grow – although not always successfully.

Forms of synchronization: timelines and memorials

In their by now classic definition, the American sociologists Pitirim A. Sorokin and Robert K. Merton contend that "[a]ll time systems might be reduced to providing means for synchronising and coordinating the activities and observations of the constituents of the group" (1937, 627). Elsewhere, I have discussed how this "work of synchronization" is not only performed by time technologies like clocks, calendars, and timelines but also by concepts, narratives, and, memorial sites (Jordheim 2014; Jordheim and Wigen 2018). Similarly, what is synchronized goes far beyond "human activities" and includes experiences, emotions, and affects as well as non-human life forms and material objects. Temporal arrangements emerge when times embodied in these various manifestations of life are synchronized and become manifest in social and political practices. Synchronization, their German counterpart Niklas Luhmann argues, "ties the medium time to forms … and tries to find suitable forms" (1990, 109–110). In this chapter, I explore these forms: on the one hand, a set of timelines, on the other hand, plans for a memorial site. And I ask what kind of synchronizing work they perform and how it contributes – positively or negatively – to restoring of the temporal fabric of Norwegian society after the 22 July attacks. I'll begin with the timelines.

Lines as temporal forms have a long and culturally diverse history. Along them, events can be plotted at regular or irregular intervals, at specific paces and rhythms, even though they belong to completely different lives, narratives, fields, cultures, or geographical areas. In her book on time as a key word in the social sciences, Barbara Adam offers the following almost poetic description of the line as a "representation of time":

> The time of the line is spatial
> The time of the line is historical
> The time of the line is projective
> The time of the line is irreversible
> The time of the line is before & after
> The time of the line is tied to a beginning.
>
> *(2004, 21)*

All these claims can and should be discussed, even contested: from the spatiality of the line, discussed in large detail by Sybille Krämer (2016, 179–192), to the idea of a beginning, called into question by the extreme success of Year Zero in Christian chronology, opening up to an endless past, at least in theory (Grafton 2003). Neither do all these claims hold true in relation to the timelines made to reconstruct the events of 22 July 2011. Still, they represent an interesting and useful starting point.

As Anthony Grafton and Daniel Rosenberg point out, the timeline is an invention of Greek and Roman antiquity, developed to record the succession of historical events (2010, 26). However, the timeline is not intrinsically linked to historical time as such. Timelines can plot other kinds of life processes than those that usually qualify as historical, they can also plot time intervals different from those that usually count as "history": hours, minutes, seconds, even nanoseconds, but also millions and billions of years. In other words, timelines are used to register time *within* historical events as well as between them. In his book about lines, the anthropologist Tim Ingold mostly avoids any discussion of the timeline, probably because for him it represents the arch-example of the linearized line of Western modernity: "Anthropologists", he claims, "have the habit of insisting that there is something essentially linear about the way people in Western societies comprehend the passage of history, generations, and time" (2007, 2). If we consider the prevalence of timelines in coping with an event such as the 22 July attacks, there seems to be a lot of truth to that anthropological assumption. Ingold himself is interested in another kind of timeline, characterized by being non-linear, in the Western sense, or even multilinear. For this reason, he makes the genealogical line, or rather the genealogical tree, his point of reference. Time's linearity, he claims, "is of a particular kind": "not the kind of line that goes from point to point connecting up a succession of present instants arrayed diachronically". Rather, he continues quoting Bergson, "[d]uration is the continuous progress of the past which gnaws into the future and which swells as it advances". This continuous present of the past is what necessitates "the work of memory" (Ingold 2007, 122). In this way Ingold is able to link the seemingly neutral and objective timeline, on which discrete, controllable, and always already synchronized time is documented, to the highly moral and politicized question of memory and indeed heritage.

Heritage is a fairly recent name for this continuous progress or, indeed, swelling, of the past into the future (Harrison 2012, 68–94). At present, "heritage" is probably the concept of historicization most able to assemble and align various lifetimes. In modernity, Luhmann writes, "there is no supersynchronization, which could bring order to the act [*practice, work*] of synchronization itself" (1990, 109–110). This used to be the role of "the concept of destiny" and later, we could add, the concept of "progress". Today, "heritage" might be the closest thing to a candidate for a new "supersynchronizer", in Luhmann's sense, which synchronizes multiple pasts by projecting them into a single, more or less uniform future. In view of the ubiquitous use of timelines to come to terms with the terrorist attacks, it is striking how the first attempt at memorializing the events at Utøya, moving beyond the timeline, took the shape of a wound (see also Storm, chapter 3) – something which intentionally breaks with any kind of linearity. Just as striking, however, was the very public failure of this project, which I am going to return to toward the end of the chapter. In both cases, the tension between chronological and phenomenological time emerges as an important point of contention in coping with the events.

Timelines after 22 July

Looking through the reports, evaluations, and other presentations that in various ways attempt to make sense of the events of 22 July 2011, there is no way to ignore the omnipresence of timelines. Press coverage, governmental evaluations, and scientific analyses alike feature visualizations of the event as points on a line, marking successive moments. It might vary which

FIGURE 11.1 Timeline at the 22 July Information Center in the Government Quarter, Oslo, Norway

Photo: Ana Rita Ferreira/22. juli-senteret.

moments these are, at what intervals they occur, and how they relate to the events as a whole. Far from offering a complete survey of all the timelines produced in response to the 22 July events, I will limit myself to offering some examples, based on certain analytical parameters. In every case, I will focus on how the timeline reconstructs, synchronizes, and in a certain sense *mends* a time that appears shattered.

The first striking observation concerns the range and variety of the timelines that have been produced. A rather neutral way of organizing them is on a scale from the most general to the most specific. At one end of the scale, there are the timelines mounted on the walls of the 22 July Information Centre in the Government Quarter, in Hegnhuset at Utøya, as well as the ones found on Wikipedia (n.d.) and on the website of the Norwegian Broadcasting Corporation (NRK), which all cover the totality of the event from several different angles. At the other end, we find innumerable specialized timelines published in official reports and white papers as well as in the press. Among them are the three timelines included in the Norwegian government's official evaluation report, produced by the so called "22 July Commission", headed by Alexandra Bech Gjørv, and published on 13 August 2012: a timeline of police response times (NOU 2012, 119), a timeline of the boats used to transport the Emergency Response Unit (*Beredskapstroppen*) (NOU 2012, 139), and, finally, a timeline for a project to improve the security at the Government Quarter in downtown Oslo (NOU 2012, 445). These are timelines of three well-documented and much-debated failures: the police responded too late, the boat used by the Emergency Response Unit almost sank, and the street that should have been closed to improve security remained open. Another set of related timelines can be found in the report containing the self-evaluation of the Norwegian police of their efforts, national and local, to stop the terrorist, which famously gave a more favourable view of the involvement by law enforcement than the later report from the Gjørv Commission. The Sønderland Committee Report includes several rather scant timelines, reconstructing local and national alarm procedures (Politidirektoratet 2012, 44), first warnings and witness reports (ibid, 67), the use of the police helicopter (ibid, 70), the mobilization of a military helicopter and later a medical emergency helicopter from Rygge airfield (ibid, 70–71), communication with actors the police were supposed to cooperate with (*samvirkeaktører*) (ibid, 75), requests for assistance from other actors (ibid, 77–78), and other similarly specific moments and processes. Since they isolate very limited series of events, these timelines are not very informative with regard to why it took the police so long to respond. By contrast, the report from the Oslo police department features several integrated timelines in which different kinds of events are combined under headings like "Most important events" in Oslo and at Utøya: helicopter resources, security and object protection, information and media handling etc. (Oslo politidistrikt 2012). A more synchronized timeline seems to have a far greater explanatory value than those timelines that only document events connected to a specific aim, like getting a helicopter in the air.

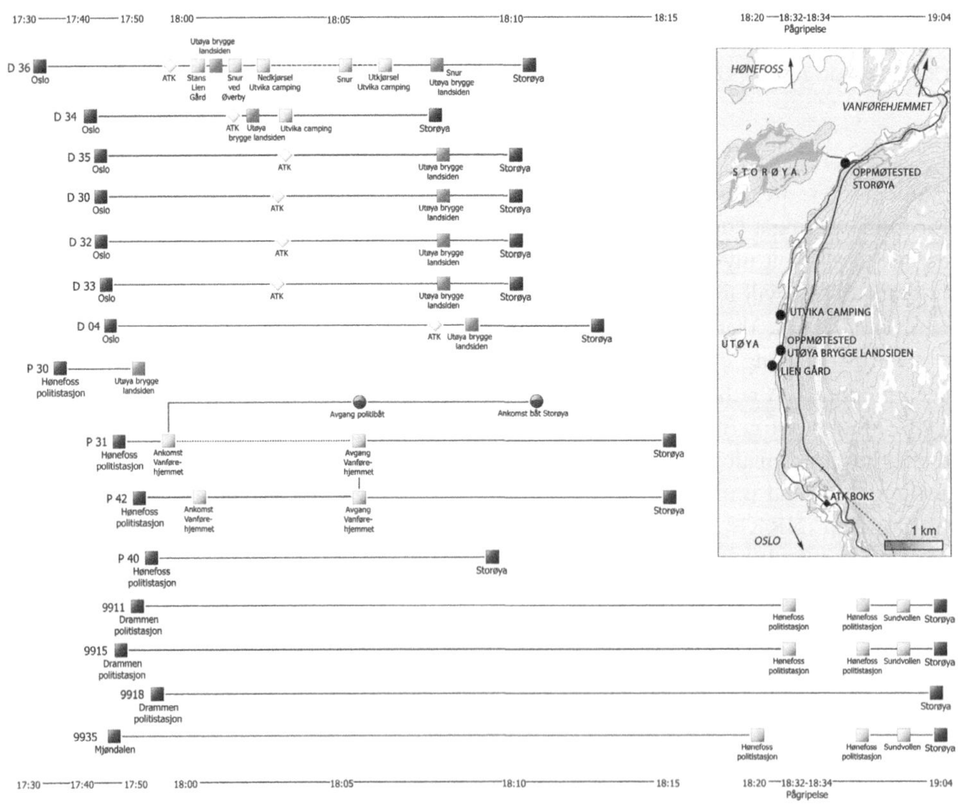

FIGURE 11.2 Timelines in the report by the Gjørv Commission (NOU 2012)

Another set of texts that use timelines extensively are the articles published in academic journals dealing especially with trauma and emergency medicine (Gaarder et al. 2012; Sollid et al. 2012; Waage, Poole and Thorgersen 2013). One important use of timelines here is to document at what time and in what numbers victims from the shootings were admitted to hospital. In a text accompanying the timeline mapping the number of patients triaged and admitted at Ringerike hospital, the authors conclude that "[t]he main surge of patients occurred between 19.40 and 20.40 hours when the hospital received 22 patients in 1 hour" (Waage 2013, 1200). In this way, the timeline is used to establish a quantitative correspondence between a time interval and a number of patients arriving from Utøya and thus turn the event into an object of quantitative research. The timelines are also effective in documenting the workload of the doctors and nurses who were working or on call that day. In this way, they are also included in the narrative of events taking place on this fateful day in Norwegian history.

The press has systematically used timelines as a tool to make the terrorist attack comprehensible to their readers. In addition to more general timelines, the newspaper *Verdens Gang* published a timeline of the events leading up to the attack, plotting the life of the perpetrator, from his birth to the moment the bomb went off at the Government Quarter, illustrated by private pictures (Figure 11.3). Beginning with the aftermath of the event, a timeline produced by NRK documents the events connected to the planning of memorial sites both at Utøya and in the centre of Oslo, including an artist competition, the selection of a winner, protest from the future neighbours etc. Finally, the most comprehensive and technologically advanced timeline is found on the NRK website (Figure 11.4) (NRK 2012). By clicking on one of the headings at the top, the user can choose between three different timelines, labelled "Anders Behring Breivik", "24 hours of terror", and "The Aftermath". For the last two, the user also has the possibility of choosing between three "filters", presenting events connected to the perpetrator and to the police respectively, as well as other events. In the default mode, all kinds of events are weaved together, with the effect that various kinds of simultaneities between different lifetimes come into view: While the former Prime Minister of Norway is speaking at Utøya, Breivik parks his escape car in a parking lot close to the Government Quarter. Most striking and indeed heartbreaking to read is the long list of often rather trivial incidents that delay the arrival of the police and the Emergency Response Unit at Utøya, when the perpetrator is already walking around killing people on the island. Indeed, this might very well be the most powerful effect of all the timelines produced for the attack: to plot the seconds, minutes, and hours that were wasted on inefficient communication and command routines after Breivik had started his killing spree on the island until he was finally apprehended by the police.

Timelines between diachronic causation and synchronic simultaneity

According to my analysis, the timelines documenting the events of 22 July can be said to mend or reassemble shattered time in two separate ways, or rather in two separate directions: diachronically and synchronically. On the one hand, events, including the cold-blooded murders of 77 people, are reconnected into diachronic temporal series, which re-establish time in accordance with principles of succession and causation. These timelines are designed to answer two questions that were on the lips of every Norwegian and certainly many others in the days after the attacks: what really happened, and how could it have been stopped? In other words, timelines were made for the specific purpose to document why it took the police emergency units so long to get to the island and stop the terrorist. Timelines plotting the mobilization

BIL | BIL OG MOTOR | FOTBALL | NYHETER | NYTTE | RAMPELYS | SPORT | TV-GUIDE | VG+ | VGD | VÆR

Anders Behring Breivik - fra fødsel til 22. juli

TEKST: Marianne Vikås FOTO: VG/Privat
GRAFIKK: Tom Byermoen og Dan Kåre Engebretsen

En splittet familie, en brokete oppvekst, turbulente ungdomsår, økonomiske problemer og mislykkede forretninger. Her er historien om vestkantgutten Anders som ble terrorist.

GÅ TIL TIÅR: 1970-1980, 1979, 1980-1990, 1980, 1983, 1986, 1990-2000, 1992, 1994, 1995, 1996, 1997, 1998, 1999, 2000-2010, 2000, 2001, 2003, 2005, 2006, 2007, 2009, 2010-2011

1970-1980

1979
Tirsdag 13. februar ser Anders Behring Breivik dagens lys i Oslo. Faren er til stede under fødselen.

1979
Anders' foreldre tar ut separasjon, ifølge faren. Da har de vært gift i cirka ett år. Anders blir boende hos moren, som har en seks år gammel datter fra før, i dette borettslaget på Nedre Silkestrå på Skøyen.

1980-1990

Mot midten av 80-tallet
Anders går i Vigelandsparken barnehage på Frogner i Oslo.

1983
17. januar blir foreldrene til Anders formelt skilt

1983
ANDERS 4 ÅR.
En psykolog kobles inn for å observere samspillet mellom Anders, halvsøsteren og moren over tid. Dette ender i en rapport der psykologen hevder at Anders utsettes for omsorgssvikt og anbefaler at han flyttes fra hjemmet. Det skjer ikke.

1983
Faren (bildet) til Anders går rettens vei for å få omsorg for sønnen. Ifølge saksdokumenter ber han om midlertidig forføyning – han vil ha Anders umiddelbart.& Det blir avholdt et rettsforberedende møte i august, men innen saken kommer opp for retten skal kravet ha blitt frafalt.

1983
Faren til Anders utnevnes til ministerråd for den norske OECD-delegasjonen i Paris av kong Olav.

1986
7 ÅR

FIGURE 11.3 The timeline from the newspaper *Verdens Gang* (*VG*)

of helicopters and boats, first sightings of the perpetrator, as well as first reports of shooting at Utøya, are designed to account for the amount of time that passed: What caused the delays? What faulty procedure or human error was responsible? But timelines also map synchronic simultaneities. Several of the 22 July timelines document what happened elsewhere while the Utøya killings took place, at the communication central for emergency medicine (AMK), at the police headquarters, at Rygge airfield, even at the dock on the mainland, while the police were trying to get people across to the island. Timelines both in the 22 July Information Center and at Utøya also contain text messages in which teenagers try to get in touch with

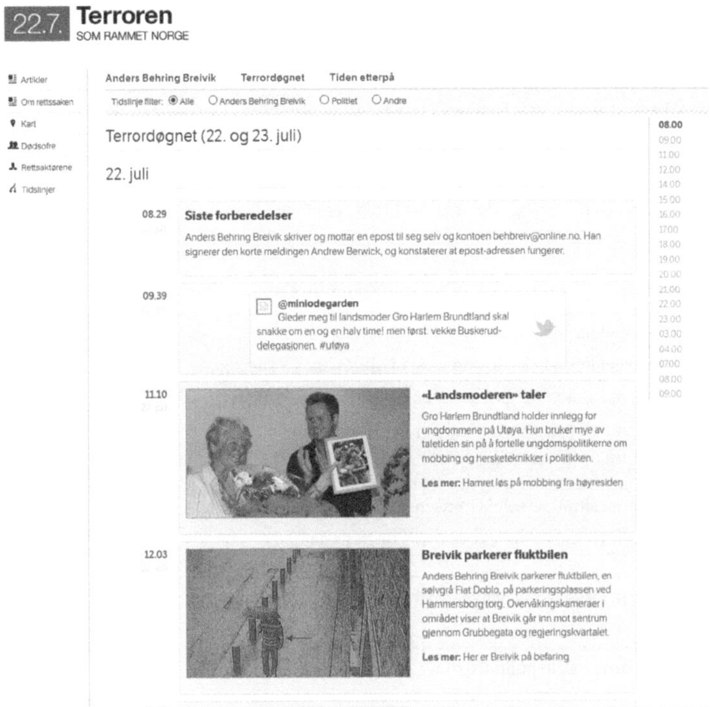

FIGURE 11.4 The timeline from the Norwegian Broadcasting Corporation

their parents to warn them about what is going on. In this way, the chronological simultaneity visualized by the timeline is expanded to include increasingly more people and places, lined up in parallel at any chronologically defined moment.

In reality there is no limit to what and who can be included in the timeline. The time of the timeline is universal, which means that anyone who has a physical or emotional relationship to the terrorist attacks is prompted to ask themselves where they were and what they were doing during those desperate 85 minutes between 17:08 and 18:33, when 69 young boys and girls were shot to death and many more wounded. In his path-breaking book on "imaginary communities", Benedict Anderson discusses how the date on the front page of the newspaper creates a feeling of community between all readers, even though they have never met or seen each other (2006, 30–36). This effect of synchronization is even more pronounced when the media in question are not physically distributed newspapers but phone calls, text messages, public warnings, and, after a while, live news coverage in television and radio. Across the world, people can tell exactly what they were doing when they first heard about the planes crashing into World Trade Center. Across Norway, people can offer similar information about their whereabouts and activities when the news about the terror attacks reached them. For most Norwegians, synchronizing with the events at Utøya happened *post festum*. By aligning their own individual timelines with the timelines produced by Wikipedia, newspapers and television channels, people make the terrorist attacks part of their own lives. In other words, they use their own lives as the temporal form into which the time-shattering events of 22 July can be placed. In this way, extreme events like the terrorist attacks at Utøya, which seem

Tidslinjen under viser tidspunkt for noen av de oppdragene som "stjal" ressurser fra innsatsen og samtidig bidro til å skape et uoversiktlig situasjonsbilde 22.7.11:

- Poster markert med grått: Henvendelser i forbindelse med eksplosjonen. Kvittert ut løpende pr. telefon.
- Poster med sluttidspunkt: Henvendelser som det ble bruk ressurser på.
- Øvrige poster: Henvendelser det ikke ble brukt ressurser på grunnet manglende kapasitet.

Start kl.	Hendelse	Slutt kl.
15.25	Patrulje melder om eksplosjon i Skippergata.	
15.25	Patrulje melder om eksplosjon inne i Folkets hus i Youngsgate.	
15.26	Melding om at det har gått av en bombe i Christian Kroghs gate.	
15.27	Melding inn til politiet om eksplosjon i Ibsentunnelen.	
15.28	Melding fra AMK om at det er mange skadde etter at noe har falt fra taket på VG-bygget.	
15.28	Melding om såret person i Grensen 10.	
15.29	Melding om bombe som har gått av i Rosteds gate.	

[115] PBS I, pkt. 6.1, jf. PBS II og Politidirektoratets rundskriv 2010/12.

Evaluering 22.7. 2011 - Oslo politidistriktet 58

15.29	Patrulje melder om at en restaurant er sprengt i Møllergata 16.	
15.30	Melding om eksplosjon i Grensen 3.	
15.31	Melding om skadet person med glassbiter i hodet etter en eksplosjon i Torggata 5.	
15.33	Melding om mange skadde på Hammerborgtorg.	
15.35	Vaktsentralen ved Oslo sentralbanestasjon ringer og spør om de skal evakuere Oslo S.	
15.35	Melding fra Wilses gate 4 B om at noen har skutt ut rutene i nr. 6.	
15.37	Melding om trafikkulykke i Sinsenkrysset.	
15.39	Patrulje til Storingsgata 14. Melding om eksplosjon og at et byggestillas hadde kollapset. Patruljen fikk opplysninger fra vitner som hevdet at det var en bombe. Patruljen fikk raskt avklart at det var et arbeidsuhell uten skadde.	16.00
15.40	Melding om eksplosjon i Fredensborgveien 22.	
15.40	Melder hevdet at han hadde blitt frastjålet lommebok og at noen forfulgte ham. Han hevdet at han befant seg ved Ankerbrua og at han slåss med vedkommende.	
15.41	Patrulje med viktig melding om en Mercedes som står forlatt i Ibsentunnelen.	
15.46	En vekter opplyser til en patrulje at det skal være to udetonerte bomber igjen. Dette resulterer i evakuering av innsatsområdet.	15.54
16.08	Patrulje melder om en hensatt Volkswagen Caravelle i Teatergata.	
16.11	Telefon fra vitne som har fulgt etter en bil som kjørte fra Regjeringskvartalet rett før eksplosjonen.	
16.14	Telefon fra vitne som har sett en utenlandsk mann løpe opp Brugata rett etter eksplosjonen bærende på en pose.	
16.16	Patrulje melder om to mistenkelige kofferter utenfor TV2s lokaler.	21.49
16.18	Samme patrulje melder om to forlatte kolli i Karl Johans gate utenfor TV 2, mulig etter sigøynere.	21.49

FIGURE 11.5 The timeline from the 22/7 report by the Oslo police department (Oslo Politidistrikt 2012)

to take place in a different time altogether, are reintegrated with the times of people's lives, alongside trivial events like working, shopping, cooking, or hanging out with friends. We are forced to accept them as part of our own reality, at least in a temporal sense – as part of our own lifetimes.

Timelines should not only be understood as diachronic series of successive homogenous presents but also as possible cross-sections into the flow of time, which cuts through and thus aligns multiple and often very different trajectories or narratives. Any moment on any of the 22 July timelines contains endless lives and lifetimes, endless diachronic trajectories, from the past, through the present, and into the future, aligned by the reference to universal, absolute, chronological time. Not all of these lifetimes are embodied primarily by humans. Some are assemblages of human and non-human actors, such as helicopters, boats, and cars, as well as emergency warning procedures like the "national alarm [*riksalarm*]". In these temporal assemblages, human intentions are thwarted by material and administrative shortcomings.

As already noted, many of the 22 July timelines are part of contested and controversial attempts to explain the almost three-hour delay between the explosion in the Government Quarter (15.25) and the arrest of the perpetrator at Utøya (18.34), discussed in-depth by the report from the Gjørv Commission (NOU 2012). But the timelines also bring out other contentious issues, like the pre- and post-history of the attack as well as the selection of events included. Timelines present their arguments mainly in two ways, by the choice of beginning and ending and by the choice of temporal intervals. The 17-meter-long timeline at the 22 July Information Centre (Figure 11.1) begins with the explosion, or rather with the moment the terrorist parks his car outside the Government Quarter, and ends with a picture taken of the island Utøya from the mainland at 22.23. In other words, it presents the terrorist attack from the perspective of the victims. No attempts are made at explaining the attack by showing the events that led up to it, such as the preparations of the terrorist or the lack of security both at the Government Quarter and at Utøya. According to the historian Tor Einar Fagerland, who was academically responsible for the contents of the exhibition at the Centre, the intention was "to take people back to the moment when Norway was hit by terror" (2015). He does not comment on the specific use of a timeline to achieve this effect, but the choice of time period fits precisely with this intention. In this way, "the moment Norway was hit by terror" is defined and delimited.

By contrast, both the Gjørv Commission, the newspaper *VG* (2011), and the national broadcasting company NRK made timelines documenting periods that precede and succeed the event itself, documenting the prehistory and the aftermath respectively. Included in the Gjørv Commission report is a timeline for the so-called "security project", tasked to assess and improve the security of the Government Quarter, in part by closing the road that the terrorist entered in his van. According to the timeline, this project started in fall 2003 and ended when the bomb went off on 22 July 2011 (NOU 2012, 445). Between these dates, the timeline lists a series of reports, plans, and decisions, which are not carried out or are delayed. Based on this timeline, the Commission concludes that it was not the planning or the financing but the "implementation" that failed (NOU 2012, 444).

Another prehistory is told in a timeline produced by *VG*, which recounts in detail and with a large number of private photos the life of the terrorist Anders Behring Breivik, from his birth on 13 January 1979 until he commits the extremely brutal and ruthless acts of terrorism on 22 July 2011 at the age of 32. A similar trajectory is mapped out by NRK as part of the multi-purpose and multi-temporal timeline mentioned previously (Figure 11.4). Whereas both timelines recount the prehistory of the event by means of Breivik's biography, the *VG* timeline is clearly more private and personal (Figure 11.3), especially in the use of photos. The NRK timeline instead chooses to document systematically the material surroundings of Breivik's life, the houses he lived in, and the public institutions he visited, as well as the receipts for ammunition, fertilizer, and other things he needed to carry out the attack. Both timelines document the parents' divorce, the encounters with child welfare services, the lack of contact with his father, and so on. In this sense, both timelines seem to argue that the causes for the attack can be found in the life of the perpetrator and thus opt for a psychological, rather than a political, explanation for the act, corresponding to the first of the two psychiatric evaluations made in connection with the trial (Melle 2013). As in the Gjørv Commission report, it is clear that timelines are not just pieces of objective documentation but actually arguments in a struggle to understand and explain the single most brutal terrorist attack carried out on Norwegian soil.

What also separates the timelines of the prehistory from the timelines of the event itself are the temporal intervals. While the timelines of the terrorist's life as well as the "security project" extend over years, even decades, the timelines of the attack span only a few hours. Time seems to be accelerating when more and more fateful incidents "stack up". Every minute becomes significant, measured against the loss of life at Utøya. At several occasions, the minute rhythm of the police timelines is still not fine grained enough to separate the different incidents and organize them in relation to each other in terms of before and after. Instead we are confronted with simultaneity not of events belonging to different trajectories but of events that find themselves in a causal relationship to each other, which the timeline is unable to account for. At 17.29, the head of operations at Northern Buskerud police district received orders to distribute weapons to the officers (*bevæpningsordre*), but at the same time, the emergency central received yet another report from a civilian about shots fired at Utøya. In this way, timelines illustrate how there are moments when time is drawn together and events "stack up" in a way that makes decision-making much harder. For understanding the event, the latter aspect is just as important as the former. A crisis is a moment in history when time accelerates, things happen at the same time, rather than in succession, and the normal procedures for planning and deciding are suspended (Jordheim and Wigen 2018).

In general, timelines can be said to have two functions, which partly contradict each other: on the one hand, they reconstruct logically the succession of events, highlighting relationships of cause and effect, planning and implementation, attempt and failure; on the other hand, they return us to the opacity, ambiguity, and confusion of specific moments, when the future that is represented by the timeline still had not emerged on the horizon of the historical actors. In this way, timelines play an important role in the reconstitution of continuous, collective time, after it has been shattered by a traumatic event like the terrorist attacks of 22 July.

Mobilizing the times of nature: Dahlberg's "memory wound"

In the final part of this chapter, I want to expand the discussion of timelines to look at another, very different attempt at reconstituting the fabric of time shattered by the events of 22 July: by shifting them into a context of memory and heritage. The NRK webpage presents a timeline designed to document the aftermath of the event, dominated by the process to select a memorial site and erect a memorial for the victims. This timeline documents the attempt to shift what happened on 22 July from past events to future memory and to processes of grief, remembrance, and forgetting. One way to think about the temporal work performed by the planned memorial site for the victims of the 22 July attacks is in terms of an expansion of the timeline, taking us into the future. Then again, the timelines were themselves always already about the future: One way to conceptualize the synchronizing, time-mending work of the timelines is as an attempt to imagine a future that continues beyond the mass killings, which reconnects the time before with what comes after. Even though most of the timelines end with the arrest of the perpetrator by Norwegian police, the format of the line also points beyond that moment, into a future that has been disconnected, broken off from the past. The plans for a memorial site and thus for future heritagization of the 22 July attacks represent an attempt at reconnecting past, present, and future and thus re-establishing a continuous timeline.

This work of memorialization and heritagization has received its own timeline documenting the official initiatives to establish a memorial site for the 22 July events, or actually two sites, one in the vicinity of Utøya, at a place called Sørbråten, and another at the Government Quarter in downtown Oslo (NRK 2017). This particular timeline begins with the attack and ends on 15 June 2017, when the plans for Sørbråten are shelved and replaced by plans for a future memorial located at the harbour across from Utøya. On this timeline, there are several moments and indeed several documents that deserve closer scrutiny (Hjorth 2018). For the purpose of this chapter, I will enter in the middle and ask what kind of temporal work was – and indeed still is – performed by the memorial site planned for the island. The work in question was designed by the Swedish artist Jonas Dahlberg and entitled "Memory Wound" (Figure 11.6). For almost three and a half years, this artwork was paraded around the Norwegian public sphere as the official memorial of the victims of the terrorist attack, until 21 June 2017, when the government announced that it would not be built.[5]

Before I go into this story, this unsuccessful attempt at mending shattered time, however, I need to remind us that the timeline for the work to build a memorial site at Utøya by no means represents the only relevant aftermath of the attack. In addition to the struggles of survivors to find a way to go on with their lives and the struggles of families and small communities to cope with their losses, the events of 22 July also have a long-lasting effect on Norwegian politics. These effects mainly come to the fore in the conflicts over statements by far-right politicians, evoking the role of the social-democratic Labour Party (*Arbeiderpartiet*) as target and victim of the terrorist attacks (NRK 2019). A comprehensive timeline of what the Germans call *Vergangenheitsbewältigung*, or "past mastering", in Meskell's words, would have to include both these equally important but contrasting trajectories (Meskell 2002; Reichel 2007).

Dahlberg's "Memory Wound" is made up by a cut through a peninsula covered by pines, forming a channel that should allow the water of the Tyrifjorden Lake to pass through, and revealing a smooth, almost mirror-like granite rock face, in which names of the victims are engraved and can be read by the visitors from a balcony built into the rock on the opposite side. This was how the Norwegian government, represented by KORO – Public Art Norway – and *Statsbygg* – Norwegian Directorate of Public Construction and Property, planned to commemorate the 77 people murdered on 22 July 2011 (KORO n.d). Furthermore, this was how the Norwegian people were told that the memory of the victims would be carried into eternity – or at least for as long as the rocks, water, and trees were able to sustain

FIGURE 11.6 Jonas Dahlberg, "Memory Wound"

Source: Courtesy of Jonas Dahlberg Studio.

it. Today, we know otherwise; the site has been moved from Sørbråten to the harbour facing Utøya, and the contract with the artist has been annulled. Dahlberg's "Memory Wound" will never be realized, at least not there. Questions about whether it was a good or a bad work of art, whether it was plagiarized, or trivial, and whether it would actually cause psycho-social traumata among the neighbours, who had also witnessed the terrorist attack and taken part in the rescue of survivors, will never be answered, or only theoretically. From now on, what characterizes "Memory Wound" is that it never came to be.

But for a while, its absence will be felt. In aesthetic terms, Dahlberg's artwork might be construed as the absence of absence, since the work was originally set to represent the absence of 77 people and thus continued a tradition of memorial aesthetics going back to the 1980s and working with representations of absence. In this context, however, I am less interested in aesthetic and theoretical speculation on what is absent. What concerns me is presence. For three and a half years, probably until Erik Poppe's film *Utøya* was screened in February 2018, Dahlberg's "Memory Wound" was the most prominent public representation of one of the most brutal and shocking events in Norwegian history. In spite of it never being built, Dahlberg's work represents a manifest attempt at integrating different forms of time in order to create a long-term memory of a highly traumatic event and thus enter it into Norwegian national heritage.

Again, it might be useful to refer to the timelines discussed in the first part of this chapter as an analytical tool. What the multiplicity of timelines, as well as the controversies around them, make clear, however, is that the task of reconstituting one, singular collective time, into which the event of 22 July can be fitted, is almost impossible. The role of the timelines is rather to document a plurality of diachronic connections and synchronic simultaneities, which in the future might form a new web or fabric of time, into which those afflicted by the events can place their own life stories in order to come to terms with grief, fear, and anger. Dahlberg's "Memory Wound", on the other hand, is more ambitious. And maybe it has to be, in order to fulfil the function of aiding all present and future Norwegians and indeed many other visitors to fit the events of 22 July into a continuous, relatively stable, and collectively accessible, even shared, time. To achieve this, Dahlberg could not choose any single one of the times represented in the timelines discussed previously, since they are all politicized, contested, and controversial in different ways. In general, Dahlberg's design can be understood as an attempt to find another form of time, which Norwegians and others are likely to accept as uncontroversial, shared, and with almost universal qualities. It also cannot be the absolute, mathematical time of the timeline itself, before it is filled with events and lives. "Empty, homogenous time", as Walter Benjamin famously called it (1961, 276), cannot take on the function of symbolizing people's grief, fear, and anger or serve as a vehicle for individual and collective memory work.

To mediate this tension, between the universal and the empty, Dahlberg turns to nature, to natural time, which is equally valid and accessible to all humans. In the zoning plan for the area (*Forslag til statlig reguleringsplan med bestemmelser for Nasjonalt minnested 22.juli*), published in 2014, we find the following description of the planned memorial site:

> The concept for the memorial site at Sørbråten involves a cut through the tip of the peninsula and hence also in nature itself. It recreates the physical experience of something being taken away, and reflects the sudden and permanent loss of those who died.
>
> *(Statsbygg 2014, 20)*

I have already pointed out how the design for the memorial site at Sørbråten draws on an aesthetics of rupture and absence, which came to dominate designs for memorials for similar traumatic events from the 1980s onward (Krankenhagen 2001). But more than that, the planned memorial features the presence of specific ideas about nature and natural time: what it means to perform a cut "in nature itself", as described in the zoning plan. The description continues:

> The cut is going to be a three-and a half meter wide excavation that cuts from the top of the cape at Sørbråten to below the water line, where it is fitted to the bottom of the lake. The empty space created in the landscapes makes it impossible to reach the point of the cape.
>
> *(ibid)*

In this way, the excavation, that is, the incision into "nature itself", cuts through both space and time, making certain experiences both spatially and temporally inaccessible. The events of 22 July are reintroduced into the flow of time and history by a gesture that, far from bridging the gap and mending the deep cracks in the temporal fabric of society, stylizes and absolutizes the gap itself. Time is reconstituted as an aestheticized, almost transcendent moment, abstracted from the multiple times of political and social reality. Furthermore, the cut itself emerges as an unbridgeable gap, from which present time can neither reconnect with the past nor restart into the future. Inside the cut, another kind of presence is evoked, which is no longer a part of "nature itself": the presence of the dead:

> The visitors find themselves at one side of a water channel created by the cut. Right across the channel, in the flat vertical rock face on the other side, the names of the deceased are engraved. It will be up to the bereaved to decide whether they want to include the name of the deceased.
>
> *(ibid)*

Names of victims engraved into a flat, polished rock face echo the black granite wall of Maya Lin's Vietnam Veterans Memorial in Washington (Scruggs and Swerdlow 1992) as well as the borders of the recessed pools making up Michael Arad's and Peter Walker's memorial for the victims of the 11 September terrorist attacks, aptly named *Reflecting Absence* (Blais and Rasic 2011). What distinguishes Dahlberg's "Memory Wound" from the two others is the reference to nature. The granite rock faces of the two US memorials make no claim to being, or even representing, nature; on the contrary, they are selected for their material qualities, their hardness and shininess, not their ability to symbolize, or even "be", nature. The granite rock face in Dahlberg's design for Sørbråten is the result of a cut "in nature itself" and carries this symbolism with it.

More than anything, Dahlberg's design testifies to the artist's belief that the time of nature is more stable and predictable, less contested and controversial than the times of social, political, and technological processes, and hence more suitable for representing and safeguarding the memories of the 22 July events for both the foreseeable and the unforeseeable future. In Western historiography, similar ideas have been formulated by the French historian of the Annales school Fernand Braudel, the Polish cultural historian Krzysztof Pomian, and the German social and intellectual historian Reinhart Koselleck. All three of them base their historical studies on a stratigraphic theory of multi-layered historical time, according to which the oldest, most long-term, and slowest-moving time can be found in the bottom layer, whereas the time of short-term events and rapid changes constitutes the top layer (Braudel 1958; Koselleck 2000; Pomian 1984. For further discussion, see Jordheim 2014). The excavation at Sørbråten wants to reach this deep, hidden layer, for which Braudel coined the term *la longue durée*, and mobilize it for the memorialization of an event which was very much part of event-history, *histoire evenementielle* (Braudel 1958).

Even though this stratigraphic theory of history has many problems and weaknesses, Dahlberg's design to merge short-term event-time with long-term nature-time proved sufficient to convince both the jury and the art critics (Due and Bjørdal 2014). In the discussion following the announcement of the winner, however, it was soon made clear that the stability and universality associated with nature-time could not easily be mobilized for social and political purposes. Nature as a temporal unity was already challenged by the German theologian Johann Gottfried Herder in the eighteenth century, in his attack on Kant's critical philosophy: "In reality every mutable thing has its own inherent standard of time", Herder claimed, added: "this exists even if nothing else is there" (1998, 360). All his examples are from nature, including human nature: "My pulse, my step or the flight of my thoughts is not a temporal standard for others; the flow of a river, the growth of a tree is not a temporal standard for all rivers, trees and plants". And then he concludes: "In other words, there are (one can say it earnestly and courageously) in the universe at any time innumerable different times" (ibid). Taking our cue from Herder, we quickly realize that the specific time in and of nature Dahlberg targets in his design for the 22 July memorial is the time of stones and rocks, what we usually refer to as "geological time", going back hundreds of millions of years.

Heritage in geological times

The peninsula at Sørbråten, into which the "cut in nature" should have been made, dates back to the Cambro-Silurian period in geological history, encompassing the Cambrian, the Ordovician, and the Silurian, beginning 545 and ending 417 million years ago. These rock layers are part of the so-called Oslo Graben, the most famous geological formation in the southern part of Norway. However, they were formed about 100 million years earlier than the Oslo Graben itself, dating from the Late Carboniferous period (Larsen et al. 2008). This makes an important difference, the geologist Bjørn Tore Larsen points out: The peninsula is made up of sedimentary rocks, more precisely, of sandstone, the so-called Ringeriks-sandstone, or "downtonsk sandstone", which formed more than 400 million years ago from sand gathering in the rivers from the so-called Caledonian orogeny (Carstens 2016). The rocks, often in the form of flagstones, carry marks of waves and from drying. They easily crack or split up when exposed to any kind of force, Larsen emphasizes (ibid). In this sense, they differ from the rock formations that are visible above the peninsula, which are formed later by rift formations and volcanism, forming the Oslo Graben. These rocks from the Permian period are much harder and more solid than the ones found at the bottom of the basin (ibid). In other words, the kind of solid granite rock face that Dahlberg imagined that he would find when he and his team cut through the peninsula, and which would be an ideal surface for engraving the names of the victims, was never more than a fantasmatic vision, in direct conflict with the geological reality, according to geologists like Larsen and his colleagues. In his response, Svein Bjørkås, leader of KORO, the government organization responsible for the artist competition, pointed out that the realization of the design is the responsibility of the artist (ibid). Since the whole project has been aborted, we will never know how it would have turned out.

For this study of the multiple times involved in the work to reintegrate the events of 22 July in the social and historical time of Norwegian society, rock quality is not the decisive question. The point is rather to emphasize how the attempt to mobilize nature and, more specifically, deep, geological time in order to secure the stable, long-term memory of the terrorist attacks and include them in Norwegian heritage, was always bound to fail, mainly for two reasons: On the one hand, the effect of making a "cut in nature", creating a gap in the rock formation of the Sørbråten peninsula, was perceived as both too brutal and too symbolically imprecise to function as a memorial site for the victims of the attack. The protests by the neighbours, who in the end threatened to take the government to court if they went through with the construction of Dahlberg's "Memory Wound", were brought about by the feeling that the design was "far more brutal than expected". Rather than making "such a brutal wound in nature", and generally, "more wounds, … deeper, more traumas", we need to "heal the wounds" that are already there, according to one of the neighbours interviewed for the report about the psychosocial effects of the planned memorial (Michel and Hellevik 2015, 32).

Possibly what both Dahlberg and those who selected him and his design for the job forgot to take into account was the intense and even intimate relationship many Norwegians have with nature, especially those living at fairly remote places like Utstranda. In comparison with the high-speed, accelerating life in the cities, in environments where things constantly move and change, rural Norway is characterized by stability and cyclical, long-term rhythms. To cut an opening in a peninsula, which has been there for thousands of years, only changing according to natural processes like seasons and growth, might make a more violent impression than similar interventions in a cityscape, in which rapid change is the norm.

On the other hand, the design raises a more theoretical question, which has to do with how we handle living in a multi-temporal world, like the one described by Herder, also encompassing the long times and durations, the slow rhythms of nature, recently being mobilized by the concept of the Anthropocene. A similar question is raised by Peter Schjeldal in a piece in *The New Yorker*: "What had the land done to deserve being fissured?" he asks. Why this "offense to blameless nature" (2017)? Schjeldal's question is the right one, only it has more to do with time, than with blame. The rocks making up the peninsula have been there, exposed to uplifting and erosion, for around 400 million years. What does it mean when we – Norwegians in the second millennium AD – decide to intervene in, change, cut, some would even say destroy, this landscape so dramatically? How – if at all – can it make sense to commemorate the greatest crime committed on Norwegian soil since World War II, that is, during the last 60 years, by mobilizing 400 million years of geological time? Of course, this tension between long-term natural history and short-term human history is made even more tangible by the other context, in which humans appear as a geological force or agent, changing the earth beyond the point of no return: the Anthropocene. At first glance, the 22 July terrorist attacks and the slow destruction of the global climate do not seem to be connected; however, in both cases, the main victims of these brutal destructions are generations of young people deemed to take over when the present adult generation leaves the wheel, that is, if they manage to survive for so long.

Again, different times become entangled with each other at a moment when we need to keep them apart – when we need to be able to define what is nature, what is history, what is politics, what is responsibility, what is guilt, what is hope. As Herder pointed out, there are in the world innumerable different times, but they do not come as neat timelines, clearly separated or separable from each other. What the case of 22 July in Norwegian collective memory demonstrates is that times come in bundles (Bjordal 2018), entangled and infiltrated with each other, and that every attempt to separate and control them, either by means of timelines or memorials, will by necessity be controversial and contested.

Acknowledgments

This chapter was written at the Centre of Advanced Study (CAS), Oslo, during the stay of the InSync research group, August 2018–July 2019. I would like to thank CAS for funding and hosting the group and Geoffrey Bowker, Espen Ytreberg, Lucian Hölscher, and Erica Baffelli for reading and commenting on the manuscript. I have also had inspiring discussions with Mattias Ekman about the central arguments and ideas, and he has helped me in locating material and literature.

Notes

1 The original quote from Act 1, Scene 5, goes as follows: "The time is out of joint; O cursed spite,/ That ever I was born to set it right!" Famously, Jacques Derrida used this quote both in *Spectres of Marx* (1993) and in a later article "The Time Is Out of Joint" (1995).

2 "Impact events", in Fuchs' terms, may "shatter the material and symbolic worlds that we inhabit" (2011, 11).
3 By "lifetimes", I understand not only, not even primarily, the biographical and generational times of human life but, in a much broader sense, how scales of time and scales of life converge to form temporal arrangements at work in human societies, more precisely, how biological, geological, and cosmological time scales combine with social and political concepts to form temporal arrangements governing human life, how these arrangements change and come in conflict with each other. See also Lifetimes project website, https://temporalities.no/.
4 The most systematic and thought-provoking effort to place Breivik's actions in the context of right-wing extremism, racism, and ideologies of white supremacy can be found in the book-length study by Sindre Bangstad: *Anders Breivik and the Rise of Islamophobia* (2014).
5 In his short film about the process, *Notes on a Memorial* (2018), the artist Jonas Dahlberg comments that the visual representation of the plans for the memorial, showing the cut through the peninsula and the sleek granite rock on either side, was probably visually too strong and too suggestive, causing people to respond more to the image than the actual plans for the memorial. By contrast, the draft for the memorial at Utøyakaia recently published by *Statsbygg* – Norwegian Directorate of Public Construction and Property contains no similarly striking visual representations (Manthey Kula and Bureau Bas Smets with M. Ekman. 2019. *Nasjonalt minnested ved Utøyakaia*).

References

Adam, B. 2004. *Time*. Cambridge: Polity.

Alexander, J. C. 2004. "Toward a Theory of Cultural Trauma." In *Cultural Trauma and Collective Identity*, edited by J. C. Alexander, P. Sztompka, R. Eyerman, N. Smelser and B. Giesen, 1–30. Berkeley, CA: University of California Press.

Alexander, J. C. 2012. *Trauma: A Social Theory*. Cambridge: Polity Press.

Anderson, B. 2006. *Imagined Communities* (rev. edn). London: Verso.

Assmann, A. 2013. *Ist die Zeit aus den Fugen? Aufstieg und Fall des Zeitregimes der Moderne*. Munich: Carl Hanser Verlag.

Bal, M. 1999. "Introduction." In *Acts of Memory: Cultural Recall in the Present*, edited by M. Bal, J. V. Crewe and L. Spitzer, viii–xvii. London: University of New England Press.

Bangstad, S. 2014. *Anders Breivik and the Rise of Islamophobia*. London: Zed Books.

Benjamin, W. 1961. *Illuminationen*. Frankfurt am Main: Suhrkamp.

Bjordal, S. H. 2018. "Tidene på sidene: Om tekster som tidsknipper og stavkirker i og ute av synk." *Tidsskrift for kulturforskning* 17(2): 5–26.

Blais, A. and Rasic, L. 2011. *A Place of Remembrance: Official Book of the National September 11 Memorial*. Washington, DC: National Geographic Society.

Braudel, F. 1958. "Historie et sciences sociales: La longue durée." *Annales ESC* 18(4): 725–753.

Carstens, H. 2016. "Stridens Odde." *GEO365.no*, 30 May. Accessed 19 November 2019. www.geo365.no/bergindustri/stridens-odde/.

Delanty, G. 2017. *The European Heritage: A Critical Re-Interpretation*. London: Routledge.

Derrida, J. 1995. "Time Out of Joint." In *Deconstruction Is/In America: A New Sense of the Political*, edited by A. Haverkamp, 14–38. New York: New York University Press.

Due, A. and Bjørdal, S. 2014. "Han skal lage minnesmerkene etter 22.juli." *Dagbladet*, 27 February. Accessed 10 December 2019. www.dagbladet.no/kultur/han-skal-lage-minnesmerkene-etter-22-juli/61440905.

Fagerland, T. E. 2015. "Kronikk av faglig ansvarlig ved 22. juli-senteret: Vi tas tilbake til øyeblikket da terroren rammet Norge." *Aftenposten*, 15 June. Accessed 10 December 2019. www.aftenposten.no/meninger/kronikk/i/8VeE/Kronikk-av-faglig-ansvarlig-ved-22-juli-senteret-Vi-tas-tilbake-til-oyeblikket-da-terroren-rammet-Norge.

Felman, S. and Laub, D. 1992. *Testimony: Crises of Witnessing in Literature Psychoanalysis and History*. New York: Routledge.

Fuchs, A. 2011. *After the Dresden Bombing: Pathways of Memory, 1945 to the Present*. London: Palgrave Macmillan.

Gaarder, C., Jorgensen, J., Kolstadbraaten, K. M., Isaksen, K. S., Skattum, J., Rimstad, R., Gundem, T. et al. 2012. "The Twin Terrorist Attacks in Norway on July 22, 2011: The Trauma Center Response." *Journal of Trauma and Acute Care Surgery* 73(1): 269–275.
Grafton, A. 2003. "Dating History: The Renaissance and the Reformation of Chronology." *Daedalus* 132(2): 74–85.
Grafton, A. and Rosenberg, D. 2010. *Cartographies of Time*. Princeton, NJ: Princeton Architectural Press.
Harrison, R. 2012. *Heritage: Critical Approaches*. London: Routledge.
Hartog, F. 2015. *Regimes of Historicity: Presentism and Experiences of Time*. Translated by Saskia Brown. New York: Columbia University Press.
Herder, J. G. 1998. "Eine Metakritik zur Kritik der reinen Vernunft." In *Schriften zur Literatur und Philosophie 1792–1800*, 303–640. Berlin: Deutscher Klassiker Verlag.
Hirschberger, G. 2018. "Collective Trauma and the Social Construction of Meaning." *Frontiers in Psychology* 9: 1441.
Hjorth, I. 2018. "Hvorfor minnesteder? En undersøkelse av den minnepolitiske håndteringen av 22.juli-terroren." *Nordisk kulturpolitisk tidsskrift* 21(2): 246–268.
Hoy, D. C. 2012. *The Time of Our Lives: A Critical History of Temporality*. Boston, MA: MIT Press.
Ingold, T. 2007. *Lines: A Brief History*. London: Routledge.
Jordheim, H. 2014. "Multiple Times and the Work of Synchronization." *History and Theory* 53: 498–518.
Jordheim, H. and Wigen, E. 2018. "Conceptual Synchronisation: From Progress to Crisis." *Millennium: Journal of International Studies* 46(3): 421–439.
KORO. n.d. *Minnesteder etter 22. Juli*. Accessed 19 November 2019. https://koro.no/prosjekter/minnesteder-etter-22-juli/.
Koselleck, R. 1985. *Futures Past: On the Semantics of Historical Time*. Cambridge, MA: MIT Press.
Koselleck, R. 2000. *Zeitschichten: Studien zur Historik*. Frankfurt am Main: Suhrkamp.
Koselleck, R. 2011. "Introduction and Prefaces to the *Geschichtliche Grundbegriffe*." *Contributions to the History of Concepts* 6(1): 1–37.
Krämer, S. 2016. *Figuration, Anschauung, Erkenntnis: Grundlinien einer Diagrammatologie*. Frankfurt am Main: Suhrkamp.
Krankenhagen, S. 2001. *Auschwitz darstellen. Ästhetische Positionen zwischen Adorno, Spielberg und Walser*. Cologne: Böhlau.
Larsen, B. T., Olaussen, S., Sundvoll, B. and Heeremans, M. 2008. "The Permo-Carboniferous Oslo Rift Through Six Stages and 65 Million Years." *Episodes* 31(1): 52–58.
Latour, B. 2007. *Reassembling the Social: An Introduction to Actor-Network Theory*. Oxford: Oxford University Press.
Laub, D. 2003. "September 11, 2001: An Event without a Voice." In *Trauma at Home*, edited by J. Greenberg, 204–215. Lincoln, NE: University of Nebraska Press.
Luhmann, N. 1990. *Soziologische Aufklärung 5: Konstruktivistische Perspektiven*. Berlin: Springer.
MacDonald, S. 2008. *Difficult Heritage: Negotiating the Nazi Past in Nuremberg and Beyond*. New York: Routledge.
Manthey Kula and Bureau Bas Smets with M. Ekman. 2019. *Nasjonalt minnested ved Utøyakaia: Skisseprosjekt for Statsbygg, 30. juni*. Accessed 10 December 2019. www.statsbygg.no/Files/prosjekter/utoyakaiaVurdering/planforslag/Vedlegg14_Illustrasjonshefte_skisseprosjekt.pdf.
Melle, I. 2013. "The Breivik Case and What Psychiatrists Can Learn from It." *World Psychiatry* 12(1): 16–21.
Meskell, L. 2002. "Negative Heritage and Past Mastering in Archaeology." *Anthropological Quarterly* 75(3): 557–574.
Michel, P.-O. and Hellevik, P. 2015. *Psykososiale konsekvenser av nasjonalt minnesmerke i Hole kommune*. Rapport 3/2015. Oslo: Nasjonalt kunnskapssenter om vold og traumatisk stress A/S.
Morris, D. J. 2015. *The Evil Hours: A Biography of Post-Traumatic Stress Disorder*. New York: Eamon Dolan.

NOU. 2012. *Rapport fra 22. juli kommisjonen*. Oslo: NOU 14/2012. Accessed 19 November 2019. www.regjeringen.no/contentassets/bb3dc76229c64735b4f6eb4dbfcdbfe8/no/pdfs/nou201220120014000dddpdfs.pdf.

NRK. 2012. *22.7 Terroren som rammet Norge* (*Multi-stranded Timeline for the Terror Attack*). Accessed 10 December 2019. www.nrk.no/227/fakta/tidslinje/tidenetter/.

NRK. 2017. *Tidslinje minnesmerke for 22. Juli* (*Timeline for the Construction of a 22 July Memorial*). Accessed 10 December 2019. www.nrk.no/tidslinje-minnesmerke-for-22.-juli-1.13558940.

NRK. 2019. *Striden om Sylvi Listhaug*. Accessed 19 November 2019. www.nrk.no/nyheter/striden-om-sylvi-listhaug-1.13965363.

Oslo Politidistrikt. 2012. *Evaluering 22/7*. 27 February (rev. edn). Oslo: Oslo Politidistrikt.

Politidirektoratet. 2012. *22. juli 2011. Evaluering av politiets innsats*, POD-publikasjon 2012/03. Oslo: Politidirektoratet.

Pomian, K. 1984. *L'ordre du temps*. Paris: Gallimard.

Reichel, P. 2007. *Vergangenheitsbewältigung in Deutschland: Die Auseinandersetzung mit der NS-Diktatur in Politik und Justiz*. Munich: Beck.

Schjeldahl, P. 2017. "Did a Cancelled Memorial to Norway's Utøya Massacre Go Too Far?" *The New Yorker*, 25 July. Accessed 10 December 2019. www.newyorker.com/culture/cultural-comment/did-a-cancelled-memorial-to-norways-utoya-massacre-go-too-far.

Scruggs, J. C. and Swerdlow, J. 1992. *The Vietnam Veterans Memorial: To Heal the Nation*. New York: HarperCollins.

Smail, D. L. and Shyrock, A. 2011. "Introduction." In *Deep History: The Architecture of Past and Present*, edited by D. L. Smail and A. Shyrock, 3–20. Berkeley, CA: The University of California Press.

Sollid, S. J. M., Rimstad, R., Rehn, M., Nakstad, A. R., Tomlinson, A.-E., Strand, T., Heimdal, H. J. et al. 2012. "Oslo Government District Bombing and Utøya Island Shooting July 22, 2011: The Immediate Prehospital Emergency Medical Service Response." *Scandinavian Journal of Trauma, Resuscitation and Emergency Medicine* 20(3): 1–12.

Sorokin, P. A. and Merton, R. K. 1937. "Social Time: A Methodological and Functional Analysis." *The American Journal of Sociology* 42(5): 615–629.

Statsbygg. 2014. *Forslag til statlig reguleringsplan med bestemmelser for Nasjonalt minnested 22.juli – Sørbråten*. Planbeskrivelse med illustrasjoner. Del av gnr. 233, bnr. 2 og 6 i Hole kommune.

VG. 2011. *Anders Behring Breivik – fra fødsel til 22. juli* (VG timeline). Accessed 10 December 2019. www.vg.no/nyheter/innenriks/22-juli/tidslinjeabb.

Waage, S., Poole, J. C. and Thorgersen, E. B. 2013. "Rural Hospital Mass Casualty Response to a Terrorist Shooting Spree." *British Journal of Surgery* 100(9): 1198–1204.

Wikipedia. n.d. *Tidslinje for terrorangrepene i Norge 2011*. Accessed 22 October 2019. https://no.wikipedia.org/wiki/Tidslinje_for_terrorangrepene_i_Norge_2011.

Wilson, R. 2009. "History, Memory, Heritage." *International Journal of Heritage Studies* 15(4): 374–378.

Zerubavel, E. 2003. *Time Maps: Collective Memory and the Social Shape of the Past*. Chicago, IL: The University of Chicago Press.

12

THE REMEMBRANCE OF THINGS

The industrial heritage of mining and the ecology of memory

Timothy LeCain

It is widely understood that individual things or environments can be spurs to our memories. An artifact from the past, whether deliberately meant to be a "souvenir" or not, can have the power to take us back to an earlier time, just as a visit to a particular place may bring feelings and thoughts of the past unbidden. Despite the extraordinarily evocative power of such things and places, however, we often tend to think of them as largely passive signifiers, placing the locus of memory itself solely in an abstract brain that is largely distinct from its environment. Likewise, we assume that the value of a souvenir or place lies in its relative fixity, believing that their unchanging nature stands in contrast to our own mercurial lives. However, recent insights into the nature of memory and cognition suggest a far more fluid and dynamic understanding of the role of things in memory. In this chapter, I will draw on the history of several former industrial mining sites to examine the ways in which things interact with human brains and bodies to constitute what we might aptly term an "ecology of memory". I argue that human cognition and memory must be understood as extensive with the material environment, suggesting that we think and remember *through* material things. In this sense, to remember is literally to re-*member* our extended selves by reincorporating the scattered material pieces of our individual and societal past being (Olsen 2003). Further, I will examine the dynamic ecological nature of the things re-membered, as no artifact or environment remains truly static. Rather, the power of things and places in evoking memories stems in part from their dynamic ecological nature, including the tendency for human-made structures and artifacts to degrade, change, and even disappear. The experience of re-membering an abandoned industrial mining site thus offers us an opportunity to escape our anthropocentric concepts of memory and self and to recognize the ways in which ecological processes might give rise to new memories and insight. Even the disappearance of human-made structures and effects might, in some cases, prove more useful than their preservation, as broader and less anthropocentric patterns and insights emerge about the power of things themselves to shape human history.

Re-membering the past

What does it mean to have a memory? An ability to remember the past, whether recent or long passed, is obviously essential to our individual and collective identities as human beings.

Yet this quotidian phenomenon is typically little understood or even considered by most historians and humanists. We often tend to assume that our memories are highly immaterial and abstract phenomena, fleeting flashes of images or sensations that reside somewhere in our brains, or an even more ineffable "spirit" yet nonetheless constituting our essential identities. However, in recent years, new insights into the physiological functions of the brain have made our understanding of memory increasingly solid and materially rooted. At roughly the same time, a growing number of cognitive theorists have begun to argue that memories and remembering occur not just *within* our brains but rather extend outward into the countless material things and environments that are essential to sparking and sustaining memory. Memories in this extended sense truly reside in the interactions *between* humans and things.

This extensive concept of memory has long been informally recognized with the concept of the "souvenir", an artifact whose very purpose is to help someone remember and whose name is rooted in the French verb "to remember". Yet the ability of things or environments to spark memories and remembering is not confined solely to consciously recognized artifactual souvenirs. Rather, countless other material things and environments may be implicated in what we mistakenly think of as solely "our" memories – things and environments that have their own properties and trajectories that are not fully controlled or even understood by humans. Such a view has profound implications for the way we understand the nature of both history and humanity, as it suggests our memories, and thus our identities, must respond and change in relation to an exterior material world that is constantly in flux. A scarred and poisoned industrial landscape like those created by large-scale copper mining, for example, might well be understood as carrying a type of "memory" of past events in its material nature. However, this memory is no more static and fixed than a brain-based memory. Rather, anthropogenic structures, land forms, and remains are constantly changing as they interact with the broader ecology and various human attempts to remediate or preserve what are sometimes viewed as historically significant sites. Visitors to such places thus experience what might aptly be termed an "ecology of memory", in which human memory, history, and identity are deeply entangled with non-human ecological processes.

As these ecological and human efforts change and reshape the material remains of industrial copper mines, they must also change the nature of the memories experienced by tourists who visit the sites. Typically this process of change and degradation of a site is assumed to occasion a *loss* of memory and history – an entirely negative process of forgetting. However, in this chapter, I want to suggest that an ecologically rooted understanding of memory might instead reveal a more positive process in which the power of material things to assert themselves point us towards a deeper and less anthropocentric understanding of our industrial past and heritage: a profound process of re-*membering* that reveals the many ways in which material things create and sustain human thought, memory, and history.

The materiality of extended memories

Despite considerable progress in recent decades, the precise biochemical nature of human memory remains somewhat mysterious. Nonetheless, contemporary neuroscientists do largely agree that memories arise from material changes in the brain that result from individual experiences and thoughts. Neuroscientists refer to these physiological carriers of memory as "engrams", and they believe they are created by the biochemical strengthening of synaptic connections between populations of neurons in the brain. When we remember something, we reactivate these neural connections and patterns in a way that echoes the original experience, thought, or idea that formed the engram in the first place. This subsequent experience of a synaptically "stored" memory is inevitably incomplete and imperfect, a reflection of the initial experience rather than an entirely faithful reproduction. Further, no matter how strongly embedded a memory might be, over time, the synaptic connections will inevitably weaken or change to at least some degree.

This unavoidable weakening or fading of memories has long been assumed to be an entirely negative process that reduces our ability to understand and act effectively within our environments. Obviously this is true to some degree – I cannot learn from and respond to a previously encountered danger if I cannot even remember the initial experience. However, today, some theorists posit that the brain not only passively "forgets" as synaptic connections degrade or evolve but is rather actively engaged in erasing or weakening memories, both to permit the formation of new memories but also because some level of imprecision in our memories seems to facilitate the recognition of useful experiential patterns. A memory with too much specific detail, they suggest, can make it difficult to spot deeper commonalities that might prove more useful for understanding and reacting to the world around us. A very precise memory of being bitten by a brown Doberman in New York's Central Park in February, for example, might be less useful than a more vague and thus more generalizable memory that suggests we should watch out not just for brown Dobermans in Central Park in February but perhaps for any unfamiliar dog regardless of setting (Siegfried 2019).

The matter of memories is further complicated by the realization that our memories also often depend heavily upon *external* physical artifacts and environments to sustain or spark the biochemical engrams in our brains. Perhaps the most famous example of this is the involuntary flood of memories Proust experienced on tasting the madeleine cookies of his childhood (Proust 1913). All of us can easily recall examples when until then largely forgotten (or at least neglected) memories were brought forth on the return to a childhood house or town or an environment like a forest or seashore, thus suggesting that our memories are often deeply rooted in things and places – so much so that it is not unreasonable to ask whether the memory itself is best understood as existing not solely in the internal brain but also in the external environment in which that brain attempts to remember and understand. This understanding of memory bears some resemblance to the concept of "extended mind" or "extended cognition" as formulated in recent years by cognitive scientists and theorists like Andy Clark. Human cognitive abilities, Clark argues, can be distributed in a network of external props and aids like computers, files, texts, and maps, aspects of our material surroundings without which some fundamental part of what we consider our intelligence would vanish (2008b, 15). Indeed, Clark points towards what we might term an ecology of cognition and memory when he suggests the extended mind "drives home the degree to which environmental engineering is also self-engineering". In changing our material environment, Clark suggests, we also reconfigure "our minds and our capacities of thought and reason" – and presumably the material nature of memory and our capacities for remembering (2008a, xxviii).

The prominent psychologist Mihaly Czikszentmihalyi has also argued for many decades that material artifacts clearly play a role in human identity and memory. Czikszentmihalyi is best known today for naming and explaining the psychological concept of "flow", those periods of intense concentration and involvement in a task when we exist purely in the moment. However, Czikszentmihalyi was equally interested in the power of things to shape our identities. In their 1981 book, *The Meaning of Things: Domestic Symbols and the Self*, Czikszentmihalyi and his co-author Eugene Rochberg-Halton argued that material things are not just tools for survival or comfort but rather are the means through which we come to be who we are. "To understand what people are and what they might become", the authors conclude, "one must understand what goes on between people and things". The sense of any individual self, they write, is "to a large extent a reflection of the things with which he interacts" (Csikszentmihalyi and Rochberg-Halton 1987, 1). In contrast to the post-modernist and anthropocentric tendency to see things and artifacts as essentially empty signifiers whose meanings are largely or entirely created by humans and their cultures, Czikszentmihalyi and Rochberg-Halton argued that things were the anchors around which a stable human identity might be constructed:

> [M]en and women make order in their selves (i.e. "retrieve their identity") by first creating and then interacting with the material world. The nature of that transaction will determine, to a great extent, the kind of person that emerges. Thus the things that are around us are inseparable from who we are. The material objects we use are not just tools we can pick up and discard at our convenience; they constitute the framework of experience that gives order to our otherwise shapeless selves.
>
> *(1987, 16)*

Indeed, as many other thinkers have recognized, this is precisely why institutions like the military or religious orders that seek to foster new identities and values often insist that initiates bring few material possessions from their past lives (Randall 2016).

While their focus is not on memory per se, Czikszentmihalyi and Rochberg-Halton nonetheless make it clear that this creation and stabilization of identity through things is tightly related to memory. When an object or a smell (which is, of course, also a material thing in chemical form) brings back feelings and memories, they are not just sparks for an entirely internal process of an isolated mind. Rather, things possess a material nature, varying widely in stability and endurance but nonetheless existing outside of any human mind that might engage with its properties and potentialities. The authors offer the example of a lawyer whose trombone from his college years allows him to recapture and revive an earlier self. "So the trombone helps both focus attention", they write, "and vividly brings back old memories and experiences, thus adding a sense of depth and wholeness to the self and its owner". To be sure, the trombone can be understood as a symbol of the lawyer's earlier self, yet it also "had the power to put him back in touch with himself" (Csikszentmihalyi 1993, 25). Could the lawyer tap into the material traces of the synaptic engrams in his brains to recall some, perhaps even many, aspects of his earlier life even in the absence of the trombone? Obviously. Yet it seems equally obvious that the lawyer's embodied sense of his past trombone-playing self could be even more powerfully remembered when he picks up the instrument itself, feels its cold metallic heft, places his lips on the mouthpiece, and begins to play. The extended historical entity that had been the "young man + trombone" would be captured most fully, or at least uniquely, when engaging with the instrument itself. It thus seems evident that these powerful memories resided not solely in the lawyer's brain but also in a material thing that possesses its own independent existence and unique historical course. Absent the trombone (or perhaps one similar to it), the lawyer could never again recapture those specific aspects of his past self in such a powerfully visceral way.

Yet if memories do indeed reside not solely in our internal brains but also in external things that persist from the past, how then should we think about our contemporary heritage of toxic industrial waste sites? The ability of a specific trombone or other artifact to spark memories and feelings in an individual who used and preserved it is clearly different than the typical visitor's experience of a historic industrial site. Many visitors to these sites have little or no personal experience of the place as it existed in the past, approaching it not so much as a part of their personal memories but rather as a remnant of a shared regional or national memory or heritage. Likewise, a visitor typically experiences not a single discrete artifact, like a trombone, but rather a complex landscape that mixes various artifacts with landforms, sky, water, and countless other things. Indeed, the focus of Czikszentmihalyi and Rochberg-Halton's work was entirely on the domestic things we preserve (and often curate) in our homes, not the messy multitude of things to be found in an external anthropogenic landscape. Nonetheless, the authors do point towards a broader environmental dynamism when they suggest the concept of an "ecology of signs" in which even domestic things have their own dynamic natures and interactions, suggesting the domestic environment of a trombone is perhaps not so very different than the external landscape of a former industrial site (Csikszentmihalyi and Rochberg-Halton 1987, 17).

Even more importantly for my purposes here, Czikszentmihalyi and Rochberg-Halton squarely challenge the modernist anthropocentric assumption that the material world – whether domestic or otherwise – is little more than a passive stage on which the independent human subject is entirely free to create and recreate their identities. Anticipating the more recent scholarly turn to materiality, the authors suggest their valorization of domestic things is not new but rather is a revival of the once-dominant human belief that the external natural world and the universe broadly was a "cosmos", a living space akin to and intimately connected to human beings, who were just one expression of it. With the advent of modernist thinking, they argue, humans sought to "depersonalize the cosmos and reconceive it as a Newtonian machine or a Cartesian "apparatus". The active and creative human individual was thus placed apart from and in control of a passive material world of non-human things and organisms ruled by mechanistic laws that could be freely manipulated without concern that they would fundamentally alter humans themselves. To return to the previous example, in the modernist world, the trombone was reduced to nothing more than an empty symbol, while all of the lawyer's power to remember the past was isolated in an ephemeral brain or spirit (Csikszentmihalyi and Rochberg-Halton 1987, 12).

In all this, Czikszentmihalyi and Rochberg-Halton might point us towards some novel ways in which we can analyze the human interaction with historic landscapes as an ecologically creative and dynamic process while also moving us beyond conventional anthropocentric obsessions. Even more than a relatively protected and curated domestic space, industrial landscapes carry material memories that are both enduring and in flux as they respond to environmental factors that we would conventionally recognize as ecological in nature. In truth, these ecological factors are also at work in a domestic environment. Yet they move so slowly that they are often difficult to identify and measure. In an external landscape of substantial geographical size, however, powerful ecological forces of wind, water, and geochemistry interact with anthropogenic actions and artifacts. These rapidly changing ecological landscapes also raise intriguing questions about how material remains may fade and even disappear – a process historians and historic preservationists have typically understood as occasioning a loss of potential memories and histories about the site. Yet as discussed earlier, if our ability to recognize important patterns and generalities depends in part on the degradation and imprecision of our *internal* memories, might it be that the decay of *external* carriers of memory also allows us to recognize useful historical patterns and lessons?

The denial of things

As an example of the potential value of re-membering the past, consider the extraordinary industrial landscape to be found in the southwestern part of the western American state of Montana. In 1882, in the high northern Rocky Mountain town of Butte, an Irish immigrant turned mining entrepreneur named Marcus Daly first discovered what would eventually prove to be one of the biggest copper deposits in the world. Recognizing that Thomas Edison's contemporaneous innovations in electric lighting were already creating a growing demand for copper wire, Daly and his financial backers bet heavily on creating a gigantic new mining operation, which became the Anaconda Copper Mining Company. For decades the Anaconda pushed ever deeper down into the mineral-laden rocks that lay underneath what came to be called the Richest Hill on Earth, eventually excavating some 16,000 km of underground passages. However, the resulting ore was of an increasingly low grade, containing less than 5% copper and large amounts of undesirable contaminants like sulfur and arsenic. To winnow out the small amounts of valuable copper from this complex ore, the company built a giant smelter and concentrating works about 30 km to the northwest of the Butte mines. Christened Anaconda in honor of the mining company that created it, the town was by the turn of the century home to the Washoe smelter, one of the largest and most technologically advanced concentrating and smelting operations in the world (LeCain 2009).

As the Anaconda's Washoe smelter purified the copper ore, it inevitably generated immense amounts of pollutants, which were either released into the air in the form of vast clouds of sulfur and arsenical oxides or stored on site in gigantic tailings ponds, which frequently leaked into the neighboring creeks and downstream into the nearby Clark Fork River. In the early twentieth century, farmers and ranchers in the surrounding Deer Lodge Valley tried to force the Anaconda to limit the amount of sulfur and arsenic it released into the atmosphere. The farmers made a convincing case that the sulfur dioxide smoke killed their crops and that the arsenic settled out on the range grasses, where it sickened and often killed the cattle and horses who ate it. After years of litigation and broken promises, in 1919, the Anaconda erected a massive new 178-meter-high smokestack at the smelter along with new technologies designed to capture significant amounts of the arsenic and other pollutants. These pollution control technologies worked reasonably well, yet their good effects were often diminished by subsequent increases in the amount of low-grade ore being processed by the Washoe mill. Moreover, the Anaconda ultimately ended up storing much of the metallic dust it captured on site, along with immense amounts of heavy-metal-laden tailings in the ironically named Opportunity Tailings Ponds (LeCain 2009).

A similar story can be told for the copper mines in Butte where the dangerous complex copper ore had originated and where another technological innovation fundamentally changed the material nature of the site. By the early 1950s, the underground mines in Butte were becoming increasingly unprofitable. The grade of ore continued to drop, while the costs of mining soared as shafts and tunnels began to reach depths of nearly 2 kilometers, where extreme temperatures and poor ventilation demanded costly life support systems to keep the miners alive. To remain profitable, in 1955, the Anaconda management decided to commence open pit mining operations that gradually replaced all the underground mining. Christened the Berkeley Pit, the operation depended on the use of immense electric shovels and dump trucks to rapidly dig up and move the low-grade ore for shipment to the Washoe smelter in Anaconda (LeCain 2009).

With these powerful new technologies at work, the Anaconda managers now predicted an almost limitless future for mining in Butte, proclaiming that "the full geological limits of its resources are still to be determined" (LeCain 2009, 179). The Anaconda also began to promote the rapidly growing Berkeley Pit as a tourist destination. In one 1957 advertisement in the popular *Saturday Review* magazine, the company urged tourists to come "see America the BOUNTIFUL", a place where giant shovels and trucks were providing the foundations of "the nation's economic strength" through "seemingly inexhaustible mineral wealth" (LeCain 2009, 188–189). In all these confident proclamations it is easy to discern the broader promise of modernity itself: that through the expansive use of science and technology, humans could create a nearly infinite supply of resources to fuel a rapidly expanding mass consumer economy built on affordable cars, houses, and refrigerators, all of them depending on abundant and cheap copper. The Anaconda, with its gigantic Berkeley Pit and high-tech Washoe smelter, in effect promised that Americans could now fully control the natural world that had once controlled them. For the first time, humans would be masters of their own fate, utterly liberated from the demands and limits of a material world, which, as discussed earlier, had already been rendered lifeless and passive by modernism and stripped of any intimate connection to the nature of human beings themselves. In this grand narrative of human self-creation, the material nature of the copper ore itself, and more broadly of all non-human things and landscapes, were rendered largely irrelevant. Henceforth, the Anaconda seemed to promise, humans would determine their own fates and thus their own memories and histories.

The persistence of things

Perhaps it is not surprising in retrospect that the Anaconda's confident promises of human liberation from base materiality proved fleeting. Faced with competition from richer and cheaper open-pit copper mines around the world, as well as declining ore grades and rising costs in their Butte-Anaconda operations, in 1982, Anaconda's corporate successor, ARCO, decided to permanently close the Berkeley Pit. Ironically, the pit that had previously inspired exuberant promises of "inexhaustible mineral wealth" had failed to even survive a mere quarter-century. ARCO then all but guaranteed that the pit would never again be reopened when managers decided to turn off the deep underground pumps that had kept the previously high levels of groundwater from flooding the mines for almost a century. It was not long before a small but rapidly growing puddle of water appeared at the bottom of the pit as the groundwater began to rise to its pre-mining levels. In the years to follow, toxic water that had been steeped in thousands of kilometers of underground passages began steadily filling the pit, creating a vast environmental disaster. Recognizing that it would take billions of dollars to stop the polluted water from spreading and to remediate the area, in 1983 the US Environmental Protection Agency designated the Berkeley Pit – as well as the Anaconda smelter and some 190 kilometers of the Clark Fork River – as a national Superfund site, making it the largest such site in the nation by total area (LeCain 2009).

Today, the Berkeley Pit "lake" is by a large margin the most-visited tourist destination in Butte. Every year, thousands of visitors, many no doubt on their way to Yellowstone, Glacier, or other more conventionally attractive destinations, take the time to pull off the interstate highway and drive up to the Berkeley Pit. For a small fee, they can walk through a narrow tunnel and out onto a wooden viewing platform perched above the flooded pit. The Berkeley Pit "lake" is at this writing about 275 meters deep, filling roughly half the total depth of the pit with its dark waters. The water is so acidic that it easily dissolves heavy metals like arsenic, cadmium, and of course copper, out of the fractured rock and holds these poisons in solution so that they do not settle out. Left on its own, this sulfuric acid water with its suspended toxins would eventually fill the pit and flood over the edges, contaminating both surface and groundwater and poisoning a vast area that is home to tens of thousands of people, as well as countless species of fish, birds, and mammals who live in and consume the water. Following years of study, the Environmental Protection Agency ultimately determined there was no economically feasible means of truly cleaning up the Berkeley Pit water. Instead, the lake is maintained at roughly its current level by pumping and treating millions of liters each day to remove the heavy metal pollutants. The purified water is then released into surface creeks, while the tons of heavy metal tailings removed are added to the already immense tailing piles perched just above the pit – the remains of an earlier concentration plant. While some future technological breakthrough might make it possible to profitably and sustainably extract the potentially valuable minerals in the water for shipment and use elsewhere, at least for the foreseeable future, the only solution will be to treat the water and simply add yet even more toxic waste to the environmentally devastated area. Eventually, the water will finally dissolve out most of the available toxic metals and once again become clean, but only over the course of geological time scales. If humans wish to live in the area and protect the broader ecosystem, the pit water will have to be pumped and treated for thousands of years to come (LeCain 2009).

As for the Washoe smelter, shortly after it was shuttered in 1981, ARCO disassembled, demolished, and carted away the entire vast industrial facility. The remains of more than a century of human construction and operation were thus largely erased from the landscape in just a few years, leaving little sign that the site was once one of the largest industrial operations in the world. Thanks to the efforts of some local citizens, the giant 178-meter smelter stack, which had been both a blessing and a curse to the area farmers, was spared. In 1986, the state of Montana created the Anaconda Smoke Stack State Park to preserve and interpret the giant stack. But since the actual site of the stack is difficult to reach and still potentially toxic, the tourists who bother to make the trip to Anaconda will discover they cannot actually visit the smokestack itself. The stack is instead memorialized from a viewing area on the edge of town where visitors can see the towering brick structure about 2 kilometers to the southeast. In the modest park itself, the state built two low concentric brick walls that represent the exterior and interior dimensions of the stack. A number of stone-mounted plaques also provide a history of the stack's construction and its operations.

Yet if most of the more obviously anthropogenic remains of the smelter like buildings and foundations have been removed, the Anaconda did leave behind other types of material memories that cannot easily be carted away and forgotten. At the Opportunity Tailings Ponds just east of where the smelter once stood, some 140 million cubic meters of toxic tailings remain, confined by low earthen dams that stretch for several kilometers. The earlier attempts to reduce the release of the air pollutants that had once poisoned the nearby farmers' crops and livestock were also informally remembered in the form of about 190,000 cubic meters of arsenic-impregnated heavy metal dust that had been removed from the smoke over the course of half a century. Ironically, the toxicity of this captured dust is now even greater than it would have been had it been released into the air as smoke, as the highly selective pollution controls effectively concentrated it in highly purified form. In subsequent years, wind and water picked up and eroded these piles of toxic dust and spread it around the neighboring homes in the Deer Lodge Valley. In the early 1980s, state health officials found that children in the nearby community of Mill Creek had dangerously elevated levels of arsenic in their urine. Having again determined there was no effective way in which to clean up the community, in 1987, the EPA forced the evacuation all the residents of Mill Creek, leveled the houses, and stripped off a thick layer of contaminated soil for storage in the Opportunity Tailings Ponds piles (LeCain 2009, 104). As with the toxic water in the Berkeley Pit, the EPA ultimately concluded there was no economically feasible means of truly removing the tailings from the site, which in any event would simply mean relocating the pollutants to some other area. The agency instead opted to try to contain the pollutants in place, capping the tailings and dust with a thick layer of fertile soil planted with hardy grasses and carefully monitoring the movement of surface and ground water through the piles so as to limit toxic run-off as much as possible.

"Here", as one chronicler of the human efforts to deal with the Anaconda pollutants puts it, "was a splendid example of how embarrassingly indestructible matter can sometimes be" (LeCain 2009, 102). While the company and the EPA could (and did) effectively remove much of the more obviously human-built structures in Anaconda and Butte, the persistence of other things and the material memories they carried were not so easily erased. Today, not even half a century after the Anaconda's massive mines and smelters shut down, it is the *non-human* materialities that truly persist and that will very likely persist for many centuries to come.

The re-membering of Butte and Anaconda

How, then, should we read and understand the complex industrial heritage of Butte and Anaconda, one where much of the more readily identifiable evidence of human history has been erased, while many less clearly anthropogenic remains endure? Typically, the erasure of the structural and architectural remains of a historic site has been understood as a negative and even tragic outcome. Historians and preservations alike have long assumed that a clear and useful understanding of the past is best achieved by preserving as many of the material (and of course textual) remnants of that past as possible. To lose historical artifacts and buildings is to lose evidence that might provide critical insights into our past. On a more practical level, preservationists also recognize that historic buildings and landscapes can often attract tourists and bring vital economic growth to former industrial sites that frequently suffer from economic stagnation, which is certainly the case at Anaconda and, to a lesser degree, Butte.

Obviously there is much of value in this conventional approach to preserving an industrial site, as we can clearly discover important historical insights through the buildings and artifacts that persist into the present. However, in focusing on such human-made structures and artifacts, there is also a perhaps unavoidable tendency to keep the level of analysis primarily on human ideas, institutions, and systems. In other words, traditional historical and preservationist approaches tend to be unapologetically anthropocentric or, perhaps more accurately, *unconsciously* anthropocentric. Modernist assumptions about the centrality of humans have become so pronounced that other, less anthropocentric approaches are rarely even considered or, if considered, quickly dismissed as nonsensical, deterministic, and even anti-humanist.

Yet a close reading of the material remains in Butte and Anaconda suggests that the erasure of the more obviously anthropogenic artifacts from the site also offers an opportunity for some very different, though perhaps no less valuable, material memories to express themselves. In the erasure of some elements of material memory at these sites, in their partial simplification, we might glimpse larger and more broadly useful patterns of human interactions with the material world. As noted earlier, some contemporary scientific students of human memory argue that the process of forgetting is not merely a passive process in which synaptic engrams degrade but rather an active process in which the brain deliberately avoids overly detailed memories that make it difficult to recognize broadly useful patterns. Might we also understand the material memories at Butte and Anaconda in a somewhat analogous manner?

If so, this is certainly not the interpretive approach offered in the plaques and other historical materials to be found at the two sites today. To the contrary, the current interpretive material not surprisingly remains resolutely modernist and anthropocentric in spirit, most especially at the Anaconda site. Far from seeing the erasure of the obvious anthropogenic remains as an opportunity to discuss the power and persistence of other non-human things and memories, the sites strive instead to resurrect these vanished historic structures and events through texts, photos, and quasi-recreations. The interpretive approach thus embraces the modernist assumption that abundant historical details about past human choices and actions best explain the historical trajectory of these sites and their larger social, political, and philosophical meanings.

Indeed, the first sign the visitor encounters at the Anaconda Smoke Stack State Park immediately strikes this anthropocentric tone by noting that its principal purpose is simply "to provide visitors an idea of the magnitude of the Washoe Smelter Stack structure". Because visitors cannot go to the actual stack itself, the designers have created a simulacrum of the structure with two low concentric brick walls that provide some sense of the immense width of the stack. While distance inevitably diminishes the effect, the visitor's view of the actual stack itself also conveys some sense of its sheer mass and height. The designers of the site also elected to fill the center of the brick circles with the same black slag that towers in high hills near the stack, providing a useful suggestion of the persistence of this glass-like silica waste material that was formed over decades of smelting.

But if there are some nods to the persistent materiality of these non-human things, the texts and images offered at the park by and large suggest the stack should be seen as a (literally) towering exemplar of human ingenuity and labor. Much is made of the size of the stack and the herculean human efforts needed to build it, implicitly transforming a device designed to capture toxic copper pollutants into a monument to human inventiveness and determination. To be sure, the stack's importance in reducing pollution from the smelter is noted, yet the powerful material demands of the copper mineral itself – its propensity to bind with arsenic, sulfur, and other pollutants – are distinctly secondary to the celebratory story of human accomplishment. Likewise, the interpretive material uses historic photos in an attempt to resurrect the demolished Washoe buildings and structures. As with the stack, the gargantuan size of the smelter itself is offered up as further evidence of human ingenuity, skill, and power. Interior photos of the operations drive the point home, emphasizing the courageous accomplishments of the everyday laborers who built the stack and operated the smelter, often in very dangerous conditions.

It is hardly surprising, of course, that the people of Anaconda who sought to save the stack from demolition should want to tell a celebratory story of human ingenuity, perseverance, and bravery. We would expect nothing less, and though historians and preservationists might quibble over details of emphasis, most would by and large agree that this is indeed the truly important story to be told. From a broader material standpoint, however, it is striking that the park's attempt to *resurrect* the vanished historic smelter and the individuals who worked there is offered up in the midst of a present-day landscape where the enduring environmental effects of historic industrialization is far more obvious. A different history of the Washoe smelter needs no resurrection, because it persists to this day in the giant piles of tailings and slag and the hills denuded of trees and other vegetation. Yet these obvious non-human (or at least not entirely human) material memories receive scant attention – to the contrary, the site's interpretive material seems to encourage the visitor to erase the present-day historical materialities in favor of creating a celebratory human history based on vanished structures and artifacts that can only be recreated through photos and words.

My point here is not that the Anaconda Smoke Stack Park would do better to offer an equally conventionally modernist story of environmental decline and disaster. If the site were to be redesigned by an environmental group, for example, it might seek instead to excoriate the Anaconda Company with a detailed account of corporate callousness, the ill effects of the toxic smelter smoke, and the farmers' long struggle to fight it – which is the perspective most often offered in contemporary history books. Again, there would be much of value in this, and the slag and tailings pollution would surely receive due attention. However, to adopt such a contemporary environmentalist interpretation would still maintain a focus on the human story, simply replacing the previous narrative of human ingenuity and triumph with one of foolishness and destruction. But whether the interpretive slant is celebratory or critical, such anthropocentric narratives would leave the power of the copper ore itself on the margins. The nature of sulfide copper deposits, their association with large amounts of sulfur and arsenic that are, once liberated, extraordinarily difficult to control and contain, are largely ignored in favor of telling stories about humans. The copper never emerges as anything more than a bit player (LeCain 2017a).

Despite the current interpretive marginalization of copper and other non-human actors, I suspect a visitor to the site cannot help but experience at least a fleeting and perhaps inchoate sense of the material memories carried by these toxic remains as they gaze at the immense mountains of black slag that dwarf even the stack itself. Here the material memories are not solely about human triumphs and tragedies but rather about the agency of the copper itself which forced humans to grapple with poisons if they wanted the benefits of the red metal. Further, in contrast to the site's comforting anthropocentric interpretation, which suggests the past is long gone and harmless, these less anthropocentric material memories continue to play out over ecological and geological time scales. Well after the smokestack itself may have toppled and the interpretive site eroded to dust, these historical materialities will persist, continually bringing the past into the present. It does not seem unreasonable, then, to ask: What are the truly important memories and lessons to be learned at the site? The transient details of a particular set of human actions over the course of a few decades – whether those be of workers or corporate executives, or triumph or failure – or the much vaster material memories unknowingly and largely unconsciously released by the simple human decisions to pursue the Butte copper deposits in order to achieve the dream of national electrification?

More broadly, might such a materialist and post-anthropocentric reading also help us to recognize important new patterns that transcend the specific details of corporate and individual human stories made at Anaconda that historians and preservations have traditionally valorized? To return to our earlier discussion of how the erasure of memories can help teach us to be cautious not just of one specific dog but rather *all* dogs, might the erasure of the anthropocentric material memories at Anaconda be useful? This would obviously be a deeply unorthodox approach, but the evidence suggests it has some merit. Consider, for example, the denuded hills surrounding the Ashio copper smelter in the mountains north of Tokyo, some 8,000 kilometers from Montana. As I have argued elsewhere, there was a striking convergence between the seemingly disparate mining operations in Ashio and Anaconda, which developed over roughly the same historic period (LeCain 2017b). Despite pronounced differences in culture and geography, as well as unique historical trajectories, the similar material and ecological memories to be found today at the two sites suggest a broader set of patterns and commonalities. The sulfur dioxide generated by both smelters had very similar ecological effects, for example, killing nearly all trees and other vegetation and leaving similarly barren and erosion-prone hillsides that remain the dominant material reality to this day. Likewise, at Ashio, huge piles of slag and toxic tailings continue to be a serious problem decades after the mining operations shut down. An observer of both sites who was entirely ignorant of their specific anthropocentric sociocultural histories might thus be tempted to conclude that what really mattered was not the details of human history per se but rather the sheer material power of the copper ore itself and its closely related mining and smelting techniques.

This ecology of re-membering would thus push the human from the center of the historical narrative to instead examine how the powerful affordances and demands offered by copper swept over diverse human nations, cultures, and societies. The broader pattern to be discerned might emphasize not the many detailed contingencies of solely human differences and choices but rather the less anthropocentric insight that humans are shaped by the material things they become entangled with. Or, perhaps more accurately, that these things simply constitute who they become. In sum, a re-membering of the Ashio and Anaconda sites that focuses on the denuded hills, piles of slag, and tailings reveals the value of moving beyond divergent human histories to instead grasp that *all* sulfide coppers have the potential to bite back when humans try to use them. Thus the erasure of the uniquely human buildings and artifacts at these sites might actually be a useful process, as this "forgetting" permits other material memories to emerge more clearly.

Perhaps because the flooded pit itself is such an overwhelming material presence, the interpretive approach offered by the Berkeley Pit is less resolutely anthropocentric in tone, even hinting at a sort of technological determinism that minimizes the important of human actors and stories. A visitor to the stand will find four informational placards, two of which are devoted to the history of mining at the site and two to the subsequent flooding and ongoing efforts to control the toxic waters. Of the first pair, one placard places the open pit mine in a historical lineage beginning with small-scale placer mining in Butte in 1864, suggesting this was then followed by a logical technological progression to conventional underground mining, large-scale subsurface block cave mining and ultimately the shift to the open pit in 1955. In some sense, the Berkeley Pit interpretive material does suggest an almost inevitable progression of events which largely eliminates the human decisions that historians and preservations typically prefer to focus on. Yet, on the other hand, the site provides no real sense of the national and global demand for copper, nor of the power of the metal to transform nations and cultures. Thus the pit interpretation is oddly sterile, giving no real agency to *either* human or non-human actors. Given the political controversies surrounding the Berkeley Pit's creation and ongoing cleanup, this otherwise surprising minimization of human agency in favor of a crude technological determinism is perhaps understandable: It avoids assigning blame to any specific corporate, governmental, or other anthropocentric actor who might even today complain or threaten lawsuits. From the perspective of a material ecology of re-membering, however, this unexpectedly flat interpretive approach to the Berkeley Pit might have an unintended benefit. By minimizing the human story, the viewing stand placards permit the visitor to experience the pit without being compelled to view it through the lens of a tidy anthropocentric history. Left alone to simply gaze out at the vast pit and its dark poisonous lake, the visitor is instead free to sense the immense power of the material world around them – of the sheer destructive and constructive abilities of copper ore, hydrocarbon-fueled technology, and electricity itself – and by comparison the rather feeble human attempts to direct and control these.

Not at all coincidentally, such a manner of re-membering the Berkeley Pit could offer a message diametrically opposed to the one the Anaconda had proposed in the 1950s. In contrast to the modernist celebration of "America the Bountiful" and human mastery of a clearly separate material nature, the Berkeley Pit today might cultivate a post-anthropocentric sense of the deep human entanglement with a powerful and sometimes dangerous world that we must at length learn to approach with a measure of humble awe and respect. The modernist conceit of stability and understanding is powerfully challenged by the sheer material presence of the Berkeley Pit and its deep toxic lake, as the mind swoons at the incommensurate scale of ecological and geological change that it cannot fully grasp yet perhaps senses is fundamentally foreign to contemporary conceptions of the human. Such an ecology of memory and re-membering can thus be a reminder of human impermanence and limitation, a call for a return to a more humble understanding of the human place on a planet that, far from being a passively compliant human servant, is alive with both opportunity and danger, as well as countless unruly memories that intrude unbidden on the confident humanism of the modern age. The real value of places like the Berkeley Pit and the Anaconda smoke stack resides in their ability to destabilize our contemporary certainties about the nature of humans and the humanities. Their persistent material memories, seemingly unwanted and ignored, might gently but insistently push us to recall an animistic understanding of the world which was once common among our species. At these places, the thin veneer of modernity is briefly stripped away, and we glimpse a world that makes us as much as we make it and perhaps even *is* us. In such moments of vertigo and confusion, there lies the hope of new memories and new thoughts, a chance to escape our corrosive modernist illusions and re-member the dynamic world of non-human things that make us human.

Conclusion

It may strike many as nonsensical and even anti-intellectual to suggest that *less* detailed information about the past could in any sense actually be *more* useful. And obviously we can imagine many instances in which preserving as much detailed material or textual sources of information about the past is invaluable. However, this conventional anthropocentric perspective is also well understood and arguably even hegemonic, having become so deeply embedded in our thinking that it is in little danger of ever seriously being eclipsed. My intent here is not to abandon or replace this dominant approach so much as to carve out an additional parallel space in which the loss of anthropocentric historical artifacts permits other less anthropocentric material stories to be more easily and clearly recognized. This is precisely the value in the ecological re-membering of places like the Anaconda smelter and the Butte Berkeley Pit. In these landscapes, we can sense the vague outlines of the heretical: a humanism that rejects the axiomatic centrality of humans in order to better grasp the many ways in which other organisms and things have independent trajectories and histories apart from us but are simultaneously critical to *making* us in all our dimensions. In this light, the persistence of an embarrassment of artifactual riches from the past might be seen as encouraging the historian's largely unquestioned conceit that the clearest insights into the past will emerge from ever more finely grained accounts of human thoughts and decisions, as if the sheer volume of minutiae could in itself resurrect the past. Yet a re-membering of that past reminds us that some, perhaps even much, of this granular detail might in fact be mere noise, the sound and fury that continuously nurtures and flatters the vanities of a deeply self-obsessed species. In the stillness of the Berkeley Pit and the Anaconda Smokestack, where the human voices have grown distant and faded, perhaps we can hear the faint whispers of a radically different understanding: to understand ourselves, we must first understand the material world of unruly things that create us.

References

Clark, A. 2008a. *Supersizing the Mind: Embodiment, Action, and Cognitive Extension.* Oxford: Oxford University Press.

Clark, A. 2008b. "Where Brain, Body and World Collide." In *Material Agency*, edited by L. Malafouris and C. Knappett. New York: Springer.

Csikszentmihalyi, M. 1993. "Why We Need Things." In *History from Things: Essays on Material Culture*, edited by S. Lubar and W. D. Kingery, 20–29. Washington, DC: Smithsonian Institution Press.

Csikszentmihalyi, M. and Rochberg-Halton, E. 1987. *The Meaning of Things: Domestic Symbols and the Self.* Cambridge: Cambridge University Press.

LeCain, T. J. 2009. *Mass Destruction: The Men and Giant Mines That Wired America and Scarred the Planet.* New Brunswick, NJ: Rutgers University Press.

LeCain, T. J. 2017a. "Copper and Longhorns: Material and Human Power in Montana's Smelter Smoke War, 1860–1910." In *Mining North America, 1522–2012: An Environmental History*, edited by J. McNeill and G. Vrtis, 166–190. Berkeley, CA: University of California Press.

LeCain, T. J. 2017b. *The Matter of History: How Things Create the Past.* Cambridge: Cambridge University Press.

Olsen, B. 2003. "Material Culture After Text: Re-Membering Things." *Norwegian Archaeological Review* 36(2): 87–104.

Proust, M. 1913 [2002]. *À La Recherche du Temps Perdu.* Paris: Gallimard.

Randall, L. 2016. "For the Love of Stuff." *Aeon*, 3 August. Accessed 26 May 2021. https://aeon.co/essays/why-i-love-my-possessions-as-a-mirror-and-a-gallery-of-me.

Siegfried, T. 2019. "A Leaky Memory May Be a Good Thing." *The Atlantic*, 14 January. Accessed 26 May 2021. www.theatlantic.com/science/archive/2019/01/how-active-forgetting-makes-memory-more-efficient/580273/.

13

INTERSTITIAL HERITAGE

Industrienatur and ecologies of memory

Torgeir Rinke Bangstad

Our physical surroundings constitute the perhaps most intuitive and corporeal sense of the continuation of the past in the present. A bewildering array of histories and material traces of past times intersect in the places we pass through. This is particularly apparent in industrial areas that have been profoundly altered by the terraforming, extractive industries that perforated the ground and amassed large spoil tips of waste materials on the surface. In the German Ruhr region, the coal and steel industries of the nineteenth and twentieth centuries contributed to the shaping of an entirely new landscape that will outlast the long economic restructuring in the former industrial heartland of Germany. The Prosper-Haniel colliery in Bottrop was the last hard coal mine in the Ruhr region to shut down in December 2018, and although the Ruhr region is recovering from the loss of some of its largest employers, the remediation of the area will require large investments and careful monitoring for many years to come. Billions of euros have already been invested in the Emscher region in northern Ruhr to foster an ecological, urban revitalization.

Ruhr's industrial heritage is still apparent in preserved steel plants, pithead towers, gasometers and nineteenth-century working-class neighbourhoods. In contrast to the large re-cultivated spoil tips and listed heritage monuments that are repurposed and used for recreation, the interstitial brownfields adjacent to former production sites have evolved more or less on their own accord to form novel urban ecosystems. In this chapter, the interstices between a consciously designed heritage space and a spontaneous memory ecology will be explored in more detail. The German term *Industrienatur*, which forms the backbone of this case study, recognizes the specific ecological conditions in post-industrial, anthropogenic settings and identifies some key characteristics of environments that are formed by humans and yet continue to transform beyond human control. In Ruhr's *Industrienatur*, the history of industrial production is embedded in a substrate mosaic of dry and warm anthropogenic "technosols" with low levels of nutrients. Although botanists started exploring so-called industrial wastelands as early as the 1930s, the wider recognition of brownfields as ecologically and socially important, biodiverse urban habitats is more recent (Dettmar 1994; Kowarik 1993).

In terms of conservation, the ecology of brownfields seems difficult to reconcile with ruralist ideals aiming to return a landscape to a previous state, *status quo ante*, through costly restoration or remediation measures. The ecological value of *Industrienatur* is not tied to strict geographical and temporal thresholds that designate nature as a realm "out there" or "back when". As a conceptual frame and a specific ecology, Ruhr's *Industrienatur* also offers ways to rethink agency in the production of heritage beyond the predominant cultural frameworks of memory and deliberate acts of remembrance. The transient ecologies of *Industrienatur* articulate both specific production histories and ongoing transformations in the urban environment and might for these reasons work against the assumed inseparability of cultural memory and material persistence (Desilvey 2017). While it recalls movements, material flows and cultural histories of the modern industrial complex, it does so in ways which do not confine with the static memory object. Rather, in the brownfields, the ecological succession reflects the ebb and flow of industries, as well as cycles of construction, destruction, idleness in the built environment. Disturbance in this context is also a condition for life, but this does not imply that "nature" is posited as a redemptive force which heals all wounds in landscapes marked by industrial exploits (see Storm, chapter 3, for an in-depth analysis of the wound metaphor).

Both socially and ecologically, it makes sense to speak of the "half-life of de-industrialization" (Linkon 2018; cf. Berger and High 2019). The aftermath of mining closures, the decline of jobs in manufacturing and working-class people's experience of loss, precarity and resentment lingers on to make deindustrialization not an event of the past but an "active and significant part of the present" (Linkon 2018, 2). Industrial decline engenders a range of long-term political affects that spread far and wide, and in this ambiguous memory terrain, the ecological focus may be seen as politically dubious and shallow – a greenwashed industrial heritage which engenders forgetfulness (cf. Barndt 2010b; Chan 2009; Eiringhaus 2020). I believe nevertheless that *Industrienatur* serves a critical purpose in recognizing a new and ultimately more-than-human heritage where ecological processes may also be generative of forms of memory. This is also an important motivation for the present exploration of memory media which sustain and redistribute the material, ecological and cultural record of industry. One important assumption in this chapter is that mediation is not a unidirectional cultural process but also concerns the in-between, the media or "milieu" (Stengers 2005) where anthropogenic substances, building materials and ruderal plant species interrelate. The milieu, as Stengers notes, has a double meaning in French, referring both to the "between" and the "surroundings" (op.cit. 187). This is also an important property of the term "medium", which reappears in an elemental (i.e. earth, water, fire and air) notion of media (Peters 2015).

In contrast to the idea of nature as ahistorical and timeless (Barad 2003), it is the shifting memory ecology of *Industrienatur* and its interstitial brownfields that are the subjects of this chapter. Drawing on post-humanist and new materialist notions of memory, the capacity to retain, reshape and redistribute the memory of past events will here be treated as extensive to biotic, chemical and geological agents (see also LeCain, chapter 12). In an attempt to theorize how material memory applies in such contexts, the study will draw particularly on ecological perspectives from memory and media studies. Further, the chapter explores the sense in which the brownfield biotopes act as a specific interface where ecological and cultural processes meet and where the memory of the industrial past is carried forward via routes of mediation that are not uniquely driven by intentional cultural recollection and memory practices. The case of the industrial heritage site of Kokerei Hansa in Dortmund figures centrally in this attempt to theorize memory media beyond the static and archival "imprints-on-a-substrate paradigm" (Burton 2008, 322).

FIGURE 13.1 View toward the rumbustious *Industrienatur* at the northern end of Kokerei Hansa, Dortmund, Germany

Photo: T. R. Bangstad.

The case study investigates volatile relations at Kokerei Hansa as an assemblage of anthropogenic landscapes, material decay and plant species that reflect the dynamic and changing character of an industrial environment. Ruderal, "industrial" vegetation grows in a historically specific, anthropogenic soil and becomes an active participant in a distributed ecology of memory. Kokerei Hansa, a heritage site I worked with years ago (Bangstad 2014a), was later characterized by biologists as a diverse and highly dynamic brownfield habitat that changes so quickly that ecological surveys only manage to capture a narrow spatiotemporal section (Gausmann et al. 2016). The temporary nature of ruderal species in these brownfield ecologies invites a rethinking of the spatiotemporal co-ordinates of conservation work and heritage concepts. These ruderal ecologies are part of the wider dynamic of urban areas, and they are succeeded by other forms of vegetation. *Industrienatur* may, moreover, serve as a catalyst for rethinking approaches to conservation "after nature", where life seems "too strange to be human", "too feral to be pure" and "too social and multiple to be objectified" (Lorimer 2015, 179). *Industrienatur* belongs to the messy entanglements of natureculture (Haraway 2003) where ecological relations are formed equally by a human, technological environment, as well as biophysical and geophysical processes. At these profoundly mixed heritage sites, it is difficult to think of nature and culture as separate domains.

The concluding part of this chapter discusses how far an ontological or "elemental" (Peters 2015) understanding of media yields a credible post-humanist rendering of memory and heritage processes. I argue that organic and inorganic materials of different kinds are involved in the mediation of memory in ways that require us to rethink the anthropocentric framing of memory processes. Attempts at diagnosing the present sense of historical rupture and upheaval are often inclined to suppress myriad continuities and entanglements across domains of culture and nature (Latour 1993). This is the difference between historical post-humanism and post-humanist historiography. The first traces the advent of the posthuman as if it was an historical event on a linear timeline, while the latter attunes historical inquiry itself to a multitude of always already more-than-human genealogies (Krieg 2018). One ambition of post-humanist memory studies, ostensibly prompted by the loss of nature as modernity's imagined "outside" (Latour 2009), is to treat natural and human history as unfolding within the same frame (Rigney 2017). An ecological understanding of memory, which stresses that acts of recollection are embedded in and conditioned by the environment and different material prostheses, offers one such account of the entangled relations with other-than-human entities that are involved even in the shaping of cultural memory. This understanding of memory is not confined to the remembering human subject in the humanist tradition, where memory has often been equated with selfhood and identity. A post-humanist understanding of memory includes a wide range of memory media that are not exclusively made by humans for humans. *Industrienatur* renders the connections between media ecology and post-humanist memory more apparent; here, the landscape is a physical record of human activity, but it is also memory that is being transformed through ecological succession and geomorphological processes beyond our control. The deep entanglements of anthropogenic materials and the natural processes require a transversal approach that is capable of recognizing the memory function of "media in the wild" (Knittel and Driscoll 2017, 381). This is why I suggest that Kokerei Hansa offers a fertile ground for rethinking memory, particularly in light of the growing awareness of processes of inscription that are not authored or solicited by human beings alone (Craps et al. 2018, 507). Thus, in this chapter, I turn from the historical question of "what" is remembered to the performative one of "how" remembering is realized through different memory media.

(Media) ecology after nature

While ecology is most commonly associated with natural ecosystems, the concept has spread far and wide in recent decades to describe a wider range of systems where different beings are enrolled in a "sticky web of connections" (Bennett 2004, 365). Classrooms and cities are ecologies, Levi Bryant and Eileen Joy (2014) argue, in a similar way as a forest in Alaska or a coral reef. Ecology is the system of relations between all kinds of "lively" things such as microbes, rocks, power lines, tectonic plates, animals and plastic (Bryant and Joy 2014, ix). And, accordingly, an ecological approach does not have to limit its purview to the domain of nature. In fact, as argued by Timothy Morton, "Nature" has been detrimental to ecological thinking because it posits a distanced elsewhere, an "ethereal beyond" (2015, 115) which surrounds us like an abstract, container-like entity. For Morton, ecology without nature implies a form of coexistence of humans with nonhuman lifeforms and non-living entities, including toxic substances, nuclear radiation and massively distributed "hyperobjects" like global warming that we cannot detach ourselves from (2015).

In media studies, the term ecology has often been used as a shorthand for complex, dynamic systems in different attempts to theorize how media environments engender historically specific knowledge structures. Different media form infrastructures for experience and understanding by circumscribing the range of what is "seeable" and "sayable" (Hansen and

Mitchell 2010, vii). For Marshall McLuhan (1964) and many subsequent media theorists with an interest in formal properties of different media, the cultural agency of a given media technology is not primarily determined by the content it communicates, and, hence, the cultural level of sign and meaning is not dissociable from the specific material properties of media technologies. Different technological infrastructures, be they stone tools or smart phones, are not simply vessels for content but also ontological shifters (Peters 2015). The materiality of media (e.g. light-weight, mass-produced paper as opposed to parchment or the relative durability of the written word as opposed to ephemerality of oral communication) may be understood as an important vector for social and political change. In contrast to oral culture, writing enabled the political practice and metaphysics of inalterability (Peters 2015, 288).

Neil Postman, who first coined the term media ecology in 1968, employed an analogy of the petri dish to frame a key concern in media ecology as the investigation of how different media conditioned the growth of human cultures (Postman 2000). In a manner similar to Postman's petri dish analogy, contemporary media theorists have also conceptualized medium as an environment, where technics always "contaminates the logic of the living" (Hansen 2006, 299). Technics, in this sense, does not discriminate between Palaeolithic flint tools and modern smart phones, and it is particularly through Bernhard Stiegler's (1998) theory of the co-originarity of technics and human that much recent media theory understands the relation between organisms and the environment as profoundly mediatic, a "concrete actualization of the living via exteriorization in an environment, in a medium" (Hansen 2006, 301). Marshall McLuhan understood media as prosthetics for *human* beings, but with the rise of post-humanist research in media studies, the perspective is broadened beyond environments designed by and for humans.

In his investigation of the ancient "elemental" understanding of media, John Durham Peters claims that the technological and the ecological are not separate and that processes of mediation always involve a tight-knit ensemble of nature and culture, human and nonhuman. The entire history of shaping and being shaped by environments may be seen in terms of processes of mediation, and Peters argues that while elements are not automatically media in themselves, they are so "for certain species in certain ways with certain techniques" (2015, 49). Different forms of media structure the actions and movements of different beings in an environment by acting as interfaces that extend certain capacities and restrict others. Peters describes media expansively as "devices and crafts by which humans and things, animals and data, hold together in time and space" (2015, 18).

It is often in the sense of holding together across time that we understand the memory media of heritage. Buildings, sites, objects and landscapes but also traditions are valued because of their ways of holding together, their capacity to persist over time, outlive historical ruptures and yield meaning over generations. The media of heritage are often associated with materials and practices that convey a specific and localized cultural meaning or natural condition that may, could, or should, be preserved and passed on to future generations. Different material media, such as Plato's well-known example of the wax tablet, are capable of retaining past impressions for a long time and serve the purpose of recovering information at a later point. In cultural memory studies, it is also recognized that the specific properties of different memory media shape practices of collective remembrance differently (Erll 2010). All kinds of materials may be constitutive in enabling mnemonic communities. The German media theorist Astrid Erll (2004) includes a wide range of materials such as bodies, landscapes, stones and rivers in her list of different memory media, but their recognition as mediators is understood strictly by virtue of their cultural work and by being socially recognized as such by the community in question (2004, 18). This leaves a lot to be desired in terms of fully understanding the autopoietic chemical, mineral and bacterial memories that inhabit and shape a wide range of material bodies including the human body.

The value and challenge of recent work in the tradition of media ecology in relation to heritage and cultural memory, is that it conceives of media in the broadest possible sense as relays that engender connections, contract forces and channels energies. Media theorist Jussi Parikka has developed more-than-human notions of media as environments, including, for instance, those formed by animals and insects through interactional sensing, movement and memory of their surroundings (2010). In relation to our current concern with memory media, Parikka reminds us that fundamental processes of transmissions, recording and connecting can be found in several places other than the specifically *human* media environments (2010, xiv). "Environment" in this tradition of media ecology is no longer the stable and timeless backdrop that we know from reified nature as the eternal and ethereal outside (Morton 2007, 2013, 2015); it can also include the deep histories of media materials, such as minerals that have been sedimented for millions of years only to become part of today's information technologies or e-waste which releases toxins when electronic components are mined for valuable metals or rare earth minerals in scrapyards (Parikka 2012).

Nanomedicine, biotechnology and a wide range of biomedia have also dissolved the distinct boundaries of informational and biological realms, and yet memory in the humanities is often restricted to questions of human identity, a core constituent of the remembering, human subject. In the following case study, I will explore what happens when the post-industrial ecology is recognized for its mnemonic capacity, in which the capacity to remember past events is distributed across humans and nonhumans who pass through or inhabit a place for some time.

The starting point of the following exploration of memory in and of a specific physical, anthropogenic environment is, first, the claim that even *cultural* memory has hardly "ever been comfortably human" (Vermeulen 2017, 384) and, second, that a study of cultural memory and cultural heritage may benefit from a consideration of the work of more-than-human agents in enabling forms of memory that do not stop with stable, material imprints. Natural processes resemble the mnemonic gesture of retaining but also of re-channelling and transforming the memory of past events. This corresponds with the transversal understanding of media ecologies and their constituent elements as effective across the artificial/natural-divide. This is why it makes sense, even within cultural memory studies, to explore the ecology of memory beyond a confined human realm.

Brownfields and their unruly heritage

The Ruhr coal field supplied the region with a valuable resource which provided jobs for up to half a million people and fuel for the German economic miracle in the post-World War II period. The coal-bearing layer formed in the Carboniferous extending back more than 350 million years, was the motor of a massive anthropogenic transformation in the area. The flat plains in the Westphalian Lowland were turned into a comprehensive industrial landscape from the mid-nineteenth century onward. The Ruhr area in the federal state of North Rhine Westphalia is named after the river Ruhr, which is a tributary of the Rhine. The two other major rivers in the Ruhr region are the Lippe and the Emscher. The river Emscher runs about 80 kilometres east-west through the coalfields from Dortmund to Duisburg, where it flows into the Rhine. Emscher was for a long time known as the biggest open sewer in Europe, collecting discharge from steel mills and factories as well as wastewater from two million households, turning it into a thick, reeking slurry. In a mining landscape perforated with tunnels from underground operations, the construction of a conventional underground pipe system was impossible, as the subsidence would have caused pipes to crack and burst. Still, today, depressions and hollows accumulate rainwater that threatens to flood the land. The enduring material legacy of the now-bust coal mining operations requires pumping stations to run round the clock, all year through, for eternity, to prevent surface flooding. These tasks are referred to as "Ewigkeitsaufgaben" and rely on sophisticated monitoring systems to manage and control the distinctive afterlives of the coal mining industry – its "unruly heritage" (Olsen and Pétursdóttir 2016). Environmental engineering and sunrise technologies offer jobs for new generations of workers who are "making a living in the present by cleaning up the mess of the past" (Evans 2004, 84; cf. Hospers 2004).

As the space-intensive industries in the Ruhr region contracted, large tracts of derelict land became considered a political problem that could be remedied through restoration or redevelopment. Brownfields are commonly known as "urban wastelands", "derelict" or "abandoned areas", and are often found in the fringe areas that have fallen by the wayside of the economically functional, urban fabric. In recent years, brownfields have increasingly been recognized as important habitats for rare plant or insect species (Robinson and Lundholm 2012). In 2005, the site of an old, abandoned oil refinery at Canvey Wick in Essex became the first brownfield in the United Kingdom to be legally protected as a nature reserve providing habitat for endangered invertebrate species (Lorimer 2015, 165). As brownfields are shaped both by the spatial dynamism of the wider urban environment and the spontaneous vegetation that colonize soils that resemble natural habitats, they are also areas that require a rethinking of some core assumptions in conservation of natural and cultural heritage. As the name suggests, *Industrienatur* is a fundamentally hybrid entity, where natural processes are entwined in disturbed anthropogenic soils and where the mobility and flux makes it difficult to envision these ecosystems as discrete, bounded sites. The wild, urban ecology of roadside verges, rooftops, landfills, deserted construction sites and railroads is an affront to the territorial imagination that has supported the idea of nature reserves as "bounded containers for Nature" (Lorimer 2015, 164). The desire to design public, green areas or re-cultivate lands from brownfields may in fact undermine the unmanaged and spontaneous transformation that allows the industrial flora to thrive (Dettmar 1999).

Ruderal plants are opportunistic species that may settle in open and barren brownfields, disturbed, often dry, nutrition-poor habitats. Ruderal, from the Latin "ruderalis", meaning rubble, is a term used by ecologists to describe the vegetation on disturbed lands (Desilvey 2017, 85), in leftover or in-between spaces (Storm 2014, 108; cf. Storm, chapter 3), and the term already figured in nineteenth-century botanical surveys of roadsides, railway tracks and waste deposit areas (Lachmund 2003). Brownfields are often composed of a dense mosaic of different habitats that reflect the industrial activities that took place at a site in a fairly recent past. Plants that grow there may indicate, for instance, whether a site was used for coal mining or steel production: "The vegetal realm also has real 'mining or steel industry indexes'" (Dettmar 1999, 68). As indexes of industrial activity, enabled by unmanaged but materially persistent pasts, ruderal plants are interstitial also in the temporal sense. They arrive after the disturbance has taken place and are often succeeded after some years by more persistent beings. Ruderal systems unfold in a temporary, in-between-stage that is impossible to maintain through static preservation of given areas.

The biological diversity engendered through spontaneous vegetation in brownfield settings is maintained by spatial dynamism such as periodic rebuilding, demolition and redevelopment, and in more stable surroundings, the biological diversity tends to decrease with time (Kattwinkel, Biedermann and Kleyer 2011, 2336). This has given rise to the notion of "temporary preservation" (ibid), which reflects the symbiotic relation between ruderal plants and urban transformation. In contrast to isolated islands, which allowed nineteenth-century ecologists to study the interaction between animals, insects and plants in their natural, non-human habitats, the urban ecology also implicates us and upsets notions of nature as elsewhere. As Jamie Lorimer has claimed, urban ecologies and ruderal sites also afford new forms of environmental responsibility by enabling experimental approaches that are less obsessed with the territorial borders of a national flora and the threats posed by nonnative, invasive species (Lorimer 2015, 165, 169).

Finally, brownfield ecologies are valuable in terms of broadening the notion of memory in relation to industrial heritage. Industrial decline does not only register in abandoned sites, in idle blast furnaces, boarded-up buildings or unemployment figures; it also yields distinctive ecological processes that, in their own fashion, carry the memories of human activity and separate stages of mobility, economic downturn, abandonment and reuse. The lack of demand for commercial real estate or a halt in production may allow for sites to remain unused for a period of time and for the bare ground to be colonized by ruderal plants. At first, the lack of, or thin layer of, fertile soil discourages the growth of grass, hawthorn, bindweed and ivy, but after a while, lichens and mosses enter, and the soil thickens enough and finally to accommodate larger trees that stay put (Hunter 2014). In recognizing the transitory character of the ruderal ecology, and not least the temporary status of the encounter with such places that we pass through, plants emerge as media that shape both a physical and mnemonic environment but without offering us the reassurance of the stable imprint.

Age-value and the transitory monument

The recognition of these peculiar hybrid landscapes as indexes of both the industrial legacy and the ongoing post-industrial regeneration was one of the results of the large-scale regional urban regeneration initiative called IBA Emscher Landscape Park (Internationale Bauausstellung Emscher Park). The IBA program came into effect in 1989 to aid the transition of the Emscher area from an industrial hub to a modern service region. This ten-year-long program, funded by both federal and state money, acted as an advisory body for the government of North Rhine Westphalia and targeted around 120 projects in landscape planning, domestic housing, heritage conservation and ecological regeneration of the Emscher river. The idea of treating the Emscher region as a massive landscape park, or a cultural landscape, was introduced as a deliberate provocation to the conservative understanding of landscape prevailing in heritage practice. Both nature conservation and cultural heritage practice in Ruhr had traditionally been at the mercy of this conservative idea of landscape, which implied that funding was often channelled into the protection of pre-industrial town centres and natural biotopes unaffected by industrial development (Ganser 1999, 14). In the cultural heritage context of Ruhr in the 1980s, this would imply that less than 5% of the region would warrant the label "cultural" (ibid).

Regional heritage initiatives had, despite this long-standing conservative bias, been able to call attention to the threat posed to industrial monuments by demolition and redevelopment (Raines 2011). Thus, in the 1970s, a few of the characteristic structures of the industrial architecture in the Ruhr, such as the coke plant Zeche Zollern in Dortmund and the workers' settlement of Eisenheim in Oberhausen, were given a lease on life through state heritage legislation. This legislation would from 1980, moreover, give a special mention of the industrial heritage. In 2001, the giant coal-mining and coke plant installations at Zollverein in Essen were listed as a UNESCO world heritage site. Kokerei Hansa in Dortmund was listed as industrial heritage by the city of Dortmund in 1998 and reopened to the public as an industrial heritage site in 1999. One of the main principles of IBA Emscher Park with regard to industrial heritage was to "give things time" (Raines 2011). This also accorded with the idea that the functionally redundant industrial structures could be released from the economic pressure of the industrial planning regime and instead be allowed to ripen over time (Bangstad 2014b; Ganser 1999, 15). The planning philosophy underpinning IBA thus signalled a clear break with the expansive logic and "clean slate" planning regime which prevailed in the industrial era.

What is particularly relevant in the context of the present study of memory is the sense in which the IBA Emscher Park recognized ephemerality as part of heritage value and memory practice: "to remember can also mean to admit transitoriness, not to create the impression that everything can be held onto" (IBA Emscher Park Positionspaper in: Raines 2011, 195). A similar point was made by the Austrian conservationist Wilfried Lipp, who in the early 1990s argued that a postmodern assessment of Alois Riegl's "age-value" was not only conditioned by the transitory character of a monument returning to nature (through decay, weathering and disintegration) but also by the historical contingencies of heritage value assessment (1993). Traditional monument protection, Lipp argued, relied on an almost autopoetically isolated realm of monument care which allowed for the heritage to stand out and stand clearly apart from the triviality of the everyday realm. A more eclectic attitude to cultural heritage evolved in the third quarter of the twentieth century, which also placed new accent on processual value, milieu and letting-be ("Seinlassen"), which Lipp recognized as central to the interweaving of the monument and the everyday: "Life world and monument have become intermingled. … The value emphases are deferred unto the profane, the frame of reference for things has changed" (1993, 9).

The sense that the cultural frame of reference for the heritage was changing in the Ruhr specifically was also a consequence of the new universities in the region that aspired to foster a new social mobility in a traditional working-class region hit hard by the coal crisis in 1957. Four new universities were established between 1962 (Ruhr-Universität Bochum) and 1972 (Essen), including Technische Universität Dortmund (1968) and Gesamthochschule Duisburg (1972), and these educational institutions emphasized equal opportunities and higher education for the masses in order to facilitate a transition from manufacturing to research. The new middle classes found self-assurance in the history of structural change which also resonated strongly with their personal experience and memory of social and cultural transformation (Goch 2002). This explains the dual identity assigned to industrial heritage in the region; it was not only a memory of the industrial past but equally a monument to the post-industrial cultural, material and social metamorphoses that were articulated in the ostensibly anti-monumental monument care. The industrial heritage of Ruhr was also a heritage of change, and this might explain why heritage sites came to act as material referents of transition rather than a commitment to hold firmly onto the past.

Industrial heritage became inextricably connected to gentrification and culture-led redevelopment (Berger and High 2019), and the idea nurtured by IBA Emscher Park was one of ecological regeneration from "the black past to a green future" (Eiringhaus 2020, 254). To its critics, the term *Industrienatur* is a neologism that primarily serves to neutralize the dichotomous relation between industry and nature and gloss over the ecological crisis in the former industrial heartland (Eiringhaus 2020, 255, 262). In this guise, the story of *Industrienatur* is seen as an empty signifier that dampens critical historical reflection rather than expressing a deteritorrialized ecology that forces us to rethink the boundaries between the natural and the cultural and between ecology as a natural science and ecology as political practice of interrelations. The heuristic value of *Industrienatur*, lies in the recognition of entangled and enduring legacies of extensive, terraforming activities that are perpetuated as new forms of vegetation in the technosol. In the *Industrienatur*, the idea of habitat as necessarily natural is challenged by a range of nonhumans who inhabit the artefactual, man-made structures and disturbed soils. If we consider life from the nonhuman point of view, we have to acknowledge that industrial sites – like the subsurface structures of decommissioned oil rigs that have become home to a variety of mussels, crabs and fish – may well be inhabitable (Jørgensen 2016, see also Breithoff, chapter 6). As a consequence, "habitat" could be dissociated from the prefix "natural" to include forms of habitation – literally "taking place" – in the artificial realm and to appreciate how unconcerned non-human animals are with this distinction between the artificial and the natural (Jørgensen 2016, 138–139).

A new naturalness?

The intermingling of the monument and the profane and the processual character of the industrial legacy is expressed at the margins of sites such as Kokerei Hansa. Here "the events of history shrivel up and become absorbed in the setting" (Benjamin 1998, 179) as a spontaneous material memory that does not stand out from its surroundings like the traditional monument. The particular texture of memory in the landscape of Kokerei Hansa does not bear the exclusive imprint of the industrial era. It is also a landscape shaped by post-industrial, structural change as well as ongoing geomorphological processes. The industrial landscape was deprived of its structural cohesion when steel works, coke plants and mines were shut down at an increasingly frequent rate from the 1970s. This rupture can be understood culturally as a dissolution of the socio-economic structure which provided the industrial landscape with a fairly stable semiotic rendering as a "landscape of work", a "place where money was earned" (Ganser 1999, 14). The painful contractions of the coal and steel industries exposed the growing problem of surplus land with greater force in the 1980s and 1990s. Already in the early 1990s, there was more than 8,000 hectares of abandoned industrial land in the Ruhr region (Kowarik 2005).

De-industrialization also enabled the formation of new habitats for ruderal, pioneer species defined as the first plants to colonize barren, nutrition-poor areas after the economic meltdown. Some have chosen to regard this optimistically as life after coal, a sign that nature is reconquering the bare and scarred landscapes of unbridled capitalist exploits, or that human-made structures are returned to nature. In his analysis of wild, urban woodlands, the German botanist Ingo Kowarik (2005) has claimed that there are two different ways to understand the idea of "naturalness". The first is based on retrospective comparisons with the historic condition of vegetation in an area, which, as time progresses, produces different degrees of deviations from a pristine state prior to human influence. The second approach, by contrast, is not based on any idyllic idea of the "original" state and offers a convincing alternative to this linear, historical rendering of naturalness. The key to the second approach, which Kowarik describes as contemporaneous, is that it begins from the current state and quality of the site regardless of whether this has been naturally or anthropogenically shaped: "From this perspective, therefore, the natural capacity for process is the central point, not a particular, retrospectively determined and often idealized, picture of nature" (2005, 19).

In this understanding, the natural is not determined historically but ecologically, based on the characteristics of the processes which involve anthropogenic substrates, untamed growth and a system's degree of self-regulation. Urban industrial woodlands, rooting in building rubble and waste materials from industry and construction, may, based on their properties rather than their origin, align with the idea of the "wild". This is a wilderness, however, that unfolds not in realms unaffected by humans but may for a period of time remain unmanaged by humans, and this is a central distinction (see also Bryant, chapter 4). The important point is that it is hard to determine where the anthropogenic ends and where the natural begins:

> Both ... clearly belong together and create, from the characteristic double nature of urban-industrial woodlands, a product that is equally natural and artificial. This bipolarity does not, however, correspond to the general image of nature. Rather it brings about confusion and insecurity in the general classification as well as in the scientific classification of these spaces.
>
> *(Kowarik 2005, 3)*

This confusion is worth noting and keeping in mind as we now turn to a closer study of Kokerei Hansa as an instance of this double nature. Instead of designating qualities along a historical continuum from the pristine to the disturbed, from the natural to the artificial, the relation between contemporaneous elements which interact and affect one another over time is the central issue (cf. Lucas 2015). The value of a material's processual aspects may be difficult to align with preservation and the idea of a memory contained in bounded objects and well-tended landscapes which reflect a specific historical, ecological condition. The processual, materialist perspective is more partial in the sense that material memory cannot recreate a "what was" but only acknowledge the continual formation of a "what becomes of what was" (Witmore 2014, 213), and it is in this formation that a variety of entities are brought into play.

The "black" and the "white" side – environmental legacies and critical consciousness

The production of coke is known for being particularly messy. Organic substances, impurities in the brown coal, are vaporized in the "coking" process to increase the heating value of coke. The pollution associated with this process made it difficult for Kokerei Hansa's heritage to be appraised exclusively as an *architectural* monument honouring the modernist design of the German architect Hellmuth von Stegemann und Stein (1892–1929) (Kastorff-Viehmann 1992). Until the coke plant closed in 1992, a large cloud would rise above the suburb of Huckarde at regular intervals, as carloads of burnt coke were drenched with 25,000 litres of water pouring down from a tall wooden quenching tower. It would take about 90 seconds for the coke to cool down to 60 degrees Celsius, and this procedure was repeated 135 times a day. The "coking" of coal at high temperatures was conducted on the aptly named "black side", while the vaporized organic elements were processed on the "white side" into by-products such as tar, benzyl, ammonium sulphate and hydrogen sulphide. The white side was where the aromatic compounds from coke gas were absorbed and distilled. During the Second World War, Soviet prisoners of war were forced to work at Kokerei Hansa, and the site was targeted by several Allied air raids towards the end of the war.

Resorting to the interpretative frame of natural regeneration and aesthetic fragmentation motifs in such contexts can easily be deemed politically dubious, as they may serve to reinforce the idea of nature as redemptory, if not forgetful. It is often assumed that the emotional appeal of industrial ruin parks might work to depoliticize memory (cf. Chan 2009). By being staged in the frame of natural and geological temporalities rather than a contemporary political and social history, industrial ruins in Germany, according to Kerstin Barndt, appear to fall back to a past untouched by the Holocaust. The natural transformation unfolding in the industrial ruins renders them "instantly ancient" (Barndt 2010a, 138). This point bears direct relevance to Kokerei Hansa. However, the point I want to explore is not how far regrowth detaches the monument from its social and historical fabric but rather how ecological and material processes enable a reframing of memory beyond the dualism of a remembering subject and a passive, material substrate. Industrial and natural processes are, as Anna Storm points

FIGURE 13.2 The "black" side of Kokerei Hansa with conveyor belt on the right and coke ovens on the left

Photo: T. R. Bangstad.

out, co-constitutors of new landscapes which make us acutely aware of the mutual imbrications of nature and culture (Storm 2014, 101).

There is a critical insight in admitting this memory ecology where the two are intimately connected and enfolded. In front of the main gates of Kokerei Hansa, visitors are introduced to the site's industrial nature. Several interpretation signs in the form of mockup steel slabs are bolted to the ground with something resembling rail lines. They provide the backstory of the site, its history, the process of coke production and a map of heritage trails, *Route Industriekultur*, which connect the regional heritage sites in a network of thematic routes. One of the steel slabs deals specifically with the theme of *Industrienatur* and landscape (Figure 13.3). The logo of Emscher Landschaftspark sits in the top right corner, and the text in German and English is underlined by the monochrome skyline featuring the characteristic silhouettes of the cooling towers, the conveyor belt and the coal bunkers on the "black side". The illustration on the sign shows large leaves protruding from the wooden quenching tower under a text which reads:

> The setting of coke ovens, towers and pipelines in the Hansa coking plant industrial monument looks rather bizarre. Indeed it has a vague resemblance to the lost worlds of the Maya and the Inca in Central and South America. A landscape metamorphosis is taking place here in the heart of Dortmund, where nature is "reconquering" the site. Flora and fauna have adapted to the changed conditions around the technical relics of industrial history, and now offer new habitats and protected areas for rare species.

In this particular story, the industrial structures are framed as ancient relics despite their relatively young age and fairly well-preserved state. The text also hints at the careful choreography required to strike a balance between two heritage domains traditionally considered separate. The words *reconquering* (rückeroberung) and, further down, *repossession* (besitzergreifen) reflects how this traditional ontological distinction is still deeply embedded in the vocabulary of heritage. Within historic preservation, the world of plants is often seen as a danger to material heritage, causing the biodeterioration of buildings through unwanted colonization (Farstadvoll 2018), and yet what also opens up here, according to the sign, is a "unique and fascinating combination" (ein einzigartiges und fazinierendes Miteinander).

Once inside, on the wall of the former locker room building, some plants from Kokerei Hansa's *Industrienatur* are suspended in the form of photos covered by a thick layer of plexiglas. The row of photos shows soapwort (*Saponaria officinalis*), purple loosestrife (*Lythrum salicaria*), Japanese knotweed (*Fallopia japonica*), wild carrot (*Daucus carota*), wall-rue (*Asplenium ruta-muraria*) and Roberts geranium (*Geranium robertianum*), which have all been found on the site. Only a small patch in front of the coke ovens has been actively planted as part of its transformation from an abandoned production plant to a heritage site. The cultivated areas amount to a mere 4% of the site's total area of 36 acres, and, apart from these, vegetation at Kokerei Hansa is left more or less to its own devices (Gausmann et al. 2016). The site's northern section is dominated by birch trees which have evolved over the course of the decades passing since its closure. This is today the most conspicuously overgrown section of the site. As part of the visitor facilities designed for the IBA Emscher Park, elevated walkways now twist their way through rusty gas pipes and lush birch trees.

FIGURE 13.3 The interpretation sign outside the main entrance of Kokerei Hansa describes the site's metamorphosis into "*Industrienatur*"

Photo: T. R. Bangstad.

FIGURE 13.4 Vegetation at Kokerei Hansa presented in the entrance of the former locker rooms

Photo: T. R. Bangstad.

The wall-rue

As already mentioned, the site's specific ecological properties have allowed for the development of a diverse local vegetation. The industrial flora that have settled along the rail tracks, in the dark coal-rich gravel and on the brick stone walls of industrial buildings are beings that reflect back on, and also perform, their own past. The bedrock they sit on is formed by a century of industry, and the conditions of their present interactions are determined by the characteristic traits of an abandonment ecology. The dark coal dust covering the ground and tinting the human-built edifices also affects the microclimate. The rays of morning sun warm the ground faster, and it also stores the heat more efficiently. In the first phase of the ecological succession on brownfields, the fallow lands are often open areas with no trees to provide shade. Coarse human-made substrates, rich in rubble, rail track gravel and brick stone fragments, also drain water quickly and thus further limit the range of plants and trees that are able to grow there. These conditions will favour trees that do well on dry, acidic soils, such as the silver birch (*Betula Pendula*), whose light seeds easily disperse with the wind and are among the first trees to sprout in disturbed contexts.

The routes of travel may also be anthropogenic, and, with the expansion of railways in the late-nineteenth century, the so-called railroad wanderers ("Eisenbahnwanderer"), plant species spreading along the railroad, became more prevalent, and new plant species may have found their way to the Ruhr with the import of goods such as textiles or raw material such as iron ore. New plant species were also carried as biotic stowaways when hay was used as packing material (Keil and Loos 2002). As more and more goods today are moved by heavy freight vehicles on highways, the road becomes a more likely pathway of dispersal for synanthropic plant species and an important corridor for migration of introduced species ("xenophytes"). In this sense, vegetation in (sub)urban, industrial contexts reflects both the culture-historically specific, often anthropogenic, "transportation medium" (Keil and Loos 2002) as well as the ecological conditions of different brownfield contexts.

The substrates at Kokerei have been described as a dense mosaic of man-made rubble, and a recent survey identified as many as 27 different substrates on the property (Gausmann et al. 2016). These consist of rocks, slag from iron-smelting processes, deposits from coal mining, building masses, basalt-rich gravel and debris from brick walls. Each one of these materials has properties that enable memory to unfold in a vegetal form. One particularly interesting plant in this regard is the wall-rue (Figure 13.5) that is found in both rural and urban contexts, on old walls as well as on limestone rocks. The wall-rue, which appears both in the photos in the locker room building (Figure 13.4) and in the table of the 482 vascular plant taxa found

FIGURE 13.5 Interstitial species in anthropogenic settings. Wall-rue on an old stone wall

Photo: T. R. Bangstad.

at Kokerei Hansa, has also been found in a shallow shaft next to the gas compressor station (Gausmann et al. 2011). Wall-rue is a calciphile, common fern that normally grows in rocky habitats like vertical walls of old mortared stone buildings or base-rich crevices in natural cliffs or outcrops. Categorized as mural vegetation, wall-rue commonly occurs in heritage contexts, both on fairly recent brick wall buildings and on old chalky, stone walls. An early record of *Asplenium ruta-muraria* identified it on the old city walls of Duisburg in 1800 (Gausmann et al. 2011), and it is also seen in the background of the painting *The Nativity with Donors and Saints Jerome and Leonard* (1510–1515) painted by the Dutch Painter Gerard David (Leech 2015). As a "cryptogam" reproducing through spores and clinging to the fabric of historic buildings, the wall-rue is the perfect stowaway carried by the wind or transfixed by the medium of oil paint in Renaissance ruin paintings.

Its occurrence on the shaft wall attests to a symbiotic relationship and an overlap between natural and cultural heritage, though possibly more of a parasitic character than the harmonious "miteinander" described on the signs at the main entrance. Along with micro-climatic conditions and sun exposure, the prevalence of mural fauna is also determined by the age, composition and condition of the mortar. As a rule, new cement mortar has a high pH-level which deters the growth of ferns (Gausmann et al. 2011), especially compared with older lime mortar, but the wall-rue is tolerant of base rich substrates and may grow on fairly modern brick walls. As the mortar ages, living conditions change and become more welcoming to ferns but also to larger plants and trees that take advantage of the new habitable spaces opened up by the decomposing binding material.

The gradually dissolving mortar in shafts and facades of brick stone buildings constitutes one type of the complex industrial biotopes found in contexts like Kokerei Hansa. As the

FIGURE 13.6 Habitable spaces open up in a decomposing brick wall building (former Sieberei-building)

Photo: T. R. Bangstad.

crevices open and seeds of larger plants adhere to the vertical walls and take root, decomposition also accelerates, and the choreography of industrial nature and monument care becomes more challenging. The tension-filled, spatiotemporal entanglements expressed in the assemblage of wall-rue, decomposing matter and a preservation ethos of granting things time relates to the sense that material memory unfolds spontaneously and in ways which exceed pre-ordained categories of the natural and cultural. This is perhaps best described as an organic memory materializing the concurrent processes of dissolution and reassembling. The coming together of the ageing mortar, the crack in the wall and the spores of the wall-rue articulate a memory in the intersection of natural processes and industrial history.

Memory media

The anthropogenic substrates contain traces of material movements which have left their mark on the land and become rechanneled in a biotic form – unauthored by humans but nevertheless carrying their faded historical signature. The common labelling of these nondescript places as "urban wastelands" or "non-places" suggests a kind of vacuity, void or nothingness that pays little credit to the actual being of the interstitial, that which literally lies "between" and where different beings extend and connect. In the wider sense, media – the in-between – is understood as an "environment for life", an originary "correlate" of the living (Hansen 2006, 302). This is evident in how tools, technical prostheses and artefactual memory aids, condition our being with and experience of the environment. The "human being invents himself in the technical by inventing the tool" (Stiegler 1998, 141). Consequently, life is mediation (Hansen 2006). This notion of media applies with equal accuracy to a Palaeolithic stone tool which presents a trace of a past need and a past operation, thus acting as an "archive" of its "projected and recollected use" (Bennett 2009, 31).

In this sense, materials escape the narrow understanding of media as vessels for transmission and become intimately tied to embodied memory, to what we remember and not least how we remember. An important aspect following from this extensive notion of media concerns how media as prostheses actively shapes the relation between an organism and its environment, and how, as an intermediary, it is never passive and idle. As we have seen with the wall-rue clinging onto the horizontal interstices of brick walls where mortar decomposes and moist and dirt gather, media allows for the interstitial to take centre stage in an "elemental" (Peters 2015) or organic understanding of medium as a "substance that something grows in, lives in, or moves through" (Cambridge Dictionary 2020). The "nothing" between absence and presence is replaced by a mediatic relation (Kittler 2009). The between, the space of mediation, is not void, it is physical, it has an existence and an extension (Phillips 2017).

But how does this relate to the conventional conception of memory media as an inscription in material form that stores an impression of past life and which continues to hold meaning and relevance for a contemporary public? This notion of memory is umbilically tied to the human realm, where the significance of materials largely derives from their formal, symbolic function in a mnemonic community. Memory, in this form, is also regarded as fundamental to the human being and a building block of individual and collective identity. In memory studies, materials often feature as carriers of memory which enable a relation between past and present, "between a remembering mind and a reminding object" (Assmann 2008, 111). In this approach, the function of materials is derivative; they carry the meaning people have invested in them, and it is primarily in this capacity that things bear relevance to memory:

> Things do not "have" a memory of their own, but they may remind us, may trigger our memory, because they carry memories which we have invested into them, things such as dishes, feasts, rites, images, stories and other texts, landscapes, and other "lieux de memoire".
>
> *(Assmann 2008, 111)*

The understanding of memory media I have relied on in this case study, however, has emphasized how materiality is a central element in memory processes and, thus, how these simultaneously involve the human and yet extend beyond it. The important property of mediation is its ongoing composition and how this allows for a continuity between entities rather than a radical disjuncture between mind and matter. Memory media, in this sense, does not entail only the things purposefully engineered as reminding objects but also raw material retentions which afford new relations and which extend into the future without facilitation of a recollecting human subject. This notion of memory media can extend to landscapes, plants and substrates, not as the passive grounds for culturally encoded and recovered memory but as the spatial extensions that connect human and non-human processes in forms of inscriptions, where the geological, the chemical, the social and the biological intertwine. This idea of mediation frames the interstitial not as emptiness and void space but as the arena where things, plants and toxic compounds forge relations and act as each other's extensions or prostheses.

FIGURE 13.7 Mnemonic environment at Kokerei Hansa industrial heritage site

Photo: T. R. Bangstad.

In being filtered through taxonomic ordering, specific nomenclatures and administrative apparatuses of preservation, spontaneous memories become remoulded into discernible, unambiguously demarcated knowledge objects. This applies both to ecosystems and forms of material culture which, in order to acquire a discernible and manageable identity as heritage, rely on the support of maps, images, documents and not least systems of comparison (which again require a common system for value assessment). Heritage systems, both natural and cultural, rely heavily on processes of translation and mediation of memory. From cultural landscapes, engravings, paintings and parchment transferred by textual interpretation, to high-resolution images saved on magnetic storage devices and made available as digital formats online, these procedures amount to a mediation of memory (cf. Olsen et al. 2012). The fundamental premise underlying such practices of translation is that the new units of knowledge are able to convey the richness, abundance and diversity of the world "out there" in a form "in here". This is relevant in this context, because biologists' identification and designation of *Industrienatur*, and "wild urban woodlands" as recent and valuable biotopes concerns a friction or distance between an institutional, archival memory and spontaneous memories deposited elsewhere in the landscape as "accidental archives" (Rigney 2017, 2015) or a memory "in the wild" (Knittel and Driscoll 2017, 381). The challenge posed by brownfield ecologies is not only that they are evanescent "by nature" or escape our desire for natural roots and origins but also that they rely on a spatial dynamism of urban built environment and of plant species that move back and forth, follow in our footsteps, their seeds sticking to the soles of our shoes, without our conscious involvement. *Industrienatur* comprises spontaneous processes where the material legacies of the human terraforming species become tangled up with ruderal species that also form soils, albeit on a different scale and at a different pace. In the midst of this highly anthropogenic, disturbed landscape where natural brown coal was heated and its organic residues removed, the belief in the distinction between nature and industry, between production landscape and external nature, in many ways culminated in the techno-scientific industrialism that engendered places like Kokerei Hansa. The concept of *Industrienatur* sheds light on the entangled legacies that comprise both persistent human-made substances, as well as short-lived, opportunistic and highly mobile plant species. The relations that are forged confuse the distinct spatial regimes which ordered the world of industry. To acknowledge that natural processes unfold in anthropogenic contexts is not the same as positing nature as the reconciliatory domain that will absorb and annul human wrongdoing. As an autopoietic memory landscape, *Industrienatur* reveals the non-human agencies at work when the firm hold on space loses, and sites are given room and time to become intermediate, connective and recollective. With regard to preservation, these sites enact a fluid memory ecology that requires a "postpreservation" ethos in heritage matters to recognize "its objects of concern as temporary arrangements of matter that shuttle between durability and vulnerability in response to social and physical forces often outside our control" (DeSilvey 2017, 8).

In the process of mediation, industrial nature as a spontaneous and "lively" thing acquires a scientific and administrative presence and consequence. However, by virtue of their unruliness, heritage objects which are distributed, complex and consist of many separate components may act as vectors of transformation in conservation practice (Domínguez Rubio 2014, 2016). It is in this sense important to recall the assumption that motivated and opened this chapter, namely that non-human entities and materials exercise a distinct gravity and agency, which feeds back into practice. As an interface between the accidental archive and the organized record, the material lifeworld of Kokerei Hansa and similar brownfields has spawned new categories and new methods within heritage management. There is now a botanical term, "industriophytes", which refers to the plant species colonizing the Ruhr after the onset of extensive industrial developments around 1840.

The spontaneous memory processes articulated in the assemblage of the mortar, the brick stone and the wall-rue, also, importantly, precede the arrival of such conceptual framing. Processes like these might inform practices of "Seinlassen" – of stepping back and letting things unfold spontaneously. This is why I think it is inadequate to approach *Industrienatur* only in terms of its aesthetic framing, political greenwashing or cultural symbolization. It is an emergent memory ecology that recollects and remediates irrespective of our taking notice of its action. Heritage studies are now beginning to realize alternatives to the anthropocentric vocabulary and framing of memory's ecologies (Desilvey 2017; Farstadvoll 2018, 2019; Fredengren 2015; Harrison 2015; Olsen and Pétursdóttir 2016; Pétursdóttir 2020). This involves recognition of the constitutive role of different forms of materials in engendering the retention of memory and how this is not always sanctioned by cultural memory or, indeed, by notions of what counts as heritage. The understanding of media as "that which connects", but also as a milieu "in which something lives", should comprise both what is consciously inscribed through mnemotechnics and, moreover, naturally occurring processes and dispersed, material retentions that are rarely recognized as media. The inspiration for this rethinking was the idea that memory cannot be treated apart from the materiality of its media and that the mediation of memory rely on the non-human domain (see also LeCain, chapter 12).

At Kokerei Hansa, the volatile relation between the anthropogenic soil, the crumbling mortar and the colonizing wall-rue which spreads its spores with the wind illustrates the scope of this more-than-human memory ensemble. De-industrialization creates a surplus of space that enables abandonment ecologies in these idle, interstitial areas. In Ruhr, the surplus of space also created a semiotic deficit, a loss of the cultural scaffold of meaning provided through the site's identity as a landscape of work. With the foresight and ingenuity of a preservation ethos of "Seinlassen", these interstitial spaces were, moreover, allowed to gradually assert their own processual value as the ruderal brownfield ecology evolved and a new understanding of the post-industrial, urban ecology developed. *Industrienatur* is the name given to ecological processes taking place in disturbed brownfield contexts where the temporal and territorial precepts of nature conservation and the original, natural condition are challenged. Plants act as memory media (*mnemophytes*) insofar as they are indexes of a landscape shaped through past events, industrial decline and new hybrid ecologies unfolding beyond the nature/culture-divide. They impress traces of past events in an ephemeral form that changes over time and disperses regardless of prescribed orders of the native or natural domain. In place of the transcendent and eternal outside nature, we are compelled to reflect on and engage the in-between and the mediatic domain where human and non-human beings live, move and interact. The hybrid, ephemeral and disturbed is vital to the ecology of industrial nature which signals the dissolution of nature and culture as separate spheres of existence and as distinct and purified heritage domains.

References

Assmann, J. 2008. "Communicative and Cultural Memory." In *Cultural Memory Studies: An International and Interdisciplinary Handbook*, edited by A. Erll and A. Nünning, 109–118. Berlin: Walter de Gruyter.

Bangstad, T. R. 2014a. *Defamiliarization, Conflict and Authenticity: Industrial Heritage and the Problem of Representation*. PhD Dissertation. Trondheim: NTNU.

Bangstad, T. R. 2014b. "Industrial Heritage and the Ideal of Presence." In *Ruin Memories: Materialities, Aesthetics and the Archaeology of the Recent Past*, edited by B. Olsen and Þ. Pétursdóttir, 92–106. London: Routledge.

Barad, K. 2003. "Posthumanist Performativity: Toward an Understanding of How Matter Comes to Matter." *Signs* 28(3): 801–831.

Barndt, K. 2010a. "Layers of Time: Industrial Ruins and Exhibitionary Temporalities." *PMLA* 125(1): 1–15.

Barndt, K. 2010b. "'Memory Traces of an Abandoned Set of Futures': Industrial Ruins in the Postindustrial Landscapes of Germany." In *Ruins of Modernity*, edited by J. Hell and A. Schönle, 270–293. Durham, NC: Duke University Press.

Benjamin, W. 1998. *The Origin of German Tragic Drama*. Translated by John Osborne. London: Verso.

Bennett, J. 2004. "The Force of Things: Steps Toward an Ecology of Matter." *Political Theory* 32(3): 347–372.

Bennett, J. 2009. *Vibrant Matter: A Political Ecology of Things*. Durham, NC: Duke University Press.

Berger, S. and High, S. 2019. "(De-)Industrial Heritage: An Introduction." *Labor* 16(1): 1–27.

Bryant, L. and Joy, E. A. 2014. "Preface: Object/Ecology." *O-Zone: A Journal of Object-Oriented Studies* 1(1): i–xiv.

Burton, J. 2008. "Bergson's Non-Archival Theory of Memory." *Memory Studies* 1(3): 321–339.

Cambridge Dictionary. 2020. "Medium." In *Cambridge Advanced Learner's Dictionary & Thesaurus*. Cambridge: Cambridge University Press. Accessed 31 July 2020. https://dictionary.cambridge.org/dictionary/english/medium.

Chan, E. C. 2009. "What Roles for Ruins? Meaning and Narrative of Industrial Ruins in Contemporary Parks." *Journal of Landscape Architecture* 4(2): 20–31.

Craps, S., Crownshaw, R., Wenzel, J., Kennedy, R., Colebrook, C. and Nardizzi, V. 2018. "Memory Studies and the Anthropocene: A Roundtable." *Memory Studies* 11(4): 498–515.

Desilvey, C. 2017. *Curated Decay: Heritage Beyond Saving*. Minneapolis, MN: University of Minnesota Press.

Dettmar, J. 1994. "Vegetation auf Industrieflächen – Bedeutung von Industriebrachen für den Naturschutz." In *Altlastensanierung: Sicherung, Sanierung und Folgenutzung kontaminierter Flächen*, edited by K.-W. Kiefer, 95–106. Berlin: Springer.

Dettmar, J. 1999. "Die Route der Industrienatur." In *Industriekultur: Mythos und Moderne im Ruhrgebiet*, edited by A. Höber and K. Ganser, 67–71. Essen: Klartext Verlag.

Domínguez Rubio, F. 2014. "Preserving the Unpreservable: Docile and Unruly Objects at MoMA." *Theory and Society* 43(6): 617–645.

Domínguez Rubio, F. 2016. "On the Discrepancy Between Objects and Things: An Ecological Approach." *Journal of Material Culture* 21(1): 59–86.

Eiringhaus, P. 2020. "Inventing the Green Metropolis Ruhr, Germany: Postindustrial Greening Narratives and Critical History Culture." In *Located Research: Regional Places, Transitions and Challenges*, edited by A. Campbell, M. Duffy and B. Edmondson, 253–277. Singapore: Springer.

Erll, A. 2004. "Medium des kollektiven Gedächtnisses: Ein (erinnerungs-)kulturwissenschaftlicher Kompaktbegriff." In *Medien des Kollektiven Gedächtnisses: Konstruktivität – Historizität – Kulturspezifität*, edited by A. Erll, A. Nünning, H. Birk, B. Neumann and P. Schmidt, 3–22. Berlin: Walter De Gruyter.

Erll, A. 2010. "Cultural Memory Studies: An Introduction." In *A Companion to Cultural Memory Studies*, edited by A. Erll, A. Nünning and S. Young, 1–18. Berlin: de Gruyter.

Evans, N. 2004. "Life after Coal." *Agenda*: 83–84.

Farstadvoll, S. 2018. "Growing Concerns: Plants and Their Roots in the Past." *Journal of Contemporary Archaeology* 5(2): 172–193.

Farstadvoll, S. 2019. *A Speculative Archaeology of Excess: Exploring the Afterlife of a Derelict Landscape Garden*. PhD Dissertation. Tromsø: UiT – The Arctic University of Norway.

Fredengren, C. 2015. "Nature: Cultures: Heritage, Sustainability and Feminist Posthumanism." *Current Swedish Archaeology* 23: 109–123.

Ganser, K. 1999. "Von der Industrilandschaft zur Kulturlandschaft." In *Industriekultur: Mythos und Moderne im Ruhrgebiet*, edited by K. Ganser and A. Höber, 11–15. Essen: Klartext Verlag.

Gausmann, P., Keil, P., Büscher, D. and Loos, H. G. 2016. "Flora und Vegetation der ehemaligen Zeche und Kokerei 'Hansa' in Dortmund-Huckarde im östlichen Ruhrgebiet (Nordrhein-Westfalen)." *Dortmunder Beiträge zur Landeskunde* 47: 45–104.

Gausmann, P., Keil, P., Fuchs, R., Sarazin, A. and Büscher, D. 2011. "Eine Bemerkenswerte Farnflora an Mauern der ehemaligen Kokerei Hansa (Dortmund-Huckarde) im Östlichen Ruhrgebiet." *Floristische Rundbriefe – Zeitschrift für floristische Geobotanik, Populationsbiologie und Taxonomie* 45(44): 71–83.

Goch, S. 2002. "Betterment Without Airs: Social, Cultural, and Political Consequences of the Deindustrialization in the Ruhr." In *De-industrialization: Social, Cultural, and Political Aspects*, edited by B. Altena and M. Van Der Linden, 87–111. Cambridge: Cambridge University Press.

Hansen, M. B. N. 2006. "Media Theory." *Theory, Culture & Society* 23(2–3): 297–306.

Hansen, M. B. N. and Mitchell, W. J. T. 2010. "Introduction." In *Critical Terms for Media Studies*, edited by W. J. T. Mitchell and M. B. N. Hansen, vii–xxii. Chicago, IL: The University of Chicago Press.

Haraway, D. 2003. *The Companion Species Manifesto: Dogs, People and Significant Otherness*. Chicago, IL: Prickly Paradigm Press.

Harrison, R. 2015. "Beyond 'Natural' and 'Cultural' Heritage: Toward an Ontological Politics of Heritage in the Age of Anthropocene." *Heritage & Society* 8(1): 24–42.

Hospers, G.-J. 2004. "Restructuring Europe's Rustbelt." *Intereconomics* 39(3): 147–156.

Hunter, P. 2014. "Brown Is the New Green." *EMBO Reports* 15(12): 1238–1242.

Jørgensen, D. 2016. "Artifacts and Habitats." In *The Routledge Companion to the Environmental Humanities*, edited by U. K. Heise, J. Christensen and M. Niemann, 138–143. London: Routledge.

Kastorff-Viehmann, R. 1992. "Die Kokerei Hansa Dortmund-Huckarde als Denkmal." *Deutsche Kunst und Denkmalpflege* 50(2): 148–157.

Kattwinkel, M., Biedermann, R. and Kleyer, M. 2011. "Temporary Conservation for Urban Biodiversity." *Biological Conservation* 144(9): 2335–2343.

Keil, P. and Loos, H. G. 2002. "Dynamik der Ephemerophytenflora im Ruhrgebiet – unerwünschter Ausbreitungspool oder Florenbereicherung?" *Neobiota* 1: 37–49.

Kittler, F. 2009. "Towards an Ontology of Media." *Theory, Culture and & Society* 26(2–3): 23–31.

Knittel, S. C. and Driscoll, K. 2017. "Introduction: Memory after Humanism." *Parallax* 23(4): 379–383.

Kowarik, I. 1993. "Stadtbrachen als Niemandsländer, Naturschutzgebiete oder Gartenkunstwerke der Zukunft?" *Geobotanische Kolloquien* 9: 3–24.

Kowarik, I. 2005. "Wild Urban Woodlands: Towards a Conceptual Framework." In *Wild Urban Woodlands: New Perspectives for Urban Forestry*, edited by S. Körner and I. Kowarik, 1–32. Berlin: Springer Verlag.

Krieg, C. P. 2018. "'By Means Other Than Life': Literature as Posthuman Memory." *Con Texte* 2(1): 22–27.

Lachmund, J. 2003. "Exploring the City of Rubble: Botanical Fieldwork in Bombed Cities in Germany after World War II." *Osiris* 18: 234–254.

Latour, B. 1993. *We Have Never Been Modern*. New York: Harvester Wheatsheaf.

Latour, B. 2009. "Spheres and Networks: Two Ways to Reinterpret Globalization." *Harvard Design Magazine* 7(30): 138–144.

Leech, C. 2015. "Cryptogams in the Nooks and Crannies." *Metropolitan Museum of Art*. Accessed 14 January 2020. www.metmuseum.org/blogs/in-season/2015/cryptogams.

Linkon, S. L. 2018. *The Half-Life of Deindustrialization: Working-Class Writing about Economic Restructuring*. Ann Arbor, MI: University of Michigan Press.

Lipp, W. 1993. "Vom modernen zum postmodernen Denkmalkultus? Aspekte zur Reparaturgesellschaft." In *Vom modernen zum postmodernen Denkmalkultus? Denkmalpflege am Ende des 20. Jahrhunderts. Arbeitsheft 69, Bayerisches Landesamt für Denkmalpflege*, edited by W. Lipp and M. Petzet, 6–12. Munich: Lipp Verlag.

Lorimer, J. 2015. *Wildlife in the Anthropocene: Conservation after Nature*. Minneapolis, MN: University of Minnesota Press.

Lucas, G. 2015. "Archaeology and Contemporaneity." *Archaeological Dialogues* 22(1): 1–15.

McLuhan, M. 1964. *Understanding Media: The Extensions of Man*. New York: McGraw-Hill.

Morton, T. 2007. *Ecology without Nature: Rethinking Environmental Aesthetics*. Cambridge, MA: Harvard University Press.

Morton, T. 2013. *Realist Magic: Objects, Ontology, Causality*. Ann Arbor, MI: Open Humanities Press.

Morton, T. 2015. *Hyperobjects: Philosophy and Ecology after the End of the World*. Minneapolis, MN: University of Minnesota Press.

Olsen, B. and Pétursdóttir, Þ. 2016. "Unruly Heritage: Tracing Legacies in the Anthropocene." *Arkæologisk Forum* 35: 38–45.

Olsen, B., Shanks, M., Webmoor, T. and Witmore, C. 2012. *Archaeology: The Discipline of Things*. Berkeley, CA: University of California Press.

Parikka, J. 2010. *Insect Media: An Archaeology of Animals and Technology*. Minneapolis, MN: University of Minnesota Press.

Parikka, J. 2012. "New Materialism as Media Theory: Medianatures and Dirty Matter." *Communication and Critical/Cultural Studies* 9(1): 95–100.

Peters, J. D. 2015. *The Marvelous Clouds: Toward a Philosophy of Elemental Media*. Chicago, IL: The University of Chicago Press.

Pétursdóttir, Þ. 2020. "Anticipated Futures? Knowing the Heritage of Drift Matter." *International Journal of Heritage Studies* 26(1): 87–103.

Phillips, J. W. P. 2017. "The End of Ontology and the Future of Media Theory." *Media Theory* 1(1): 122–136.

Postman, N. 2000. "The Humanism of Media Ecology." *Proceedings of the Media Ecology Association* 1: 10–16.

Raines, A. B. 2011. "Wandel Durch (Industrie) Kultur [Change through (Industrial) Culture]: Conservation and Renewal in the Ruhrgebiet." *Planning Perspectives* 26(2): 183–207.

Rigney, A. 2015. "Things and the Archive: Scott's Materialist Legacy." *Scottish Literary Review* 7(2): 13–34.

Rigney, A. 2017. "Materiality and Memory: Objects to Ecologies. A Response to Maria Zirra." *Parallax* 23(4): 474–478.

Robinson, S. L. and Lundholm, J. T. 2012. "Ecosystem Services Provided by Urban Spontaneous Vegetation." *Urban Ecosystems* 15(3): 545–557.

Stengers, I. 2005. "Introductory Notes on an Ecology of Practices." *Cultural Studies Review* 11(1): 183–196.

Stiegler, B. 1998. *Technics and Time 1: The Fault of Epimetheus*. Stanford, CA: Stanford University Press.

Storm, A. 2014. *Post-Industrial Landscape Scars*. New York: Palgrave Macmillan.

Vermeulen, P. 2017. "Creaturely Memory: Shakespeare, the Anthropocene and the New Nomos of the Earth." *Parallax* 23(4): 384–397.

Witmore, C. 2014. "Archaeology and the New Materialisms." *Journal of Contemporary Archaeology* 1(2): 203–246.

14

MEMORY AND REDEMPTION

Lessons from a peasant ecology

Alfredo González-Ruibal

Introduction

In this chapter, I undertake a critique of the two concepts that form the basis of the present volume: "heritage" and "ecology". My aim, however, is not to engage in simple deconstruction or to propose a conservative turn. Rather, I want to unpack what I found unsatisfactory in both and take what I find valuable so as to find new ways in which to address the material legacy of the past in the present. My inspiration, and the case study that I will use throughout the present chapter, is the area where I come from, an old peasant landscape in northwest Spain that has undergone a deep and irreversible change during the last century. In the first part of the chapter, I will focus on the ecological side: the constitution of a peasant ecosystem and its deconstitution – the creation of a post-agrarian world. I will use an ecological approach to understand this landscape but then will indicate what I see as the shortcomings of this perspective. In the second part of the chapter, I will look at the legacy of the peasant world and the two main ways in which it has been appraised so far: through annihilation or through heritage – that is, as a lack or as an excess. This will allow me to undertake a critique of the limited nature of the concept of heritage. In the conclusion, I will suggest what I believe is a more powerful way of relating to the disassembled remains of past worlds, one in which Benjaminian notions of memory and redemption are deployed.

Ecology

If there is something valuable in ecological (and neomaterialist) perspectives, it is that they have made us acutely aware of the indissoluble mixture of human and non-human beings in the constitution of the world (Witmore 2014). We no longer perceive plants, animals or things as a mere backdrop of human action but all beings (human and non-humans) as part of complex ecosystems characterized by myriad relations of dependence and collaboration. The concept of ecology, as it has been popularized in recent years across the humanities and social sciences (see introduction to the volume), intends to overcome anthropocentrism and the modern dualisms that created ontological rifts through the universe. The term "society" has been increasingly perceived as partial, as it is too centred on humans, and new terms have been proposed to refer to the mixtures of human and non-human beings: networks, collectives, assemblages (De Landa 2006; Latour 1999). Here I would like to use the term "world" to refer to the relations between humans and non-humans in a specific context. With this concept, I intend to make sense not just of social and ecological relations but geological ones as well while at the same time avoiding the specific connotations of other terms. Yet this does not mean that "world" is a simple name without philosophical connotations. I assume it in the way Levi Bryant (2014, 112–114) does, that is, considering that "*World* is that which *intrudes* on the networks of meaning we throw over it; it is what Lacan sometimes refers as the Real" (his emphasis). It is not just a space of lived experience but something that exceeds such experience in its irreducible materiality.

A peasant world

It has been argued that one of the problems with the Anthropocene is that it creates a divide between moderns, who actively impact the world in which they live, and non-moderns, who are perceived as passive (Bauer and Ellis 2018). To counteract this assumption, some have argued for an Anthropocene that is equated with much of the Holocene, as this is the period when humans clearly start to have an impact in the atmosphere and the environment (Smith and Zeder 2013). While I believe that such conflation is both ontologically wrong and politically dangerous – there is something unique in the way modern capitalism depredates nature (Moore 2016), I agree that humans have been biogeochemical agents well before the Industrial Revolution, as stated by Bauer and Ellis. The world that I am going to describe here as a good example of this.

Galicia, the northwesternmost region of the Iberian Peninsula, was, until the late twentieth century, a prototypical peasant world. Its origins can be traced back to the early Iron Age, around the ninth to eighth centuries BC, when the first permanent settlements were created that shaped the land in a new way and created new sorts of relations between people, plants, animals and the Earth. The landscape that emerged then has undergone serious changes during the last three millennia: the Roman conquest (late first century BC) and the emergence of cities, the decline of cities and the collapse of Roman rule (fourth-fifth centuries AD), the arrival of American cultigens (seventeenth century), an explosive population growth (eighteenth and nineteenth centuries) and a mass migratory movement from the late nineteenth century to the late 1960s. Yet the agrarian ecosystem, characterized by specific relations between farmers, lords, land, animals, plants, spirits and gods, largely persisted. It has been only during the last 40 years that it has been thoroughly dismantled. This has not meant only the disappearance of cultural traditions, forms of sociality and preindustrial materialities but also the radical rearrangement of biological and geophysical relations.

The peasant world of Galicia was characterized by a heavy modification of the landscape: extensive deforestation, dense population density in dispersed settlements, widespread construction of agricultural terraces, intense land partitioning through smallholding and the development of a thick meshwork of paths and roads. By the end of the eighteenth century, primeval forests had virtually disappeared, and cultivated fields, pastures or scrubland covered virtually everything. It was the culmination of a process that had started during the Neolithic, 7,000 years before.

Galicia is characterized by a rugged terrain. Mountains and hills occupy most of the surface and are a challenge to intense agriculture (Figure 14.1). Thus, probably from the Iron Age

FIGURE 14.1 The peasant landscape is better preserved in the mountains of eastern Galicia, Spain – despite afforestation

onwards, farmers started modifying the land through cultivation terraces which today occupy a large percentage of the region's surface. A terrace is a good example of a premodern device through which a human collective becomes a geophysical agent: to make them, the slope is cut, levelled, and the resultant soil pushed forward, making a parapet, which is contained by a masonry wall – the stones themselves quarried from ubiquitous granite or shale outcrops. In addition to this, terraces were often irrigated through many small channels which diverted the water from nearby creeks and created a parallel hydrological network. Even when such works were not undertaken, as in some mountainous areas and in the few plains, peasants partitioned the land in many small lots with stone walls or standing stones (*chantos*). In a region where smallholding was predominant, this meant tens of thousands of kilometres of walls and a bewildering typology of enclosing systems (Lorenzo Fernández 1982, 16–21) (Figure 14.2).

The only land that was not fenced off was the commons. Communal lands survived in Galicia more than in other parts of Spain – or Western Europe. Communal lands could be *monte*, the bush, or it could be forests. *Monte* was more common, due to intensive deforestation, which was achieved through cutting down and burning the woods. This strategy, coupled with very acidic and thin soils, meant that deforested land was largely colonized by gorse (*Ulex sp.*) and heather (*Ericaceae*). Gorse became a crucial element in the peasant world (Lorenzo Fernández 1982, 179–181). Scrublands were managed to increase production and then divided among neighbours. Gorse was harvested to produce compost by mixing it with

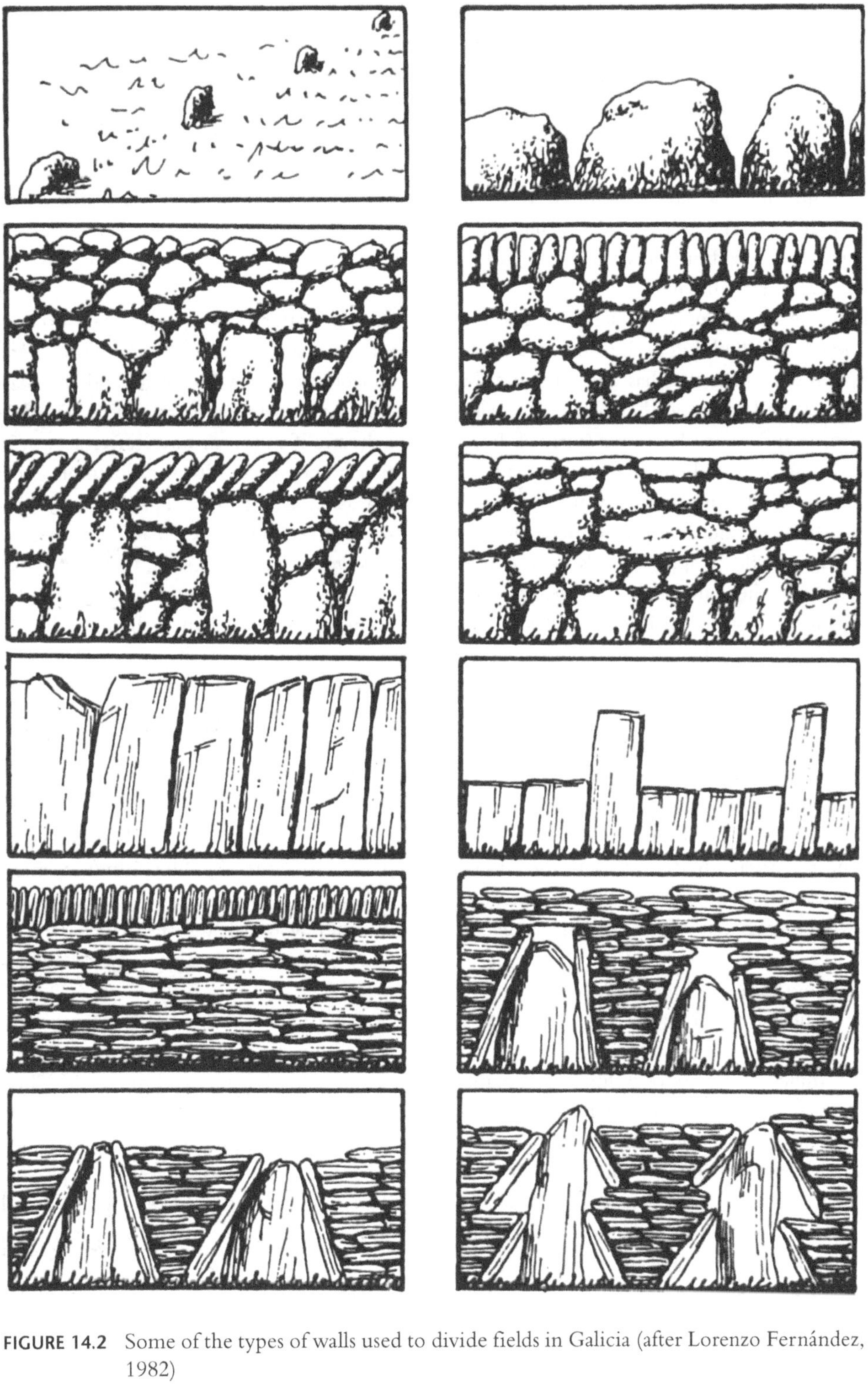

FIGURE 14.2 Some of the types of walls used to divide fields in Galicia (after Lorenzo Fernández, 1982)

cow dung in the stables, for fuel (in the absence of proper wood) and even to make tools, in the case of the greater bushes of *Ulex europaeus*. As for the few forests that were not felled, they were carefully managed. The two predominant types were *soutos* (chestnut-tree [*Castanea sativa*] groves) and *carballeiras* (oak-tree [*Quercus robur*] groves) which could be the property of a specific village or could be divided among different neighbours. Every family knew their trees, and I remember, as a teenager, going with my father to search for the family's oaks under the guidance of old neighbours. Trees often possess convoluted biographies. In the village where I presently have a house, the neighbours as a collective possess oaks in the communal land of a neighbouring village, and vice versa. Trees, thus, are not seen as a mass but as individuals – even today.

The high population densities, a scattered settlement pattern (villages are nucleated, but small), smallholding and a heavily anthropogenic landscape explain the thick meshwork of paths and roads that has characterized Galicia, probably since the Iron Age, and the many terms used to name them: *camiño*, *sendeiro*, *rúa*, *surrieira*, *carreiro*, *congostra*, *canella*, *corredoira*, *estrada* and so on. There is a spiritual dimension to the proliferation of ways: a rich lore exists associated with them, and people had frequent encounters there with supernatural beings, the most famous of which were the *Santa Compaña* ("Holy Company"), a procession of spirits that went out at night to announce a death or a disgrace. There is a material dimension to this world of spirits: roads were flanked with *cruceiros* (stone crosses) and *petos de ánimas* (altars to the souls in the purgatory). They can be regarded as the continuation of the cult to the road spirits of Roman times (*Lares Viales*), which are particularly well attested in the Roman epigraphy of Galicia (Gómez Vila 2004) and whose roots are obviously pre-Roman. The material side is extended to the non-spiritual realm. Many roads and paths were heavily constructed with stone slabs and side walls, vegetable fences (*sebes*) and ditches (*cómaros*). The pavement was interrupted by drainage ditches or irrigation channels (a compromise between the circulation of people and animals and the circulation of water). The dense hydrological system of Galicia required frequent bridges and stepping stones. Most traditional ways were so profoundly imprinted in the land that they can be considered geological phenomena, but this is particularly the case with the *congostras*, deep paths excavated by the continuous movement of the heavy Galician oxcart, made of oak wood and provided with iron-plated, massive wooden wheels (whose earliest parallels are to be found, again, in the Iron Age). In some cases, they are 3 or 4 metres deep – a permanent land feature that will take many millennia of geomorphological processes to erase.

The last element that I will mention here are rivers (Figure 14.3). As has been pointed out (Edgeworth 2011), rivers – at least many of them – are artefacts, perhaps one of the best examples of the impossibility to disentangle human and natural actors and processes. This is very evident in Galicia, where one can scarcely walk 50 metres along a river without noticing human interventions. They have been the main source of non-animal energy until well into the twentieth century. Water mills ground the cereals that made up the base of the peasant's diet, fullers were used to pound paper pulp and fabric, flax was soaked in river pools, the flow was partially diverted to irrigate neighbouring fields, oxcarts were driven into rivers in dry times to soak wheels and axle – an operation without which the cart would not move. Mills and fullers in particular radically transformed the banks with channels, sluices, ditches and floodgates. The river banks were in many cases totally engineered for several kilometres.

Terraces, paths and rivers are all good example of the geophysical character of traditional Galician peasants, their collaboration (and conflict) with plants, rocks and rain. The landscape

FIGURE 14.3 A heavily modified riverbank with bridge and paved way

that I have just described, however, was a fragile one. The fact that it was made predominantly of stone does not mean that it is perennial. It was (and is) continuously menaced by entropy: on the one hand, because of its strongly anthropogenic character – heavily dependent on humans for maintenance – and, on the other hand, because of the peculiar Galician environment, which has one of the highest precipitation rates of the Iberian Peninsula. Maintenance activities were crucial, as stone walls tend to collapse, iron rusts, wood roots and fungi and bacteria thrive, whereas the growth of vegetation is incredibly fast.

It is not surprising then, that traditional tools and several crafts in Galicia were not so much aimed at creating new things as at conserving what there was. This fact has not been properly appraised because of our blindness towards maintenance activities (Domínguez Rubio 2016; Edgeworth 2006; Montón-Subías and Sánchez-Romero 2008) but is quite obvious when one takes a look at the material inventory (Liste Fernández 1991; Lorenzo Fernández 1983). Many tools were used for repairing instruments or architectures, and peasants devoted many hours to repairing their tools (Liste Fernández 1991, 73–74). Several strategies were developed to extend the life of things: tools were placed over the hearth so that the smoke would dry them, kill bacteria and provide them with a protective patina (Liste Fernández 1991, 74). Some artefacts were oiled or kept in leather sheaths. To prevent the plough from abrading the plough or cart rudder (*timón*), it was partially covered with a leather or brush collar, called *timoeiro*. As for infrastructures, stonemasons spent more time repairing roads, bridges and terraces than building new ones. Maintenance work was often undertaken collectively. People gathered to clear paths and to repair drainage ditches. Maintaining things implied maintaining the community, both of the living and of the dead, because repairing old works meant maintaining links with past generations and reproducing and renewing their know-how.

This continuous work of maintenance transformed artefacts and architectures and provided them with new aesthetic and practical qualities. Domínguez Rubio (2016) distinguishes between objects and things. For him, objects are things endowed with a special status by virtue of material, cultural and social processes that allow them to be distinguished from other things. As such, objects are anything but given, they are fragile and temporal realities, always falling out of place. They "wear down and change, … break, malfunction and have to be constantly mended, retrofitted and repurposed" in order to keep their identity qua objects (Domínguez Rubio 2016, 60). Yet maintenance is not just about the impossible task of arresting time and maintaining the identity of the thing as an object. It can also be about producing objects out of objects. A good example is the ladder used to make haystacks or prune trees. It is only after it has been around for some time and two grooves have appeared where the clogs usually step that the ladder becomes the perfect artefact (Liste Fernández 1991, 332). The same happens with other instruments, hand planes, yokes or sickle shafts. They are the product of the accommodation of hand and wood, the exchange of chemical substances (sweat, wax, resin), the physical work of friction. This is why many well-worn, but well-maintained, artefacts were more valuable than new ones. Patina was not just the appreciation of the past or a way of establishing social distinction, as in the bourgeois world (Dawdy 2016): peasants valued patina for both aesthetic and practical reasons. Artefacts look better and *work* better when they show the mark of time and of care. This is why my great-great-grandfather's yoke is still in my house. And it could be put to use any time.

The constitution of a post-agrarian world

The decline of the peasant world started in Galicia in the late 1960s, with the last wave of mass migration and the fast modernization of the Spanish economy. In my childhood and early teenage years (1980s and early 1990s), it was already agonizing and, in material terms, largely disassembled. However, the sight of ox-carts transporting dry gorse or straw, water mills grinding corn and cows tied to the plough with elaborately carved yokes was definitely not an oddity, and vernacular architecture was still dominant. By the time I did my work on the ruins of peasant life – 1997–1999 (González-Ruibal 2003) – premodern technology had all but disappeared. Since then, it has vanished, except for the most durable elements – such as houses – which have, nonetheless, been seriously affected (González-Ruibal 2005). The collapse was both material and demographic: my area, Terra de Montes, lost 70% of its inhabitants during the last hundred years.

This provoked a drastic ecological transformation. Yet instead of the dystopic image of desertification which is how we tend to picture the end of a world, Galicia saw a dramatic increase in forested land, parallel to the shrinking of cultivated fields and pasture lands. Natural or artificial afforestation deeply changed the ecosystem: species that depended on open land and cereal cultivation, such as larks (*Alauda arvensis*) and quails (*Coturnix coturnix*), have dwindled. Instead, the populations of species that thrive in forests, such as jays (*Garrulus glandarius*), have increased. These, in turn, favour the regeneration of forests by dispersing their seeds (Frost and Rydin 2000). At the same time, the abandonment of houses has benefitted some species such as the black redstart (*Phoenicurus ochrurus*), which nest in the ruins, and ruderal plants.

The expansion of new-growth forests and eucalyptus plantations (*Eucalyptus globulus*), coupled with the collapse of the maintenance system of old ways and bridges, which depended on collective work, has severely limited human perambulation (Figure 14.4). Most of the landscape has become a no-go zone for humans, an impenetrable barrier behind which there is a thick bush, which can only be crossed through asphalted roads. This "rewilding" in turn has helped the expansion of some animals, including medium- to large-sized mammals that had disappeared or were receding, such as wild boar (*Sus scrofa*), wolf (*Canis lupus*) and otter (*Lutra lutra*). The latter, in particular, thrives in rivers that are no longer objects but things (Domínguez Rubio 2016).

The process of rewilding is also actively driven by humans through the introduction of new species for game or other purposes, like wild boars, squirrels (*Sciurus vulgaris*) and roe deer (*Capreolus capreolus*). These animals are largely incompatible with the old peasant ecosystem, and those who still cultivate patches of land complain of the destruction that they bring. The limitation of human movement, due simultaneously to the advancing forests and the end of activities that implied perambulation (hunting, fishing, milling, gorse harvesting, etc.) has expelled other beings from the landscape, apart from humans: spirits. Traditionally, the land was thick with them, both evil and benevolent: proof of that are the many crosses inscribed in doors and windows (to prevent evil from entering the home). Their disappearance is due, of course, to secularization through education but also to the end of walking. It is difficult to have supernatural encounters if one does not walk the land.

Changes are not limited to what can be seen. We have to take the entire atmosphere into account (Bille and Sørensen 2016). Thus, the soundscape has undergone serious modifications: there are no cowbells tinkling, no massive wooden cart wheels squeaking; the sound of sickles and axes has been replaced by that of electric brushcutters and chainsaws, and in some areas, human voices are rarely heard anymore. Silence, in general, is much more prevalent now than it was 50 years ago – and it can be eerie. As for light, another crucial element in

FIGURE 14.4 An old path flanked by agricultural terraces peters out in the new-growth forest

defining an atmosphere (Bille and Sørensen 2007), the situation is paradoxical. On the one hand, the lack of inhabitants has made the land darker in many places. As villages are abandoned, they are turned off, so to say. On the other hand, the arrival of electricity (which was complete by the 1970s), created an ever-illuminated landscape that contributed to its disenchantment. Thus, a neighbour told me that it was the arrival of electric light that made de Santa Compaña, the nocturnal procession of spirits, disappear – an excellent example of the agency of (electric) light (Bille and Sørensen 2007). Traditionally, outdoor illumination was something associated with supernatural powers or with specific moments of the year, such as the vigil of the Feast of Saint John, when huge bonfires were lit to celebrate the summer solstice, often on prehistoric hillforts. Light now is vulgar and profane. And finally the smell: when I was a child, my village smelled of smoke, dry gorse, straw, manure and cow dung. The cows long gone, the smell is less pungent, less characteristic.

When I speak with old neighbours, they evoke a lost cartography and sensorial world: the fields cultivated with flax, millet or barley; the winding path to the mountain pastures; the shouts of children; the crossroads and the pool where the devil dwelled. There is no idealization of the past, though. Maybe the opposite: they remember all too well the slavery of tilling a poor land, the hard work, the ubiquitous stench of manure, the hunger of the postwar years. Still, the present landscape is perceived as dystopic: the end of the world as they knew it and the emergence of an alien one – impoverished, depopulated, broken, disenchanted. What urbanites see as pastoral and picturesque is lived here as the dismantling of a solid system that managed to sustain life for centuries.

The limits of an ecological approach

I have tried to provide an ecological description of two different worlds: one peasant, the other post-agrarian. I have emphasized the manifold relations that sustained the first and that produce the second. The problem with the ecological approach, as it is often presented, is that it depicts a homogeneous mass of human beings dealing with a heterogeneous mass of non-humans (bacteria, rain or cattle). Therefore, the emphasis on human/non-human relations is making us overlook asymmetrical and dialectical relations between humans. Domínguez Rubio (2016), as we saw, talks about the discrepancy between objects and things: it surely takes an effort to maintain an object as such. But there is another discrepancy: it also takes an effort to maintain a group of people as a group, a community as a community. The more so when egalitarian values are strong. Things keep falling out of place, but so do people.

The peasant landscape that I described was a landscape in tension, where different conflicts were played out – between peasants and landlords, between peasants and the Church, between peasants themselves. The traditional world survived for a long time – longer than in other Western European regions – not just because it was sustained through maintenance activities that kept entropy at bay but also because the community was kept alive, through a strict moral economy, common celebrations, the participation in collective work (spinning, threshing, house-building) and maintenance activities (repairing terraces and paths) and through their (usually silent) opposition to the misdeeds of the Church and the landlords (Hervés et al. 1997; Scott 1985). This is a tension that medievalists have studied well. The world of the forest, writes Fossier (1996, 341–342), was a place of confrontation and rivalry: the lord tried to maintain his game reserves off the villagers' hands, but the peasant insisted on gathering wood and poaching. The ecology of the forest was not just the product of humans interacting with deer and oaks but of the tension between nobles and villains and their respective ecologies. Likewise, the *monte* in Galicia, the common land, is not primarily the result of an ecological process or of the conflicts and collaborations between humans and non-humans. It is the result of the successful struggle of the peasants against Church, aristocrats and the bourgeoisie. When one of the first modern factories was established in Galicia in 1798, its owner intended to limit common access to the forest by the local peasants: in his ecology, woodlands had to be transformed into charcoal to fuel his kilns. Four thousand people mutinied and destroyed the factory (Hervés et al. 1997, 168–169).

Surely, the peasant world started to fall apart when shared activities of making and repairing the world stopped, entropy was unrestrained and objects became things. But this process was motivated in the first place by a break in relations between people: when the community of equals was fragmented and atomized through the impact of capitalism and modernity (Alonso González and González Álvarez 2016).

Heritage

The peasant world that I have described has disappeared as a functioning entity. So where are we now, and where are we heading? It seems that we have two options: the option of amnesia and the option of heritage. I will briefly describe them here and then suggest an alternative way.

Heritage zero

I devoted part of my archaeological work on the abandonment of houses and villages to trying to understand the attitude of former Galician peasants toward the material legacy of preindustrial life (González-Ruibal 2005). The question was not so much that the peasant world had succumbed so rapidly. This could be explained due to the way in which modernity had arrived to the region: as a sudden catastrophe. Unlike northern Europe, where the introduction of modern economies, ecologies and mentalities was a relatively gradual process, in the case of Galicia and other areas of southern and eastern Europe, the modern-capitalist system prevailed through a traumatic process of mass emigration which reached its apex during the mid-twentieth century. Unlike in other areas, however, there was not simply abandonment and replacement: there was thorough destruction as well. Anything that smelled of the past was disfigured, demolished or burnt. Some areas, like the western part of A Coruña province and most of the coast, lost virtually all its vernacular architecture in a few decades (Figure 14.5). There are villages where it is impossible to spot a single building predating the 1960s – except perhaps the church and the iconic *hórreos* (granaries). The peasant past of the last 500 years was thoroughly cleansed. This is particularly striking in a culture where care, maintenance and relationality had played such a prominent role.

In my archaeological investigation of abandoned houses, it struck me that people had not just abandoned the houses themselves but complete inventories of artefacts and personal belongings, including letters and photographs. It is as if people did not care for the past anymore. As if their history belonged to another group – and in a sense, it did. The people of the plough and the stone house were regarded as remote from the present as the Iron Age people that had occupied the land two and a half millennia earlier – but with a difference. The people of the plough and the stone house were the peasants themselves before emigrating, or their parents. They were seen as primitives out of time – the colonial gaze applied to oneself. There

FIGURE 14.5 Cleansing the past: a village with all the vernacular buildings demolished or disfigured

was shame for a past that was perceived as anachronically poor and uncouth, and this explains the violence with which it was dealt. The word iconoclasm comes to mind, but the former peasants were not destroying symbols or images, or not just that.

When I talked with neighbours, they described the very material experiences that were related to physically handling instruments or dealing with raw matter (stone), natural substances (dung) or the built space. They insisted that they could tell me about such experiences, but it was not possible for me to actually realize what the world was like – and they were partially right, because it requires working and living with the actual things to really know. I say partially right because I have used sickles and hoes very often (still do), and as an archaeologist, I have had my fair share of working with manual instruments, toiling under the rain or the sun. I have also handled the old instruments, felt the unwieldiness of the plough, the weight of an oxcart wheel and the coarse texture of a masonry wall. It was the sensorial experience of the vanishing world that better conveyed their rejection of such world. They rarely used a metaphorical language. It is understandable, given the very material environment in which they were immersed and in which few things were mediated: almost every experience – walking to the weekly fair, slaughtering a cow or cutting gorse – implied a direct involvement with reality (Millán Pascual 2015, 58–59) – even spiritual encounters. Since they could not see themselves as different from the plough or the dark kitchen of the old stone house, the only way they saw to achieve separation was through the destruction of the plough and the house. Not a Cartesian process of purification but a very real process of removing the past from them, like the surgeon that removes a tumour. The destruction of the old world was not metaphorical but physical.

A peasant heritage

As part of my work with the remnants of the peasant past, I started an ethnographic collection. The idea was to curate a material assemblage that was disappearing very fast – faster than what has been called "intangible heritage", because songs, dances and legends were still transmitted and performed, but nobody was saving pots or winders from destruction. The former peasants saw the efforts deployed by my siblings, my father and myself as ridiculous, and when the exhibition was finally opened, it was nicknamed *casa da ferralla* ("house of rust"). For them, it was a place full of garbage, and many refused to visit it. What those who did visit experienced, however, was fascinating, and I will return to it later. The ethnographic collection that I helped create was one among many similar undertakings happening throughout Galicia – and Europe. Many people, often urbanites such as myself, fascinated by the vanishing peasant world tried to preserve what they could. There is a historical and an aesthetic component in such fascination, which is also behind the rehabilitation of old houses and, in some areas, the general gentrification of the countryside (Alonso González 2017). National and international governments and agencies are now bent on preserving the cultural identity of regions by curating artefacts, architectures and landscapes.

However, the only imaginable solution seems to be to freeze certain pasts (for identity or economic purposes), which is either impossible or unsatisfactory. It is impossible to maintain the peasant landscape of Galicia, which was based on ecological relations now lost. It is unsatisfactory to have the material inventory of a peasant word displayed out of place inside a museum. Of course, it is better to have a fossilized landscape and a museum than to lose everything forever, but there is something important that is irremediably lost in the transition to heritage. Some have recently argued for allowing entropy in heritage (e.g. DeSilvey 2017; Holtorf 2015; Kobiałka 2014). Entropy is indeed inherent to the organic logic of traditional material culture, as we have seen: the play between care and decay generates new realities and reveals hidden qualities in artefacts. It does have to have a place in our relationship with past artefacts. This is something that is actually being allowed in the daily practice of ethnographic museums today, where things are no longer cleaned and frozen to a pristine condition, and oils, dirt and soot are not necessarily removed: the biogeochemical processes in which they are involved are allowed to continue working. Decay certainly has a productive quality vis-à-vis things at the beginning of the process, the early phases of post-depositional transformations, in archaeological parlance. This is what most commentators document (e.g. DeSilvey 2017), but if unrestrained, it ends up leading to the virtual disappearance of an object, as archaeologists know well. This might not be problematic in some cases (mass-produced twentieth-century social housing, for instance), but it is in others. There is another problem: the concept of heritage itself. I find the idea of heritage unsatisfactory for at least three reasons: first, because it creates a divide between past and present – a passive past that is rekindled by living people (Bangstad 2011). Second, it implies a move from fact to representation: heritage is a social construct, it is an image of the past, it is a collective fiction, a playground and so on. There is not a huge difference here between conventional and critical perceptions of heritage (the former are perhaps more materialist). I am quite sceptical as to the possibilities of ontologically rematerializing heritage (see Pétursdóttir 2013 and introduction to the volume; Solli 2011).

Heritage, by definition, cannot be dissociated from representation and mediation. It is not the matter of the past that matters but how the matter of the past is apprehended in the present. If we want to transcend representation and mediation, then we need another concept. This brings me to the third problem: heritage is inevitably loaded with ideas of property, power, class and gender inequality. This is perhaps more obvious in Latin languages, in which we use words derived from *patrimonium*, that is, the wealth of a household transmitted through a patrilineage.

A way of circumventing heritage could be simply denying the divide between past and present and trying to return to roots (Alonso González 2015) or continue reproducing the past as it was, including agricultural and community practices, as some neorural communities try to do. This, however, requires denying that an irresolvable rupture has occurred, and repressing it inevitably leads to further problems. We have to embrace the negativity of the rupture (Millán Pascual 2015) and accept that returning to the peasant world cannot be done by simply restoring its materiality, not even its social relations and mode of production. It is impossible to sustain the peasant world without a peasant subjectivity (Millán Pascual 2015). And this cannot be done through simple goodwill. The situation is complex, and I do not have a solution. In the last section of this chapter, I would like to discuss some ideas to think the issue of heritage from other perspectives. For that, I suggest abandoning the word "heritage" and using "memory" instead.

Memory and redemption: the promise of things

Andreas Huyssen (2006, 8) surmises that our contemporary fixation with ruins is related to the fact that "they still seem to hold a promise that has vanished from our own age: the promise of an alternative future". Richard Wolin (1994, 217–218) extracts a similar lesson from his reading of Walter Benjamin:

> The foremost danger of modernity is that its radical disrespect for tradition runs the grave risk of totally eradicating our links with the past, thus squandering that invaluable "temporal index of redemption" [Benjamin 1968, 254] which tradition contains. There is a promise of redemption that has been sedimented in the artefacts and ruins of traditional life, frustrated aspirations for a better life that have to be preserved and redeemed.

Horkheimer and Adorno (2002, xv), in turn, argued that it was not so much the past that had to be preserved but "the redemption of past hopes".

Such redemptive qualities can perhaps be found in the work of archaeology and perhaps more clearly in its forerunner, antiquarianism. Both disciplines have been, from the beginning, a

> melancholic attempt to grasp, materialize and bring into being that which has been already lost – and thereby deny that loss. It is an active form of resistance to the march of history – it is in Freudian terms, a refusal to mourn; a refusal to acknowledge the loss and move on and it is instead a pathological melancholic denial of this obvious loss, desperately attempting to hang on to something which is no longer there.
>
> *(Buchli 2010, 204)*

It is this refusal to acknowledge loss of traditional heritage management that has been recently subjected to systematic revision by critical heritage scholars (DeSilvey 2017; Holtorf 2015; Kobiałka 2014). The desperate attempts to recover the past and preserve it are now mocked as futile and politically conservative. Although critical myself of the impossible desire of freezing time of conventional heritage management and its often artificial outcome, I would not dismiss altogether the concern with preserving objects as faithfully as possible to their original state. Today, when the fluidity or motility of things is continuously praised, it might seem strange to defend its permanence. Yet while it is true that things are always falling out of place (Domínguez Rubio 2016), it is also true that they persist and that this is one of their inherent qualities (Olsen 2010, 141, 151), perhaps the one that better defines them. Favouring persistence, then, can be a way of helping things fulfilling one of their defining qualities while at the same time being a way of showing respect for those who made them, used them and lived with them.

Walter Benjamin (1968) was one of the first to warn against the angel of progress. Instead of allowing himself to be carried away by the "storm blowing from paradise", he started to investigate the wreckages that were piling at the feet of the angel – a task that had been that of archaeologists and antiquarians before. The case of antiquarians is particularly relevant to the present argument. From the fifteenth century onwards, they engaged in a quest for traces of the classical past: statues, coins, inscriptions, manuscripts (Schnapp 1996). They excavated, collected and curated them. But Renaissance antiquarians never saw that as an end in itself. What they were trying to do was to retrieve the index of redemption of the Greek and Roman past in order to give birth to a new era. They did not intend to replicate that past as it was. They were very conscious both of the impossibility of such an enterprise and the limitations of the ancient world. Yet they saw in the collection and preservation of that world's materiality a point of departure.

This is what I suggest we do with the remains of the nonmodern worlds that modernity is abolishing at increased speed. They have to be the foundations to construct a new world, a Renaissance of sorts. One which appraises other ontologies in fragments and traces – not of sculptures and temples this time but in sickles and cultivation terraces. This is not about showcasing them, as if they were a past dead and done, nor about re-enacting history, as if by donning the robes of peasants one could become a peasant again (not a very desirable thing, anyway). The idea is to find the indexes of redemption in the physical presence of vanished worlds. Hence the great asset of things: the spontaneous memory that they carry with them that transcends any attempt at complete symbolization by humans (makers, users, consumers, inheritors of things). While former peasants complained to me of the very real weight of the plough and the hardness of working with it, the thingliness of the plough was manifesting other qualities, no less real, and revealing other possible worlds – characterized by an intimate experience of matter, an awareness of one's place in a wider meshwork of beings and by relations of solidarity and cooperation. Pétursdóttir and Olsen (2018, 104) encourage us to address the unrealized excess of things, their dark side. But we should be careful of not casting all excess as dark: it can be light as well. And while heritage can be a way of "encountering something that feels strange, distant and other" (Bangstad 2014, 102), it can also be a way of returning home – a home that we might not know existed.

I have said that many neighbours refused to visit the ethnographic collection that I helped create (Figure 14.6). But among those who entered, we could see something surprising. They are usually reluctant at the beginning or have a mocking attitude. And then they see the

FIGURE 14.6 A traditional bedroom displayed in the Centro Etnográfico de Terra de Montes

winder for weaving flax or the mallets for threshing corn and suddenly they rush to touch them, and there is an epiphany in that touch which brings back long-repressed memories (Figure 14.7). And the artefact of shame and toil reveals in a flash its redemptive qualities, the never-fulfilled hopes, and there come stories and songs and days and nights of work and feast. Retrieving the index of redemption means finding the ecology behind the object without defusing the uniqueness of the object – a terrace, a yoke, a watermill. But it also means finding its economy: the moral economy that sustained the world to which it belonged.

Conclusion

In this chapter, I have described a peasant world and its transformation into a post-agrarian one. I have referred to some of the beings involved in their constitution – stone terraces, gorse, acidic soils, ploughs and watermills, *Phoenicurus ocrurus*, ruins, eucalyptus trees. Yet I have cautioned against the temptation to present a mass of humans interacting with a heterogeneity of non-human beings. This, in my regard, is a common pitfall in current ecological approaches in the humanities. I have, instead, tried to make clear that the peasant world was created through tensions between peasants and the powerful and, above all, through a moral economy that successfully sustained the peasant world for centuries by weaving thick meshworks of human solidarity. With the introduction of capitalism and the triumph of modern individual subjectivities, the old world started to collapse. Its ruins face different options today. The two that prevail are annihilation – which has been the one favoured by the former peasants, traumatized by a difficult process of becoming modern – or preservation as heritage. One

FIGURE 14.7 Women threshing corn in the 1920s

Photo: Ruth M. Anderson.

of them negates the past; the other negates the liveliness of the past by freezing it in its purely aesthetic and historical qualities. A third option, which is strongly linked to heritage and which recruits its adherents amid the same social groups, is the attempt at reconstituting the peasant world. I have proposed a Benjaminian alternative: one that accepts loss, acknowledges the impossibility (and undesirability) of replicating the past and finds in the humble objects of nonmodern worlds indexes of redemption, flashes of spontaneous memory that illuminate our future. This understanding as vibrant, but latent, memory of what is usually understood as just (dead) heritage is not completely at odds, however, with traditional heritage management. It is not my intention to abolish heritage (an attitude redolent of the destructive drive of former peasants) or museums. Rather, my point is that conceptualizing and displaying past things as *just* heritage diminishes them and deprives them of their potential to be something else, much more interesting. Yet museums and other similar spaces can be places to discover both the richness of other materialities and their hidden potential – provided that more sensorial, corporeal interactions are allowed (Classen and Howes 2006) not as an end in themselves but as a way of disclosure (de Bolla 2012). It is not an issue of either/or but rather complementarity: heritage and memory. We need to preserve and display things, but we need them to be more than signs of a past era: they have to be powerful, to be alive, to carry a message for the future.

References

Alonso González, P. 2015. "The Heritage Machine: The Neoliberal Order and the Individualisation of Identity in Maragatería (Spain)." *Identities* 22(4): 397–415.

Alonso González, P. 2017. "Heritage and Rural Gentrification in Spain: The Case of Santiago Millas." *International Journal of Heritage Studies* 23(2): 125–140.
Alonso González, P. and González Álvarez, D. 2016. "A Contemporary Archaeology of Cultural Change in Rural North-Western Spain: From Traditional Domesticity to Postmodern Individualisation." *International Journal of Historical Archaeology* 20(1): 23–44.
Bangstad, T. R. 2011. "Routes of Industrial Heritage: On the Animation of Sedentary Objects." *Culture Unbound: Journal of Current Cultural Research* 3(3): 279–294.
Bangstad, T. R. 2014. "Industrial Heritage and the Ideal of Presence." In *Ruin Memories: Materialities, Aesthetics and the Archaeology of the Recent Past*, edited by B. Olsen and Þ. Pétursdóttir, 92–105. London: Routledge.
Bauer, A. M. and Ellis, E. C. 2018. "The Anthropocene Divide." *Current Anthropology* 59(2): 209–227.
Benjamin, W. 1968. *Illuminations*. New York: Schocken.
Bille, M. and Sørensen, T. F. 2007. "An Anthropology of Luminosity: The Agency of Light." *Journal of Material Culture* 12(3): 263–284.
Bille, M. and Sørensen, T. F., eds. 2016. *Elements of Architecture: Archaeology, Atmosphere and the Performance of Building Space*. London: Routledge.
Bryant, L. R. 2014. *Onto-Cartography*. Edinburgh: Edinburg University Press.
Buchli, V. 2010. "Memory, Melancholy and Materiality." In *Archaeology and Memory*, edited by D. Borič, 204–210. Oxford: Oxbow.
Classen, C. and Howes, D. 2006. "The Museum as Sensescape: Western Sensibilities and Indigenous Artifacts." In *Sensible Objects: Colonialism, Museums and Material Culture*, edited by E. Edwards, C. Gosden and R. Philips, 199–209. Oxford: Berg.
Dawdy, S. L. 2016. *Patina: A Profane Archaeology*. Chicago, IL: The University of Chicago Press.
de Bolla, P. 2012. "The State of Wonder." In *Museum Objects: Experiencing the Properties of Things*, edited by S. Dudley, 158–165. London: Routledge.
De Landa, M. 2006. *A New Philosophy of Society: Assemblage Theory and Social Complexity*. London: Continuum.
DeSilvey, C. 2017. *Curated Decay: Heritage Beyond Saving*. Minneapolis, MN: University of Minnesota Press.
Domínguez Rubio, F. 2016. "On the Discrepancy Between Objects and Things: An Ecological Approach." *Journal of Material Culture* 21(1): 59–86.
Edgeworth, M., eds. 2006. *Ethnographies of Archaeological Practice: Cultural Encounters Material Transformations*. Lanham, MD: AltaMira Press.
Edgeworth, M. 2011. *Fluid Pasts: Archaeology of Flow*. London: Bloomsbury Academic.
Fossier, R. 1996. *La sociedad medieval*. Barcelona: Crítica.
Frost, I. and Rydin, H. 2000. "Spatial Pattern and Size Distribution of the Animal-Dispersed Tree *Quercus robur* in Two Spruce-Dominated Forests." *Ecoscience* 7: 38–44.
Gómez Vila, J. 2004. "Dedicatorias a los Lares Viales en la provincia de Lugo." *Gallaecia* 23: 135–154.
González-Ruibal, A. 2003. *Etnoarqueología de la emigración. El fin del mundo preindustrial en Terra de Montes (Galicia)*. Pontevedra: Diputación de Pontevedra.
González-Ruibal, A. 2005. "The Need for a Decaying Past: An Archaeology of Oblivion in Contemporary Galicia (NW Spain)." *Home Cultures* 2(2): 129–152.
Hervés, H., Fernández, Á., Fernández Prieto, L., Artiaga, A. and Balboa, X. 1997. "Resistencia y Organización. La conflictividad rural en Galicia desde la crisis del Antiguo Régimen al franquismo." *Noticiario de Historia Agraria* 13: 165–191.
Holtorf, C. 2015. "Averting Loss Aversion in Cultural Heritage." *International Journal of Heritage Studies* 21(4): 405–421.
Horkheimer, M. and Adorno, T. W. 2002. *Dialectic of Enlightenment: Philosophical Fragments*, edited by G. Schmid Noerr and translated by E. Jephcott. Stanford, CA: Stanford University Press.
Huyssen, A. 2006. "Nostalgia for Ruins." *Grey Room* 23: 6–21.
Kobiałka, D. 2014. "Let Heritage Die! The Ruins of Trams at Depot no. 5 in Wrocław, Poland." *Journal of Contemporary Archaeology* 1(2): 351–368.

Latour, B. 1999. *Pandora's Hope: Essays on the Reality of Science Studies*. Cambridge, MA: Harvard University Press.
Liste Fernández, A. 1991. *Funcionalidad y estética en el Museo Etnográfico Liste*. Pontevedra: Deputación de Pontevedra.
Lorenzo Fernández, X. 1982. *A Terra*. Vigo: Galaxia.
Lorenzo Fernández, X. 1983. *Os oficios*. Vigo: Galaxia.
Millán Pascual, R. 2015. "Arqueología Negativa. Las fronteras arqueológicas del presente." *Complutum* 26(1): 49–69.
Montón-Subías, S. and Sánchez-Romero, M., eds. 2008. *Engendering Social Dynamics: The Archaeology of Maintenance Activities*. BAR International Series 1862. Oxford: Archaeopress.
Moore, J., ed. 2016. *Anthropocene or Capitalocene?: Nature, History, and the Crisis of Capitalism*. Berkeley, CA: PM Press.
Olsen, B. 2010. *In Defence of Things: Archaeology and the Ontology of Objects*. Lanham, MD: AltaMira Press.
Pétursdóttir, Þ. 2013. "Concrete Matters: Ruins of Modernity and the Things Called Heritage." *Journal of Social Archaeology* 13(1): 31–53.
Pétursdóttir, Þ. and Olsen, B. 2018. "Theory Adrift: The Matter of Archaeological Theorizing." *Journal of Social Archaeology* 18(1): 97–117.
Schnapp, A. 1996. *The Discovery of the Past*. London: British Museum.
Scott, J. C. 1985. *Weapons of the Weak: Everyday Forms of Peasant Resistance*. New Haven, CT: Yale University Press.
Smith, B. D. and Zeder, M. A. 2013. "The Onset of the Anthropocene." *Anthropocene* 4: 8–13.
Solli, B. 2011. "Some Reflections on Heritage and Archaeology in the Anthropocene." *Norwegian Archaeological Review* 44(1): 40–88.
Witmore, C. 2014. "Archaeology and the New Materialisms." *Journal of Contemporary Archaeology* 1(2): 203–246.
Wolin, R. 1994. *Walter Benjamin: An Aesthetic of Redemption*. Berkeley, CA: The University of California Press.

15

(SM)ALL THINGS REMEMBERED

Ingar Figenschau

Introduction

Memory is a difficult thing to grasp and concretize, as memory is largely perceived as something cognitive, something that is shaped in our minds. However, is memory a purely human property, a process that is managed without any kind of "non-human" features? In this chapter, I want to approach the complex process of memory through an encounter with things, through an *ecology of memory*, and take the backstairs into the world of recollection. More precisely, I want to explore memory through an archaeological study of prison camps from WWII. These marginalized, unpleasant objects and structures of war do reflect a dark and negative past, but the material also has an eerie and peculiar nature that is worth considering. Caught in ongoing processes of decay, juxtaposing the aesthetic "cultured" and the ready-ruined, it is hard to see these objects as separated from nature or from their ecological surroundings. The material relics expand by spreading into the surroundings and by forming new constellations with the vegetation, the landscape and the changing seasons, in a natural process of conservation.

We depend on things not only in our everyday tasks but also as eminent triggers for memory – a memory that can generate involuntary and unedited recollections. In this chapter, I will discuss how things and archaeological structures in overgrown prison camps in Northern Norway are experienced in their present state. I will argue that through their ecological presence and natural conservation, these things can be perceived as uncanny strangers while at the same time feeling uncomfortably close and familiar. Just as this archaeological material is partially overlooked, it also contains fragmented material memories of the trivial and "antihistoric" elements that conventional and linear history has a tendency to forget. What is left in these prison camps, and what stories can this material tell? How do these tangible stories relate to the written memories of prisoners who stayed here, and how can fragile and broken things possibly provide nuance to such first-hand accounts?

(Sm)all things forgotten

I stand in the middle of what was once a German prison camp for Soviet prisoners of war. The place, which was baptized Kitzbühel by the Austrian soldiers, is located in a desolate

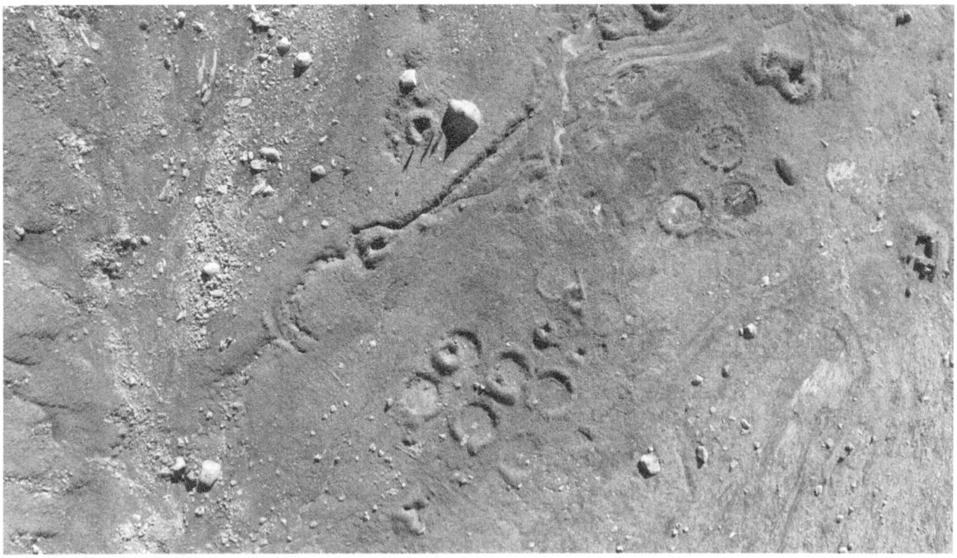

FIGURE 15.1 Kitzbühel, Austria, 2016. The PoW camps appear as circles in the landscape
Photo: I. Figenschau.

mountain landscape in Norddalen, Nord-Troms in Norway, at about 800 meters above sea level. At first glance, you cannot see them, but gradually they emerge; small groups of circular structures resiliently placed on a low ridge. Scattered around are visible remains of stoves, barbed wire, oil barrels and glass. Moving around in the area, new structures and things are constantly revealed; things and traces that have fallen into a kind of dormant state only affected by the shifting seasons.

Closing in on one of the structures, the traces of a so-called plywood tent or *Finnenzelte*, I come across a large amount of rusty tin cans. In a way, they are conspicuously strange in such a landscape but at the same time disturbingly familiar. The remains of the German field kitchen arranged among rocks and heather, stubborn remnants that involuntarily recall how disturbingly visible – and redolent – the food must have been for the many Soviet prisoners of war who starved close by throughout the winter.

Having been here several times, and confident that the place had already been exhaustively and minutely examined, two broken German helmets suddenly came to view. The vegetation and the many seasons after 1945 had camouflaged the helmets well, and at first glance, the rusty parts of the outer shell could look like stones. The contrast between the outer shell and the inner bands' weathering was striking. Unlike the soft aluminum, the hard steel no longer stood against nature's corrosion. What once protected against enemy fire had in this case given up. The aluminum, on the other hand, though somewhat grayed, looked surprisingly new. The stamp on the internal band revealed the manufacturer "Schuberth-Werk" and "Werner Zahn". The helmets were manufactured in 1938 and 1939 and marked with "66 nA, 58" and "62 nA, 57" (nA: Neues Art). The liner is a German M31, which was used in the well-known helmet M35 (Sáiz 2008, 16–18).

This was one of the first mass-produced helmets, made in the period between 1935 and 1940. The quality of the M35 was considered very good, since it was made mainly

FIGURE 15.2 Two German M35s smashed and camouflaged in a small scree in Kitzbühel

Photo: I. Figenschau.

of molybdenum-blended steel (Sáiz 2008, 22). Perhaps the helmets used in Kitzbühel were highly appreciated, a symbol of the German army, its invulnerable being, quality and function. A value that disappeared in the spring of 1945 when it was no longer needed. The helmets were broken, but if the smashing was done in delight, anger or relief, we will probably never know. Nevertheless, their camouflaged pieces were left in the deserted mountain scenery where they persist as distinctive memories of the local and worldwide events they once were part of.

A large amount of shards of both faience and porcelain lay on the surface in one of the plywood tent structures. As the sunlight reflected on the smooth enameled surfaces, they appeared more invasive than naturally at home here. Breaking the surface, our archaeological research brought forth nuances in this material as well. Many shards were made of white-painted faience specifically made for the German army, but there were also pieces of porcelain with colorful print designs. Several of these shards could be assembled into platters and cups. Among these was a smaller coffee cup and even a cake platter with sturdy color printing. The platter is produced by the Norwegian Porsgrund Porselen and is hand-painted with steel plate–printed decorations – a type of print used in the period from 1887 to about 1905. These platters were relatively expensive to produce, costly, and not everyday items. The cup had a nice floral motif with gilded edges at the top and bottom. The bottom is stamped with "Porsgrund Porselen" and a hand-painted number in brown. The stamp is green, encircled with a circular boundary, typical for the period 1913–1937 (Eek 2017).

These are just some of the things we came upon. Apart from these, representative of the finds from Kitzbühel and also Spittal – another low-lying prison camp we investigated – were shoes, bakelite equipment, sardine boxes and other canned goods, ammunition, tallow lights, ski bindings, fountain pens, medals, fragments of fabric and bottles. These finds appeared both on the surface and through the archaeological excavation. They all tell the story of everyday conditions in the camps, of military everyday life; food rations, field equipment, clothing, hygiene products and other practical paraphernalia. But also things that point to other aspects of life in the camp, such as gaming pieces, beer and wine bottles.

What does this material express? How does it remember? What can the bakelite residues of a Czechoslovakian-produced thermos; ammunition from Raufoss, Hirtberg and Saxony; porcelain from Porsgrund, Mettlach and Kronach; canned goods from France and Denmark; a fountain pen from Heidelberg; Finnish *Markka* coins; American-produced toe plates; *Bergschuhe* mountain boots; a bottle of *Haarwasser*; and a *Hitlerjugend* medal from Munich tell us? All of these objects were the result of a worldwide war, a war that brought soldiers, prisoners, structures and things to this European periphery. Although the military presence only lasted for six cold and dark months, the material traces are still there, refusing to let themselves be removed.

All of these items, manufactured in different parts of the world, were brought together in the high North and left behind, stranded. Why did they end up here, far from the well-known battlefields and the war's main arena in Central Europe? Along the Lyngen Line's inner valleys and mountainous areas, prison camps were named after German, Austrian and Swiss cities: Spittal, Mallnitz, Gastein, Kitzbühel, Kufstein, Windeck, Würgel, Seehof, Seefeld, Salzberg, Zermatt and Landech. Whether it was homesickness, a recognizable landscape or a dark sense of humor that was the reason for the naming is uncertain. However, apart from these somewhat foreign place names, little links this remote northern landscape to the war's well-documented narratives and closely researched main themes. What drove soldiers and prisoners to occupy this remote landscape in what would be the last winter of the war? What happened before they arrived at Lyngenfjorden, and where did they go, the soldiers, the officers and the prisoners, at the end of the war?

Austria, the Soviet Union and the Third Reich in Norddalen

The German *Operation Silberfuchs*, part of the German attack on the Soviet Union known as *Unternehmen Barbarossa*, was stopped on the banks of Litza and Titovka on Kola Peninsula. From the autumn of 1941, the frontline was thus characterized by locked standoff warfare (Jaklin 2006, 267–268). On 7 October 1944, the Red Army initiated an attack; an operation that resulted in German retreat and which in less than two months would bring the German army to Lyngenfjorden in North Troms, Norway, 480 km to the west. The German authorities had been prepared for such a development. The "Führer Directive" No. 50, dated 28 September 1943, proposed a complete withdrawal from Finland, Finnmark and Nord Troms to the Lyngen area. The operation was nicknamed "Operation Nordlicht" (Fossum 2014, 43) and was implemented shortly after the Soviet attack at the Litzafront. In their retreat to Lyngenfjorden, the German forces utilized scorched-earth tactics and burned down all structures – houses, barns and even churches – that could be used by the advancing Soviet army. Civilians in Finnmark and Northern parts of Troms, the northernmost regions of Norway, were forcibly evacuated under threats of fines; some resisted and hid, while others followed the evacuation order (Gyllenhaal and Gebhardt 2001, 90–91; Isachsen 2016, 147; Westrheim 1978, 55–56).

With orders from Berlin, construction of the Lyngen defense line commenced late in the fall of 1944 and should have been completed on 1 May 1945. Narrow valleys and high mountains offered topographical advantages for the establishment of the line. The line's main position was between Skibotn and Kitdal, with flank guards in Rostadalen and Bjørnefjell (Fossum 2014, 43; Sunde 2014, 62–63).

To work on the preparation of the defense line, about 10,000 Soviet but also Yugoslav and Polish prisoners of war were brought along on the withdrawal. They were set to carry out most of the heavy labor in inhumane conditions. Much of this occurred in high mountain areas in very difficult circumstances, and with few clothes and facing constant abuse, the death rate among the prisoners in the camps was high (Soleim 2009, 119).

Kitzbühel has the highest elevation of the four prison camps in Norddalen, the other three being Spittal, Mallnitz and Gastein. Both Mallnitz and Kitzbühel were investigated immediately after the war by a war crimes commission after several sources had reported war crimes in the area (Lund 2014, 71). Mallnitz was eventually referred to as the "death camp" of the Soviet prisoners of war; "many were sent there and no one left" (Gallagher 1945b). The investigations in Mallnitz revealed four mass graves with a total of 143 bodies (Jenney 2014 [1945]); in Kitzbühel, another six bodies were found. It is assumed that the construction of the Lyngen Line cost up to 1,000 human lives (Hesjedal and Andreassen 2015, 12), which is more than the present-day population in the Austrian town of Mallnitz.

Encounters with Spittal and Kitzbühel

After WWII, during and after the Cold War, the prison camps were largely omitted from historical narratives – both as a result of the changed political climate but also because of the inherently problematic and negative content of this heritage. Perhaps there was an implicit desire to neglect the "dead load of history" and let time itself bury them in eternal oblivion (cf. Arendt 1976, ix). It was not until 2010 that one felt an official change in attitude. The war remains were almost forced into the public eye due to a power line development, following which it was argued that an archaeological registration of Mallnitz and Gastein ought to be carried out (Drageset, Haugen and Hesjedal 2010). This eventually resulted in a larger archaeological registration along parts of the Lyngen Line, which led to a 2015 heritage management proposal, where a larger area of the Lyngen line was proposed listed, including the prison camps in Norddalen (Hesjedal and Andreassen 2015).

A large amount of material that had been left out of the historical narrative was suddenly given wider public recognition. Irrespective of what had been either felt, lived, written and told, or kept silent, hidden away and actively forgotten, the remains lived on as material memories of the war and were still there as I began my research at the sites of the former prison camps Spittal and Kitzbühel in 2016.

Spittal

The prison camp of Spittal, established and operated by Organization Todt, had by 1 April 1945, 232 Soviet prisoners (Hesjedal and Andreassen 2015, 122). The camp itself is located on an old river surface north of the Kitdal River. The area is relatively wide and lies between the river to the south and the mountain foot of Raikiskaisa (1369 mas) to the north. Today, the area consists of deciduous forest, rough grazing and grassy heather. Although the camp is located in an area that is easily accessible, it is well hidden by natural "conservation". Registration of the camp in 2016 revealed almost 100 structures spread over an area of approximately 900 by 300 meters. Structures range from small firing positions, cooling pits and roads to larger dry-walled structures, plywood tents and cleared areas.

Enclosed in deciduous forest and grassy heather, the stone structures appear both unfamiliar and natural. Unfamiliar by virtue of their architecture, being beautiful dry-walled structures with stairs, entrances, and foundations that provide associations to prehistoric structures of unknown origins. Natural by virtue of their adaptation to the landscape and topography and through the use of natural stone, assembled with great care – reflecting both beauty and an immovable quality approved by the passing of time.

A stone staircase constructed in front of a plywood tent testifies to an adaptation to the environment and landscape, carefully designed and constructed. Even today, the staircase is steady and solid, and it feels almost natural to utilize this when entering the structure. At the same time, one knows that these appealing structures, simple things and stones, are also part of a prison camp. My first meeting with Spittal, a camp that was assumed to have been removed and covered with layers of agricultural land, disclosed a rubbish heap of barbed wire, oven residue, shoes, bottles and other objects infilling a creek bed between two fields. The large amount of barbed wire in the ground yielded an almost metallic sound when walking on the surface. An extraordinary material and bodily reminder of what was happening here in the cold and dark winter months of 1944–1945, and of the 232 prisoners who were imprisoned in Spittal.

At the bottom of the rubbish heap, almost completely hidden underneath barbed wire, are parts of a bicycle; only the head tube and parts of the fork are visible. A white emblem labeled

FIGURE 15.3 A mountain of barbed wire percolating the ground, Spittal, Austria, 2016

Photo: I. Figenschau.

with "*Vaterland. Fr. Herfeld Söhne. Neuenrade*" stands out among the rust-colored materials. Close by, two rusty but surprisingly intact bikes hanging from a tree make a metallic sound as they rattle in the wind. The leather on the seats is long gone, but the characteristic spring feathers are left. Someone has taken the trouble to hang them up, maybe in an attempt to preserve them or simply to make them visible.

Near one of the many dry-walled structures, we come across a gas mask container, *Tragebusche*, filled to the edge with cement. The container is located in an old streambed and is probably just the tip of a waste heap. This strange composite object is also the only indication of use of cement in the area. In combination with the climatic challenges here, the absence of cement may also be explained by the Wehrmacht's command from September 1944, ordering full stop of all cement supply (Sæveraas 2016, 114). In an area further east, in a large dry-walled structure, lies a lone German helmet, another M35, slightly overgrown with lichen and moss, camouflaged as a stone. Close by, at Myrhaug, a small ridge to the west of the main camp, is a lone red cabin – the only standing structure in the area. Ironically, apart from the tin roof, the cabin is built with old roof elements from the German plywood tents. Disassembled and put together again, the elements now form a structure never intended in their fabrication. By the entrance lies a large stone slab with "1945" cut into its surface – allegedly made by Austrian soldiers.

Both objects and structures are sometimes difficult to spot and have, through their decay, grown into the natural surroundings – perhaps even more than initially intended. Spittal, which, although located in a relatively open landscape, conceals a large number of structures and things. While several of the dry-walled structures are large, they allow themselves to be

FIGURE 15.4 A German helmet lingering in the leftovers of a barrack, Spittal, 2016

Photo: S. Farstadvoll.

camouflaged in the landscape. Even the largest structure, with a measurement of approximately 7 by 15 meters and foundation walls over 1 meter high, was well hidden between birch and grass.

Kitzbühel

Kitzbühel is located in a deserted mountain landscape 800 meters above sea level, approximately 5 kilometers southeast of Spittal. The camp is confined to an area of roughly 150 by 230 meters. In the camp itself, 32 structures of plywood tents are registered, 15 of which belong to the prison camp (Hesjedal and Andreassen 2015, 146–154). The camp consists mainly of three different concentrations with circular tent structures carefully organized in the landscape. The prison camp itself is located on a crest in the terrain, exposed to wind and weather and with views to the northwest and southeast. About 100 meters northwest, the soldiers' accommodation is divided into two concentrations. One is placed on a west-facing slope, while the other is located on a drained surface. Some of the structures clearly emerge in the sparse and low vegetation like circular ramparts or embedments. They also reveal their presence through, among other things, ovens, pipe hats, coke, barbed wire, roof panels, flat-tin buckets, spades, hooks, canned goods and footwear.

Building remains and objects lie scattered in and around the structures. Like in Spittal, they appear dissonant with the landscape while at the same time merging with the terrain in a process of natural surrender. An oil barrel cut in two, handily reconstructed as an oven,[1] is conspicuous in the landscape. A copper-clad wooden box with remains of ski bindings

FIGURE 15.5 Canned food, Kitzbühel, 2017

Photo: I. Figenschau.

and a gas tank has also been left in peace and gives an impression of something partly intact. In and around the many structures, there are small pieces and fragments of glass, porcelain, faience, window glass and bakelite that seem to have been intentionally crushed and broken. There is an overwhelming material presence in this windy landscape; large unused rolls of German barbed wire; ammunition and helmets; and intimate and personal things like gaming pieces, personal clothing, a fountain pen, a lighter and a nameplate with a street address in Berlin.

It is a strange feeling to walk among the structures, sensing an aura of "prehistoric heritage" but also, among the many remains of the familiar and everyday, a special and immediate force of recognition. As material reminders, they attest not only to how the war reached far into these remote regions, but also to how this landscape is still under siege, even though the soldiers and the prisoners are long gone. Caught in a process of decay that is both alluring and repulsive, their familiarity touches us in a personal and sometimes inconvenient manner (Burström 2015, 38). In the tension between exposure, ongoing decay and natural surrender, the war remains have acquired a very distinctive potential to trigger memories and recollection.

The question is, however, how can such broken and secluded objects still hold some form of intelligible memory? Ruination and decay are often linked to something negative – information lost, fragmented and corrupted. War remains may even be particularly fragile in this respect, as they are themselves also associated with something negative, with the human suffering that took place within their confines. Nevertheless, it is precisely this negative and disturbing element that also helps to trigger memories – memories and aspects of the war that have never been conveyed in historical narratives. This is especially evident in the prison camps in Norddalen, which have hitherto largely been forgotten and omitted from the national history of war.

The strength of all the rejected objects and structures is their unedited survival that reveals and remembers what was neither documented nor preserved, that which is too banal or unpleasant. However, all these things and structures have withdrawn from human influence, and they should not be perceived as merely passive mediators of a past completed and ended. In their ruined and rusty state, they convey their own lives of use and oblivion – a memory of functional pasts mingled with "dysfunctional" or alternative presences (Olsen and Pétursdóttir 2017, 93). Hence, the question remains, how do the things and the archaeological material relate to the historical sources? What can reports from surviving prisoners from Kitzbühel communicate?

Accounts from Kitzbühel

In June 1945, an international war crimes commission conducted interviews with Soviet prisoners who had survived captivity. Most of the interviewed prisoners were housed in a provisional field hospital or at nearby camps at this time. The participants from the commission were:

> Capt. P.H. Gallagher, US, legal representative for Tromso Civil Affairs detachment together with Mr Bjorn Waarhuus, statsadvocate of Troms and Finnmark, Mr Gabriel Lund, politimeister of Troms, Mr Welja, Russian representative, Alexander Elissejew, No. 32569 of 5 Norwegian Bn and Lieut Yens Odvar Hankoy of the Norwegian army.
>
> *(Lund and Waarhuus 1945, 63)*

The situation must have been complex, with people distributed at various camps, and language difficulties probably did not make the job easier. The following examples are based on interviews from prisoners at Kitzbühel.

Witness No. 18, Michail Karakosian. Born 18 October 1913.

> He said there was a grave at this Camp and that German soldiers had told him that 250 Russians had been buried in the snow. They were not buried in the ground because the snow was so deep that it was difficult to dig through it and to make a grave in the ground. He has never seen anyone buried there but saw a large cross indicating where the grave was supposed to be. If there were bodies in this alleged grave they were buried there by some other company. He heard that those buried there died from starvation, freezing and shooting. He said there was no fuel at this Camp. He also said that he saw one Feldswebel of Company No, 1 beat four or five PW but they were not killed. The above information was obtained on June 8. Then again on June 9 I talked to this PW and found out that he had been sent to this camp on March 27 and that he left there on the 18 April to go to MALLNITZ. … He said that this large grave was just outside of the barbed wire enclosure around the camp and that there was a road running between the Camp and the grave.
>
> *(Lund and Waarhuus 1945, 83a)*

Witness No. 24, Pawel Nuchtikow. Born 23 August 1920.

> He now states that the first camp he went to was one near the end of the aerial tramway which is probably camp KITZBUHEL. He claims he went there on the 12 March and left about April 6 for camp MALLNITZ staying there until April 28 when he was sent to KVESMENES. When asked about the conflicting dates given in the previously [sic] interview he said they were wrong and that after having time to think about it he was now giving me the correct information. … While he was there he saw KNOPFLE beat a PW to death with a stick, also that ARNOLD had killed several men of this camp by dropping heavy poles on them. … He has seen other Germans beat prisoners with gun butts and stick and on several occasions stab the PW with a bayonet. He stated that the men who had died or had been killed at this camp were taken away by mules and small horses and then buried at camp MALLNITZ. He had never seen anyone buried at this camp while he was there but heard that there was a grave in the snow which [sic] 300 bodies were buried. He said this grave was marked by a large cross and a large pole which was stuck up in the snow.
>
> *(Lund and Waarhuus 1945, 89)*

Witness No. 29, Ivan Kamarow. Born 5 July 1911.

> At KITDAL Camp No. 3 – the camp at the end of the aerial tramline known as KITZBUHEL – on the 9 Jan 1945, a Russian prisoner was shot by UFFs SCHAFER of the 3rd company of 204 Bn. Orders for this company were generally given by Capt ROLLBUSCH. The witness did not see this but he knew the prisoner was shot. ROLLBUSCH was reported to be a Gestapo member according to the German soldiers. This PW knows the ranks of the other persons who mistreated the PW but does not know their names. He could identify many of them.
>
> *(Lund and Waarhuus 1945, 94)*

Witness No. 30, Alexander Komarow. Born 30 Dec 1919.

> On Jan 9th, 1945 at No. 3 Camp at KITDAL a Russian prisoner was shot by UFFs SCHAFER because he had some documents in his possession. The Germans told the PW that SCHAFER had been given orders to do this by Oberfeldwebel OTTO and that he in turn had received his orders from the Camp Commandant of the 2nd company 204 Bn. This witness helped bury the PW that had been shot. At the end of Jan 1945 in the same camp UFFs SETSCH shot a Russian prisoner. The Capt (sic) gave the order. ... This man was shot because he was ordered to lift a large stone which he could not because he was so weak from undernourishment. The PW complained about the food and Oberfeldwebel OTTO heard about it and took this information to the Camp Commandant and in about four hours SETSCH was ordered to shoot the prisoner. This witness helped bury the victim close to the victim mentioned above. Another death he knows about occurred in the middle of February 1945 at a spot between the two camps up from MALLNITZ. The men were carrying wood and one man got behind. He asked the one in charge if he could leave the wood. This witness also asked if the weak man could leave the wood there. He was told to go on and when about 60 metres (sic) away heard a shot. When the one in charge came up this witness asked why he shot the man and was told that the man would not carry the wood. The man who did the shooting is a German soldier and is now (*12th of June 1945) stationed at OVERGARD hospital. His family name is something like "IHME", he is quite tall and has a little moustache. The men doing the shooting in all three cases were from the 2nd or 3rd companies of 204 Bn. He states that the men were very poorly fed and all suffered from under-nourishment.
>
> *(Lund and Waarhuus 1945, 95)*

(Sm)all things remembered

Although the interviews with the survivors of Kitzbühel tell a specific story, about named persons, about specific dates, war crimes and an inconceivable ideology, they are largely guided by specific human aspects, aspects that try to order memories, events and actions in a logical and linear sequence – memories that will make sense in a historical narrative. The War Crimes Commission highlights the problem they encountered, shortly after the interviews were conducted:

> The information contained in these statements was all obtained through interpreters and in many instances is confusing and not in the best order. ... It was often difficult to make the witnesses answer a particular question and this accounts for the variety of matters mentioned, in no regular order.
>
> *(Lund and Waarhuus 1945, 63)*

In addition, Captain Gallagher comments: "Also the Russians are liable to make changes in their statements, if they are to repeat them, as they frequently did before" (Gallagher 1945a, 135).

One should neither ridicule the challenges facing the committee nor undermine the importance of historical documents; they contain important information about actual circumstances and war crimes. They also contain information that things cannot provide. However, there is an important feature of the interview texts and which appears to characterize the memories brought forward, which gives rise to reflection. Although the interviews to some extent follow specific questions, the answers are often marked by unrelated events, sometimes confusing and nonlinear, and sometimes they do not correspond to the questions at all. In many ways, this is a feature of the archaeological record as well: There is striking similarity between the recollection pattern presented in the interviews and how the archaeological material in the prison camps meets us. Both are fragmented, sometimes random and not always understandable (cf. Olsen and Pétursdóttir 2017, 91). Moreover, as Gallagher points out, the prisoners' statements tend to constantly change, as if they are not subject to any kind of logical, coherent or settled story – as if caught in a process of fragmentation and decay, they break and oppose the linear structure that characterizes historical narratives.

What good does it do to carry out archaeological investigations of war remains, modern ruins that are so close in time, and which are even documented in interviews with surviving witnesses? Several books about the war and the fate of prisoners have been written (e.g Figenschau 2014; Fjermeros 2013; Soleim 2009; Soleim 2016; Stokke 2008). To a large extent, we know what happened in the camps and how the German army mistreated, starved and executed Soviet prisoners during the construction of the Lyngen line. We know when the war started and when it ended – so what can archeology contribute to?

Being present in these camps, wandering between the remains of plywood tents, ovens, shoes, barbed wire, broken glass and porcelain – all these fragile things – made me realize that the war never really left us. Although the Austrian soldiers eventually returned home, although the Soviet prisoners Karakosian, Nuchtikow, Kamarow and Komarow fortunately survived Kitzbühel, the material traces of the prison camps are a viscous reminder of the war's continued presence.

Although the interviews appear somewhat fragmented, they are, by virtue of their purpose, directed by an effort to uncover war crimes and thus guided by questions focusing on the prisoners' daily struggle to survive hunger, ill-treatment and illness. The material memories, on the other hand, do not have an agenda, political, ethical or other. In fact, many of the archaeological artifacts that emerged during our investigations are remnants of the 2nd and 3rd company, 204 Battalion – traces of Knepfle, Arnold, Schäfer, Rollbusch, Otto and Setsch. Regardless of how unpleasant, wrong and irreverent the conducts of the Third Reich's Austrian soldiers, the material memory does not suppress their extended presence.

Moreover, material remains of the soldiers' life, be it cloth hangers, helmets, porcelain, checkers pieces or bottle glass, show the unsettling proximity between murder and mistreatment of detainees and the mundane, almost banal dimension of camp life – a proximity which is not conveyed through historical narratives. The material remains have no censorship or political correctness to them, no labeling of "dark", "negative" or even "evil". Rather, they show the plain presence of soldiers; playing board games by the light of a trench candle, consuming alcohol to ease the nagging distress and create distance from the war, from homesickness and war-weariness.

Hence, the afterlife of these things expresses a material resistance to the order of history and the segmentation of spheres (Olsen 2013, 204). And, perhaps, through their fragmented and messy state, as scattered and weathered bits without clear and continuous narratives, they have more resemblance to memory – and archeology – than to the linearity of historicism (Olivier 2011, 95–96). What archaeologists contribute with is therefore a different story, an angle that takes things seriously even in their fragmented state. Their nature may not make them easily accepted as cultural heritage, but it is precisely because of being trapped between modern waste and history that they create a tension that triggers critical and involuntary memories (Olsen 2013, 205).

Of course, the object matter tells us about human suffering, war and captivity, but at the same time, it takes us behind the façade, behind historical narratives. By equally embracing the banal aspects of the war, cake platters, gaming pieces, cups and personal belongings, other aspects of human presence are revealed. Not only are unexpected and involuntary memories evoked on encounter with, for example, a plate or a coffee cup from Porsgrund Porselen in a prison camp 800 meters above sea level, but they also point to different scenarios and other conditions. It is precisely these everyday and rather trivial objects that can take us behind the façade, into the almost inappropriate, the embarrassing and that which does not find the way into either historical accounts or interviews.

All these objects have endured through an ecological conservation whose corporal progression enables the material to "articulate" this specific memory of past life. Arguably, archaeology may be seen as a continuation of such memory practice, where survey, fieldwork, interpretation, post-excavation work and, certainly, this very chapter emphasize and re-member certain objects and past events. In this vein, González-Ruibal (2014, 370) describes archaeologists as postwitnesses and hyperwitnesses: "As archaeologists, we do not document the events as they

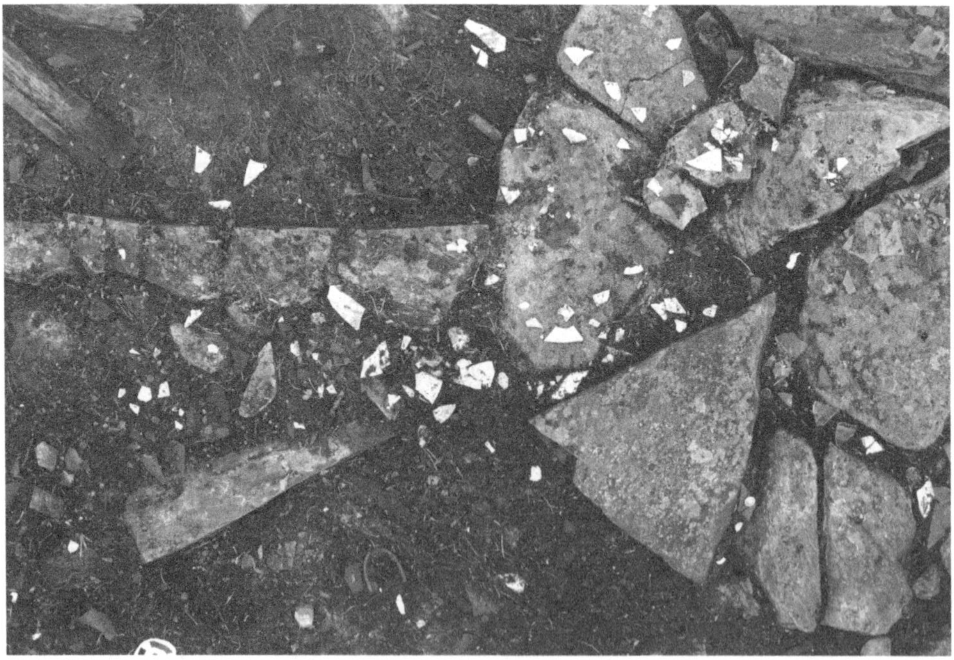

FIGURE 15.6 Kitzbühel, 2016: excavation work. Leftovers from the Austrian soldiers – porcelain from cake platters, bakelite, a fountain pen, buttons, playing pieces, a candle and a heel plate made of wrought iron

Photo: I. Figenschau.

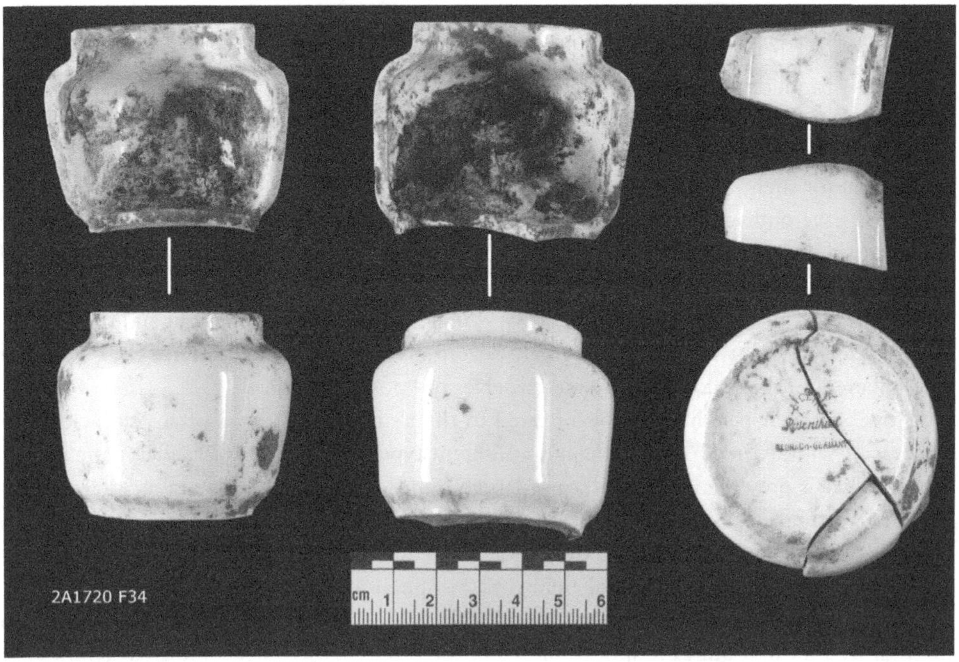

FIGURE 15.7 Post-excavation work: new care and new reminiscence. A single jar from Rosenthal with some of the original ointment still present. Spittal, 2016

Photo: I. Figenschau.

happen, but we become witnesses of a kind when we dig them up", and through engagement with objects, we gain a closer insight into passed events.

Thus, through engagement with the varied legacies of this remote war, we are dealing with several different types of memory. We have historical narratives based on a linear framework with logical and coherent representations. We have the interviews with the prisoners, which contrast such logic through being incoherent, nonlinear and fragmented and thus closely resembling the material remains. Furthermore, a contemporary archaeological survey activates yet another type of memory media which illuminates and increases the visibility of objects, including contextual finds in the field, photo documentation, 3D modeling and photogrammetry. Through post-excavation work, things are subjected to new forms of care and an interest that differs from intentional use-driven attention. Practices such as cleaning, measuring, describing, photographing, identifying and contextualizing objects are also a form of memory ecology which renders the objects proximate through various technologies and thus also presencing the past. It is through this process that we become post- and hyperwitnesses, creating new connections and new reminiscences: "We see too late, but we see more" (González-Ruibal 2014, 369–370).

Present things, past memories

Much of memory studies has so far been characterized by an intangible angle that promotes past memories as a consequence of cultural mediation with emphasis on textualization and visualization (Tamm 2013, 461), where memory is first and foremost viewed as a deliberate human process (Olsen 2010, 109). However, even if one has included a potential material dimension in memory studies through the introduction of concepts such as mnemohistory, presentism and afterlife/survival (*Nachleben*) (Tamm 2015, 9), there is little account of the active role played by things in this form of memory (Olsen and Pétursdóttir 2016, 40).

Crucial for such remembrance is the persistent quality of things; many of them are stubborn and have outlived their human counterparts. In addition, through this persistency, they have left their urgent role of being useful, to continue life in an uncensored material world unaffected by historically adapted narratives or ideologies. It is precisely through this democratic presence, through ruin and decay, that things appear in a different light and thus generate or enable alternative memories. Largely devoid of human influence and the urge for order and coherency, things do not correct themselves or "post-rationalize".

However, their peculiarity is also related to their survival, which is often "coincidental" and varies with degrees of weathering in different types of soil and external climatic conditions. What is preserved is in that sense largely random, and this happenstance factor, this peculiarity, thus creates archaeological records beyond our control. A selective recollection that we do not have the ability to override (Olsen and Pétursdóttir 2017, 93–94). This was very evident in the archaeological fieldwork at Spittal and Kitzbühel, where the aim was to try to capture an even representation of both soldiers and prisoners. Nevertheless, even though we followed our assumptions of "what was where", the objects still amassed through an obstinate and almost unruly presence in some places while gleaming only through their absence in others. The result was a great number of objects which spoke mostly of the presence of the soldiers and which in many ways reflected the random and uncensored characteristics of things.

The way things remember can be surprising and unpredictable, as in the case of the small red cabin in Spittal. The cabin was built after the war and, at first glance, also resembles any other timber cabin with a tinned roof. However, during my investigations, the war soon turned out to be an inherent part of it; the floor, the ceiling and the walls were all constructed of triangular roof units from the occupation force's plywood tents. The tents were assembled from such prefabricated elements, and due to their usefulness, these Finnish construction kits were produced in massive numbers during World War II (Pool 2016, 177; US-Army 1943, 75–79).

Here, however, the usefulness and affordance of the elements even exceeded their intended use and found – in their afterlife – new and unexpected applications. Thus, the only standing structure in the prison camp and the element that seemingly had nothing to do with the war had nevertheless prolonged its presence. The cabin was initially constructed to keep shelter during the harvest – something that was also dexterously illustrated on its walls in form of small notes and tallies for the number of hay loads, dated to 1962. In a way, the cabin thus represents the post-war normalization, where the reconstruction and re-cultivation of the area was in focus. In this process, there was also widespread reuse of military equipment, something the cabin is a good example of. The prefabricated elements have acted as practical and easily available material for building a much-needed cabin. The contrast between these complete, painted and wallpapered elements and the remains that lay spread in Kitzbühel, as rotten timber, rivets and nails, was striking. In the cabin, they were floors, walls and ceilings – they had occupied new layouts and new shapes that held a geometric attraction and aesthetics that made them come across almost as decorative elements. Thus, in its construction, an unexpected and involuntary memory has also been triggered, both in relation to a sort of

FIGURE 15.8 The red cabin, Spittal, 2017. Recycled recollections evading the fictional and linear pace of history

Photo: S. Farstadvoll.

appropriation and "conquest" of German material but equally in juxtaposition to the material still spread throughout these landscapes.

Endless endings: facing things, encountering memories

Just about all these things have participated in shaping and building these places and in the fight to survive, shelter and keep warm in these crude environs. They have witnessed many events, alarming, frightening, uninteresting and banal. But there is an ambiguity, anonymity and secrecy involved in the meeting with all these things. We will never know who chose to bring a delicate cake platter and cup from Porsgrund Porselen to this rough and remote region, who used the helmets or who designed and built the oven from an old oil barrel. These human relations have been repealed, but what is left in return is a direct and very intimate memory; a memory created in the encounter with the things themselves (Olsen and Witmore 2014, 187).

All these fragmented objects from the prison camps hold and trigger memory, though not like a passive translation of past meanings. They remain in their very own form – not complete, restored, pre-interpreted and purified but, on the contrary, in an ongoing process of change that allows us to recall their lives through use, adaptation, discard, weathering and forgetfulness. Moreover, far from shiny glass cases, illuminated cabinets, wordy accounts and written stories, this lichen-grown, overrun and crumbled heritage holds its very own nearness and intimacy. Not only does it show that the war is still present but also that it is so in a very direct and mundane way. These objects have a very direct and honest connection with the war regardless of nationality, political standpoint or pre-interpreted and fixed timeframes. They just *are*.

Note

1 See https://bit.ly/2Oh3IY3 for a 3D model of a barrel reconstructed as an oven.

References

Arendt, H. 1976. *The Origins of Totalitarianism*. New York: Houghton Mifflin Harcourt Publishing Company.

Burström, M. 2015. *Skärvor av 1900-tal. Arkeologiska essäer*. Lund: Nordic Academic Press.

Drageset, A., Haugen, A. M. and Hesjedal, A. 2010 Arkeologiske undersøkelser i 2010 i forbindelse med Kv420 kraftlinje Balsfjord Hammerfest. Registreringer langs traseen i Troms: Balsfjord, Storfjord, Gaivuotna/Kåfjord, Nordreisa og Kvænangen kommuner. *Befaringsrapport Kulturetaten – 2010*. Troms fylkeskommune.

Eek, M. 2017. Personal communication, 29 November.

Figenschau, T. 2014. *Krigen i Storfjord. Mord, mytteri, krigsforbrytelser*. Signaldalen: Tekst and Cetera.

Fjermeros, H. 2013. *Med lik i lasten. Operasjon Asfalt – de sovjetiske massegravenes skjebne i Norge*. Oslo: Scandinavian Academic Press.

Fossum, F. 2014. "Den tyske Lyngenlinja." In *Krigen i Storfjord. Mord, mytteri, krigsforbrytelser*, edited by T. Figenschau, 43–60. Signaldalen: Tekst & Cetera.

Gallagher, P. H. 1945a. *Germans Accused of War Crimes*. Not published.

Gallagher, P. H. 1945b. *Skibotn Investigation* (Kvesmenes Area). Not published.

González-Ruibal, A. 2014 "Returning to Where We Have Never Been. Excavating the Ruins of Modernity." In *Ruin Memories. Materiality, Aesthetics and the Archaeology of the Recent Past*, edited by B. Olsen and Þ. Pétursdóttir, 367–389. London: Routledge.

Gyllenhaal, L. and Gebhardt, J. F. 2001. *Slaget om Nordkalotten*. Lund: Historiska Media.

Hesjedal, A. and Andreassen, D.-M. 2015. Kulturminneundersøkelser i 2015, Krigsminnelandskap Troms. Registreringer langs Lyngenlinja i Storfjord kommune. Upublisert rapport. *Befaringsrapport Kulturetaten – 2015*. Troms fylkeskommune.

Isachsen, N. R. 2016. *Min krigsdagbok*, edited by F. Fagertun and R. Rautio. Stamsund: Orkana forlag.

Jaklin, A. 2006. *Nordfronten. Hitlers skjebneområde*. Trondheim: Gyldendal.

Jenney, E. R. 2014 [1945]. "Undersøkelser av krigsforbrytelser i Skibotn-og Kvesmenes-området. Utdrag av uformell rapport av E. Ross Jenney. Fra The National Archives, London." In *Krigen i Storfjord. Mord, mytteri, krigsforbrytelser*, edited by T. Figenschau, 144–155. Signaldalen: Tekst & Cetera.

Lund, G. 2014. "Tyske krigsforbrytelser i Storfjord: Dødsleiren i Kitdalen." In *Krigen i Storfjord. Mord, mytteri, krigsforbrytelser*, edited by T. Figenschau, 70–84. Signaldalen: Tekst & Cetera.

Lund, G. and Waarhuus, B. 1945. *Investigation of War Crimes by Germans Against Russian PW at Camp Mallnitz, Kitdal in Storfjord – Skibotn*. Not published.

Olivier, L. 2011. "The Dark Abyss of Time. Archaeology and Memory." In *Archaeology in Society*, edited by I. Hodder, R. W. Preucel and L. Silverman. London: Rowman & Littlefield.

Olsen, B. 2010. "In Defense of Things." In *Archaeology in Society*, edited by I. Hodder and R. W. Preucel. Lanham, MD: AltaMira Press.

Olsen, B. 2013. "Memory." In *The Oxford Handbook of the Archaeology of the Contemporary World*, edited by P. Graves-Brown, R. Harrison and A. Piccini, 204–218. Oxford: Oxford University Press.

Olsen, B. and Pétursdóttir, Þ. 2016. "Unruly Heritage: Tracing Legacies in the Anthropocene." *Arkæologisk Forum* 35: 38–45.

Olsen, B. and Pétursdóttir, Þ. 2017. "Sværholt: Poetics of Memory." In *Inside and Beside the Camp*, edited by J. Kaila and J. Knuutila, 88–95. Helsinki: The Academy of Fine Arts at the University of the Arts Helsinki.

Olsen, B. and Witmore, C. 2014. "Sværholt. Recovered Memories from a POW Camp in the Far North." In *Ruin Memories: Materialities, Aesthetics and the Archaeology of the Recent Past*, edited by B. Olsen and Þ. Pétursdóttir, 162–190. London: Routledge.

Pool, J. L. 2016. *Ruhetag: The Day-to-Day Life of the German Soldier in WWII: Volume II: Morale and Welfare*. Atglen: Schiffer Publishing, Ltd.

Sæveraas, T. E. 2016. *"Beton macht Geschichte". Organisation Todt og utbyggingen av Festung Norwegen*. Ph.d.-avhandling i historie. Trondheim: Norges teknisk-naturvitenskapelige universitet.

Sáiz, A. 2008. *Deutsche Soldaten*. Madrid: Andrea Press.

Soleim, M. N. 2009. *Sovjetiske krigsfanger i Norge 1941–1945. Antall, organisering og repatriering*. Oslo: Scandinavian Academic Press/Spartacus.

Soleim, M. N. 2016. *"Operasjon asfalt". Kald krig om krigsgraver*. Stamsund: Orkana Akademisk.

Stokke, M. 2008. *Sovjetiske og franske tvangsarbeidere i Norge 1942–1945. En sammenligning av arbeids-og leveforhold*. Mastergradsoppgave i historie. Bergen: Universitetet i Bergen.

Sunde, H. I. 2014. "Tyskernes Lyngenstilling. En del av krigshistorien i Norge." In *Krigen i Storfjord. Mord, mytteri, krigsforbrytelser*, edited by T. Figenschau, 61–69. Signaldalen: Tekst & Cetera.

Tamm, M. 2013. "Beyond History and Memory: New Perspectives in Memory Studies." *History Compass* 11(6): 458–473.

Tamm, M. 2015. "Introduction: Afterlife of Events. Perspectives on Mnemohistory." In *Afterlife of Events: Perspectives on Mnemohistory*, edited by M. Tamm, 1–23. Palgrave Macmillan Memory Studies. London: Palgrave Macmillan.

US-Army. 1943. *German Winter Warfare*. Special Series, No. 18, Mid 461. Washington, DC: Military Intelligence Division, War Department.

Westrheim, H. 1978. *Landet de brente. Tvangsevakueringen av Finnmark og Nord-Troms høsten 1944*. Oslo: Nordnorsk Forfatterlag – Tiden Norsk Forlag.

PART V
Entanglements

16

A POSITIVE PASSIVITY

Entropy and ecology in the ruins

Caitlin DeSilvey

Opening

In this chapter, I explore two places that have staged experiments in partial, intentional ruination. In both places, there has been some attempt to balance accommodation of gradual structural decay with a level of intervention required to sustain selective reuse and managed public access. At one site, this experiment has been running for over two decades – long enough to allow for some reflections on whether its original intentions have been realized. At the other site, intentions are largely conceptual, captured in documents and declarations of intent, though the shape of its future orientation is discernible in early actions to manage the site. These places are linked through their adoption of shared principles of openness, iteration and incompletion. They are also both experimental sites – not in the sense that they have constructed a space where they control specific processes in order test hypothetical outcomes but in an alternative sense, recently articulated by Jamie Lorimer and Clemens Driessen, that understands experiment as a "tentative procedure; a method, system of things, or course of action, adopted in uncertainty" (Lorimer and Driessen 2014, 170).

I draw on Georg Simmel's essay, "The Ruin" (1911), to frame my argument, reflecting on his analysis of the ruin as a space of tension and negotiation. Simmel discusses how ruins can be approached with a kind of active indifference, which accommodates decay and accepts vegetative and faunal colonization. In such an orientation, which Simmel describes as a "positive passivity", people become, in effect, ecological accomplices; through their curation of the ruin, they facilitate the expression of inherent tendencies of disintegration, regeneration and entropic unmaking. The sites I explore in this chapter are both radically open to the workings of entropic possibility – if entropy is understood not as a state of increasing disorder but as a measure of potential and multiplicity (DeSilvey 2017, 10). As becomes clear in the telling, however, the embrace of entropy often slips in the translation from principle to practice.

Accommodating entropy

Duisburg Nord Landschaftpark occupies a 570-acre site in Germany's Ruhr region in the far west of the country. The Ruhrgebiet, as it is known, is an area associated with intensive twentieth-century industrial production and, more recently, bold initiatives to recuperate and reinvent the extensively altered landscape as Industriekultur (usually translated as "industrial heritage") (Lange 2004). Iron production at Duisburg began in 1903 and ceased in 1985. After closure, the site was the subject of an international design competition (as part of the Emscher Park International Building Exhibition) and reopened to the public in 1994 as a landscape park, designed by architects Latz + Partner. In this most recent incarnation, Duisburg has been the subject of sustained international interest, primarily focused on the way the site's design incorporates an unprecedented openness to reuse and regeneration (both social and ecological). The other site under consideration in this chapter is a derelict Victorian estate (Kilmahew) surrounding an iconic modernist seminary (St. Peter's) 20 miles west of Glasgow, Scotland. The seminary was also abandoned in the mid-1980s, but its subsequent recycling and repurposing has been slower to take hold. In 2008, public arts organization NVA (an acronym for nacionale vitae activa, a Latin term meaning "the right to influence public affairs") began to explore options for the site's future, and in 2010, they were invited to host a debate at the Twelfth Venice Architecture Biennale. They invited Tilman Latz, a current partner in the firm that designed Duisburg Nord, to take part in the conversation. By aligning themselves with the innovative experimental practices that had been trialed at Duisburg, NVA made clear its intention to adopt strategies for the Scottish site that also privileged indeterminacy, reinvention and iterative participation. In a statement of intent published after the Biennale, NVA asserted that their plan for Kilmahew/St. Peter's would "embrace an incremental process of change" that "supports the principle of an unfinished work and accepts a level of entropy" (NVA 2011, 22).

What does it mean, in practice, to "accept a level of entropy"? In such a statement, entropy is invoked in the abstract, as a force capable of catalyzing an indiscriminate unmaking of material substance. In order to perform, however, entropy must work through a series of agents, and in a built structure, these agents come in many animal, vegetable and chemical forms. Wildlife, weeds and water initiate the transformative "micro-processes of wear and tear" that ultimately produce an entropic effect: oxidation, biodeterioration, percolation, cryoperturbation, erosion (Löfgren 2005, 53). In a counterpreservation heritage practice, any attempt to find a balance between continued process and planned use must negotiate with these agents. As will become clear in this chapter, how people carry out this negotiation – and their willingness to accept the inherent unpredictability that entropy engenders – exposes a tension between intended conceptual innovation and its realization in practical applications.

Studies of ruins and ruination (both historic and contemporary) have been concerned in various ways to call attention to the way plants generate particular aesthetic effects as they take hold in disused structures, softening the hard lines of architecture and contributing to its "pleasing decay" (Piper 1948). In such writings, plants are most often taken to be part of a generic collective of greenery, bundled into an aggregate identity. A similar lack of specificity also emerges in applied contexts when, for example, landscape architects or ecologists discuss the need for "vegetation management". The plants being managed (or, most often, removed) are rarely distinguished from one another, unless it is to demonize particular species as undesired invaders and thus further justify the need for eradication. Research on human–plant geographies has begun to explore this tension between collectivity and individuality, as well as the cultural reluctance to assign agency and sentience to individual plants (Head and Atchison 2009; Atchison and Head 2013; Pitt 2014). This work has clear relevance in the historic environment context, where attitudes toward plants are characterized by ambivalence, in turn appreciative (of their aesthetic contribution) and apprehensive (about their role in accelerating deterioration).

The two sites explored in this chapter express subtly different attitudes toward the plant species that share their sites. At Duisburg Nord, plants appear to be treated as individuals, with their own complex ecocultural histories and their own specific desires and demands. At the Scottish site, there is evidence of a more anxious and unresolved response. At base, as Michael Pollan explores in his book *Second Nature*, the story that people tell about the plants they share their space with is a story about control, about different levels of willingness to accept vegetative autonomy (Pollan 1991; Hitchings 2006). Attitudes toward herbal others can be placed on a continuum that runs from cultivation and containment to a looser form of collaboration and coexistence.

Duisburg Nord Landschaftpark

In articulating his original vision for Duisburg, Peter Latz identifies two central themes. First, he was concerned to allow the site to express its "physical nature" by accommodating ongoing "natural physical processes" such as erosion, oxidation and spontaneous succession and by creating new elements also subject to the working of such processes. The second theme he expresses as "utilisation", in that he sought to allow for the "metamorphosis of industrial structures without destroying them" (Latz 2001, 151). Many areas were to be "left to develop on their own without intensive treatment" (ibid, 153) including zones of persistent contamination, where soils and slag heaps were laced with deposits of heavy metals and hydrocarbons. In other areas (including a complex of former "sinter bunkers"), contaminated sediments were capped and sealed, and the new surface was planted with ornamental gardens. His approach sought not restoration (of the landscape to an imagined state of prior ecological equilibrium) but reconciliation between the damaged landscape and unspecified future uses. Latz states,

> The new vision should not be one of "re-cultivation," for this approach negates the qualities they currently possess and destroys them for a second time. The vision for a new landscape should seek its justification exactly within the existing forms of demolition and exhaustion.
>
> *(ibid, 158)*

Rather than extinguishing the site's "aberrant processes" (Rosenberg 2009, 216), Latz sought to work with them, to allow the discourse between old industrial forms and emergent natural process to go about, in his words, "creating values between art and nature in a way that could never be made by the artist or nature alone" (2001, 159).

In this assertion, one can hear the echo of Georg Simmel, whose 1911 essay "The Ruin" identified a tension between form and process, or "nature and spirit", in the ruined structure. Architecture, he argued, is a sublimation of the inherent properties of matter – the durability of stone, the rigidity of metal – to serve the purpose of design and function. But in the ruin, when a building is subject to forces of "weathering, erosion, faulting … and the growth of vegetation", there is a moment when a precarious aesthetic balance is achieved: "Out of what art still lives in the ruin, and what of nature already lives in it, there has emerged a new whole, a characteristic unity"(Simmel 1959 [1911], 260). Something like this "new whole", which emerges out of a relation between extinguished human function and resurgent natural process, is hinted at in Peter Latz's evocation of collaboration between the artist and nature at Duisburg (though Latz's approach arguably abandons Simmel's humanist hierarchical model to adopt a more radical symmetrical model of agency) (Hetherington 2010, 20). Russell Hitchings picks up on similar themes in his proposal that we shift from a focus on "material culture" to a study of "cultured materials", which examines the processes through which people impose their designs on physical matter and the ways in which that matter works back on these designs to express underlying "individual physical propensities" in unpredictable ways (Hitchings 2006, 368).

I visited Duisburg Nord in 2012 because I was curious to see for myself how Latz's original vision had been realized. I was particularly drawn to the site's declared and deliberate accommodation of transformative process, an approach that seemed to unhitch the work of cultural memory from an insistence on the maintenance of durable objects. The background reading I'd done in preparation for my visit had seeded a certain skepticism about what I would find there. Most academic work seemed to hinge on analysis of the site as a study in polarized states: "intervention and neglect" (Rosenberg 2009), "spontaneous and designed nature" (Keulartz 2013), "devastation and ecological reclamation" (Barndt 2010a). Kerstin Barndt had written critically of the intentional dialectical tension that Duisburg Nord cultivates, observing, "The organic appeal of this dialectical move ... smooths out the rough edges of the processes, and the overall sense is one of theatricality, in which visitors act out unscripted parts" (ibid, 281). I was led to believe that I would find a place in which process was mostly patina and where the promise of ongoing metamorphosis and dissolution was diluted through the prosaic imperatives imposed by public use of the site. In a sense, this is what I found – but I also found something else, which fits uneasily with this critical take.

At the entrance to the Duisburg Nord Landschaftpark, a row of interpretive signs marched alongside the path, steel slabs over six feet high, printed with maps and text in both German and English. One of them offered this passage by way of introduction:

> The "Sacro Bosco" (sacred grove) in the medieval Italian town of Bomarzo is a renaissance park containing statues of fantastic and fabulous animals, surrealist figures, giants and monsters. Similar associations were the inspiration for the designer Peter Latz in his work on the North Duisburg Landscape Park. The disused ironworks used to spit out huge gobbets of fire into the night sky, but the giant has now been tamed and lies exhausted. The site was formerly a no-go area for the general public, but its gates have long been thrown open to visitors. Rusty red blast furnaces and pitch-black storage towers, a labyrinth of pipelines and bunkers for coal, ore and other materials have been transformed into an exciting leisure area with attractive gardens and open viewing areas.

The areas of the park nearest to the entrance were almost completely dedicated to the repurposing of the site as an "exciting leisure area". Couples canoodled on benches in the open plaza between the hulks of the two blast furnaces; an area between the gasometer (now a diving facility) and the former casting house (a cinema and concert venue) was set up as beer garden with palm-frond kiosks, a white sand floor and an inflatable Pilsner bottle. Around the corner, I came across the *Piazza Metallica*, a grid of iron plates installed as part of the original park design and intended to express "erosion by natural physical processes" (though the space is now also used for temporary toilets and Dumpsters, which tempers its aesthetic unity somewhat) (Latz 2001, 151).

Slightly farther from the center, the uses of the site became more loose and improvisational: teenagers roamed in packs and left their traces (love locks, beer bottles) behind in odd corners and ostensibly off-limits areas. Climbers scaled the walls of the pitted concrete bunkers, and artists daubed en plein air portraits of tangled steel and brooding machinery. The mass of one blast furnace was open to allow the public to wander up through its oxidized innards to a viewing platform, 70 meters above the ground, past various interpretive panels (which few people seemed to be reading). The view from the platform was instructive; the whole horizon to the west, along the Rhine Valley, was crowded with belching smokestacks and massed industrial buildings, obviously still active and productive. In this wider context, the park came into focus as anomalous and premature, promising to heal the scars left by industry with a soothing layer of greenery and an array of diverting leisure activities. From this elevated vantage point, the dominant impression of the park itself was of a place unevenly consumed by a creeping blanket of leaves and vines, which seemed to be encroaching from the outer edges inward toward the relatively clear zone immediately surrounding the blast furnaces and the repurposed structures.

As I descended the blast furnace, I noted areas of rust and regrowth: patches of moss, tiny willow saplings rooted in pipes, flat roofs furred with green. At the base of the buildings, there was evidence of more active incursion – though primarily in inaccessible areas behind temporary fencing, where thickets of buddleia had sprung up and brambles were casting out straggling, speculative arms in search of new territory. Where the plants reached through the fences, they had often been pruned back. On one fence, a section of buddleia trunk as thick as my wrist had fused into the grid of steel, and although it had been cut off at either end, the desiccated remnant held on, witness to an attempt to control and contain. Other obvious efforts had been made to check growth and regeneration. Rotting stumps resprouted with new shoots, and some patches of ground seemed to be artificially suspended in a stage of early successional regrowth. But slightly farther from the core, the situation was more ambiguous. Clumps of ragwort were firmly rooted in cracks in a concrete wall above a well-traveled path. Ivy fingered its tough roots into ledges and loose sections of masonry, and saplings thrived along the tracks of an elevated railway. It was not entirely clear how much management was taking place, or even if anyone was paying much attention to the gradual erosion of structural integrity – and the attendant risks to public safety. There was not a total abdication of control here but something more uncertain.

The ambiguity was enhanced by the contrast between the unruly edge lands and the designed garden spaces. In one bunker garden, serried rows of lavender held in tight formation, and another was carefully planted with roses. A third contained a clipped boxwood hedge trimmed into wavy stripes, interspersed with rows of hydrangea. These gardens seemed to be erratically tended, however, and the hydrangea was rangy and unkempt, flopping over the box hedge and obscuring the clean lines of the design. In an adjacent bunker compartment, an apparently self-seeded thicket of pear and alder rose over the concrete walls. Certain species seemed to emphasize the blurry boundary between cultivation and chaos most acutely. Across the site, silver birch trees grew in incongruous places – in inch-wide gaps between concrete blocks, along elevated rail lines, clinging to narrow ledges. But they had also seeded in groves in the interstitial open spaces on the site, where they appeared to have been allowed to mature without any interference – for the most part. In one of the sunken bunkers, a gap in the wall revealed a glimpse of a spiral of short birch logs, laid onto the dipped surface of

FIGURE 16.1 *A section of buddleia trunk as thick as my wrist had fused into the grid of steel*

the ground and slowly rotting. I later learned that the birch had been cut down in the early "vegetation management" of the derelict rail lines, and rather than being chipped or disposed of off-site, the wood had been repurposed to provide an object lesson in decay and regeneration, intention and intervention.

♦

FIGURE 16.2 *Silver birch trees grew in incongruous places*

As I moved farther from the center of the site, I came across another strange zone, this one enclosed within a high brick wall. Inside, in the midst of a remnant orchard, a grid of squared beds and paths had been laid out, separated by sections of steel that resembled railroad ties. Each squared section had been filled with salvaged material composites – construction rubble, gravel, used bricks, broken tiles. The plants growing in these scrap substrates were familiar, opportunistic waste ground species: dock, bramble, willowherb, mullein. Two giant burdock plants leaned from their enclosures across a path to meet in the middle. But adjacent to these odd compartments, there were also cultivated flower beds, thick with asters, daisies and ornamental grasses.

I realized that I'd read about this space in a chapter by Elissa Rosenberg, where she describes these beds as "research plots": "True experimentation is defined by its open-ended quality, and the unpredictable results that often challenge the initial hypothesis. Here, the aesthetic of experimentation assumes, as its starting point, the dynamic flux of natural process" (2009, 216). Rosenberg points out that the garden's apparent embrace of disturbance and flux signals an intellectual alignment with postequilibrium ecological theory, in which "succession is a highly probabilistic and contingent process", not a steady progression to a stable state. As with other areas on the site, however, some intervention seemed to be required. The test plots were suspended in a state of calculated barrenness, which continually rewound the successional clock to the point at which the colonization of the industrial waste by volunteer plant species would be most apparent. Given a few years of nonintervention, the entire walled garden would have been an impenetrable thicket of burdock and bramble.

As I moved back toward the bulky steel and concrete forms of the bunkers, furnaces and rail lines, I found myself puzzling over how to integrate my experience of the apparently (if unevenly) cultivated spaces into my interpretation of the management of the site's built structures and my interest in the ways in which they were, or were not, being allowed to actively decay and disintegrate. I realized that part of the trouble came with the distinction I was used to making between architecture (as discrete and bounded) and environment (as dispersed and distributed). This site was operating through a more porous principle. I recalled a comment made by Tilman Latz in the publication that documented NVA's Venice debate:

> We should maybe overcome the notion of a building, as this always asks for function. We could rather think of it as a landscape, an open system that can transform over time and assimilate as many functions and interpretations as possible. Then a strategy of "controlled decay and growth" is maybe applicable.
>
> *(2011, 67)*

A further clue to the site's underlying philosophy was revealed when I came across an unfamiliar term on an interpretation board near a patchy meadow, in German with no English translation: *Industrienatur.* I returned to the row of translated signs flanking the main entrance and found the following definition:

> The term "industrial nature" seems at first to be self-contradictory. The ravages of heavy industrial production utterly changed much of the natural landscape in the Ruhrgebeit. Fields, meadows and farmland were replaced by colleries, coal tips, steelworks and waste tips. But even during the era of industrial production a few plant and animal species managed to take root here and there. When the industrial sites became redundant nature swiftly repossessed the derelict areas. We have given the name "industrial nature" to the particular mixture of plant and animal species which have managed to adapt to what were in places very arduous living conditions. Here you can discover rare species, a colourful mass of blossoms and bizarre forms of growth. What makes these disused

> sites so particularly beautiful and fascinating is the singular and unfamiliar link between derelict industrial sites and untamed nature.

The acknowledgement of the apparent contradiction at the core of the concept of *Industrienatur* seemed to offer a key to the philosophy that guided management of the site. Nature was framed as other – something external to culture, capable of acts of repossession and untamed intervention – but also as inextricably intertwined with industrial and cultural processes, to the point where any essential "naturalness" dissolved. In some sense, this double inflection aligns with current ecological thinking about novel ecosystems, which are defined as systems "of abiotic, biotic and social components (and their interactions) that, by virtue of human influence, differ from those that prevailed historically, having a tendency to self-organise and manifest novel qualities without intensive human management" (Hobbs, Higgs and Hall 2013, 58). The artificial ecosystem is produced by human interference and paradoxically sustained by human indifference.

At Duisburg, adoption of the concept of *Industrienatur* as an interpretive framework appeared to have allowed for the cultural valuation of ecological components that in other contexts would be considered undesirable (Storm 2014). Walking back through the site (now with the assistance of my German–English dictionary), I noticed several signs that explained the way that specific soils and microclimates support specialist plant and animals species, adapted to cope with extreme conditions, nutrient deficiency and residual contamination. One sign noted that biting stonecrop (*Sedum acre*) prefers highly compacted soils composed primarily of slag created during the smelting process. Another singled out the neophyte goldenrod (*Solidago canadiensis*), a North American immigrant that I recognized from the scrubby New England fields of my childhood. Interpretation highlighted the fact that certain plants were able to thrive on the site not in spite of but because of the way industrial process had transformed the landscape. Species that would have been classified as invasive in other contexts were distinguished through their association with the site's industrial past in an inverted value system that gave precedence to the refugees and invaders, the pioneers of a postpastoral nature (Pearce 2015).

One such species is the narrow-leaved ragwort (*Senecio inaequidens*), which has a long, spindly stalk and, as its name suggests, narrow green leaves. When I visited, its slightly scraggly yellow flowers were in full bloom. This particular species of ragwort is native (and here we encounter the embedded politics of ecological etymology) to the far south of Africa. Ragwort seeds are thought to have arrived in the area in shipments of iron ore, which were transported up the Rhine to feed the massive blast furnaces whose skeletal structures the plant has now colonized. The ragwort is one of many traveling species that have established themselves in this postindustrial landscape, finding an unlikely refuge in the scarred and contaminated soils left behind by eight decades of intensive production and processing. Here it is part of a unique recombinant ecosystem that has developed in the three decades since production ceased and other processes emerged to alter and occupy the space.

In this site, nature, as Rosenberg puts it, is "inextricably bound up with technology and shaped by social relationships and cultural memory" (Rosenberg 2009, 210). It is possible to draw a connection here between Duisburg Nord and the (admittedly less redemptive) "technonatures" that Shiloh Krupar identifies at postmilitary sites in the American West, which emerge through a "complex and dynamic co-mingling of waste and nature, the social and ecology"(Krupar 2013, 223). The tension between a framing of nature as an inherently compromised human artifact and nature as autonomous agent is central and irresolvable in such places. Materials – be they plants, steel, soil or seeds – are only ever partly cultured, and management embodies the lived contradictions of "simultaneously taking charge and ceding control" (Hitchings 2006, 377). During my self-guided tour of Duisburg Nord, I encountered an elevated meadow of blooming wild carrot, thistle and salsify, perched on the end of the rail line that once provided access to the ore bunkers from above. The meadow was in its high summer glory, abundant and sovereign. When I returned the next day, the meadow had been mown down, and workers were spreading fine gravel on the shorn ground. It had been "cultured" again to create a staging area for a mountain bike race. This pragmatic disturbance of a paradoxical ecology might be best described as entropic gardening, a variant of the "rambunctious gardening" that frames Emma Marris' discussion of novel and emerging ecosystems (2011).

It is not entirely clear to what extent the embrace of *Industrienatur* is applied to the built structures at Duisburg, although in principle, the concept would seem to encourage the blurring of the distinction between architecture and environment that I noted previously. Duisburg's core built area appears to be selectively maintained, with some remnants allowed to decay and others held together through minimal repair. Much of the critical literature about the accommodation of ruination at other sites in the Ruhr seems to emphasize the underlying artifice at work. Dan Swanton writes of the approach adopted in the redevelopment of a steel plant at nearby Dortmund:

> The invasion of plant and animal life is managed so that the ruin maintains an air of authenticity, without threatening the physical presence of the blast furnace as vast artefact. The

FIGURE 16.3 *The meadow was in its high summer glory, abundant and sovereign*

FIGURE 16.4 *When I returned the next day, the meadow had been mown down, and workers were spreading fine gravel on the shorn ground*

> chemical and biological lives of these material remains have to be controlled, and ecological processes of disintegration and regeneration truncated for these ruins to perform.
>
> *(2012, 276)*

Torgeir Rinke Bangstad offers a similar critique in his work on Kokerei Hansa (Hansa Coking Plant), pointing out that the emphasis on contemporary ecology works to excuse past industrial excess and allows managers to operate in a space that appropriates the productivity of decay and regeneration without totally surrendering to it (Bangstad 2014; see chapter 13). Other critics have suggested that there is a political and financial convenience to privileging passive renaturalization over active maintenance and repair, which can be allied with other manifestations of laissez-faire economic neoliberalism. Kerstin Barndt has also noted the way that the emphasis on ecological renewal functions to "leapfrog" from the preindustrial pastoral to the postindustrial future, eliding the difficult intervening period of military industrialism, when Ruhrgebiet production was of decisive importance in both world wars (2010b). While all of these critiques are valid, there is still a tendency within this work to frame plant communities not as collectives of independent, agentive entities but as compliant pawns, mobilized to realize the aesthetic and interpretive objectives of designers and managers. They also seem to assume that continued maintenance of the illusion will require some attempt to preserve the structural integrity and stability of the remnant structures.

But it is also possible to imagine that at Duisburg there will be an accommodation of more profound collapse and transience, as well as a willingness to collaborate with plants and other agents to interpret shared histories and craft shared futures. This experiment is well underway, and those responsible for the site appear to be practicing a selective indifference to ongoing processes of decay and dissolution. When decay is actively accommodated, Simmel writes, "such indifference is, so to speak, a positive passivity, whereby man makes himself the accomplice of nature and of one of its inherent tendencies, which is dramatically opposed to his own interests" (1959 [1911], 261). Will process be allowed to run to the point where the structures at Duisburg are too derelict to access? Could access be managed cyclically, withdrawn for a period of time while structures are unsafe, then allowed again after a new stability has been reached through partial collapse? The site seems to want to leave these questions open, to force the visitor into uncomfortable encounters with potential risk and danger. (Will this rusty step hold my weight? How secure is that chunk of concrete?) While this may be a precarious and temporary state, it is a valuable one, in that it provides an alternative to the stability and stasis we are accustomed to experiencing in heritage spaces. The aesthetic and affective potency of site operates through a disclosure of the properties and propensities of matter in the process of its unmaking. To return to Simmel, "destruction here is not something senselessly coming from the outside but rather the realization of a tendency inherent in the deepest layer of existence of the destroyed" (ibid, 263).

An attitude of "positive passivity" frames people as both "accomplices" (accommodating ongoing change) and antagonists (intervening selectively to check decay and deterioration). At Duisburg, there is no apparent attempt to resolve this tension. In the mode of accomplice, the human agent is destabilized and distributed, and the illusion of cultivation and control is abandoned to make space for a messier form of coexistence. This abandonment is not total, however, and the site does not apologize for its occasional reversion to more secure models of human subjectivity that rely on the disruption or arrest of process in the interests of material stability. In fact, much of the aesthetic potency of the site lies in the interstitial space between these modes of attention. Duisburg Nord operates through a principle of productive indeterminacy, in which the visitor is encouraged to experience (and negotiate) the tension between domesticated and wild, control and chaos, intervention and abandonment, process and preservation, intention and accident. This underlying ambiguity may be intentional and may include elements of artifice, but this too is perhaps inevitable. What the place succeeds in doing is foregrounding the impossibility of any critical clarity in the way that we approach such places. The site exposes itself as compromised and incomplete, and in this, at least, it succeeds. It is a space of possibility – possible unraveling as well as coming together – and in this, it offers one model for entropic heritage practice.

Kilmahew/St. Peter's

Is it a transferable model, though? The conversations that have taken place between the architects of the plan for Duisburg Nord and those responsible for constructing a viable future for Kilmahew/St. Peter's would seem to suggest so. But there are other indications that the Duisburg model may be too anarchic for adoption at a site where the architectural legacy is more exceptional and external expectations are more constrained. Since 2012, a group of academics, architects, activists and artists have gathered under the collective The Invisible College, working with NVA to gradually reoccupy Kilmahew/St. Peter's and carry out experimental research. The Invisible College website provides this description of the site and their work:

> The masterpiece of modernist architects Gillespie Kidd and Coia, St. Peter's Kilmahew opened in 1966 as a Roman Catholic seminary, only to be abandoned two decades later. Since then the building has fallen into spectacular ruin, celebrated by architectural pilgrims, vandals, ravers, dog walkers, and relic hunters alike. Despite being "A" listed, it has proved the wreck of numerous commercial schemes for restoration. But the history of the site is much older than the ruin. Kilmahew was originally an early Christian chapel, then a castle, and in the nineteenth century an industrialist's mansion, at the heart of a planted and managed estate. Each historical phase has left behind relics, rhododendrons, a walled garden, bridges, and burns.
>
> The woods of Kilmahew are much more than a tragic modernist ruin. They are the living document of a millennium, at least, of occupation, and, potentially, a unique "laboratory" in which, through the mechanism of The Invisible College, innovative ideas about the environment, from restoring historic buildings to managing biodiversity, may be speculated upon, experimented with, and tested.
>
> *(Invisible College n.d.)*

The potency of the seminary building complex in its current state as a "tragic", "spectacular" modernist ruin is clearly valued, and the aesthetic attraction of its extreme dereliction is well documented and discussed (Lorimer and Murray 2015). Photographs of the ruined altar, overhung with charred beams, and the ravaged, vaulted space of the former nave crowd Internet and print forums. Among those debating the future of the site, there is a clear desire to retain something of this elusive, dystopic character in any future plans. As I've already mentioned, the stated intention for the site "supports the principle of an unfinished work and accepts a level of entropy" (NVA 2011, 22). Since the early involvement of NVA, and through the Venice debate in 2010, there has been a continual return to questions about the appropriate balance between intervention and strategic neglect. Is it possible to conserve a ruin? Architectural historian Ed Hollis, in an essay prepared for the Biennale, implies that the rhetorical invocation of entropy is hollow if it does not involve an acceptance of unpredictable change. "If nothing happens St. Peter's will soon melt away, reduced by theft, arson, rhododendrons and rain" (Hollis 2011, 54). He goes on to suggest that acceptance of unpredictability can only ever be partial, and uncomfortable. Citing Ruskin ("Watch an old building with anxious care; guard it as best you may … do not care about the unsightliness of the aid"), he suggests that the future for St. Peter's might be better understood as a continual negotiation over the inclusion of unsightly aids that allow for incremental reoccupation while preserving some element of indeterminacy and openness. The embedded contradiction in the plans for the site – the promise to "save this valuable resource from irreversible ruination" and to simultaneously embrace the free play of entropy – frames an impossible proposition (NVA n.d.). Ruination, it appears, will be both retained and reversed. In this, I suppose, the parallel with Duisburg is clear. What is not clear is how this will play out on the ground. In the rest of this chapter, I try to imagine what it would look like to apply some of the principles and practices adopted at Duisburg on the grounds of a very different site. This is a speculative exercise, but it may be useful in flagging the limits, as well as the potential, in accepting "a level of entropy".

My first visit to Kilmahew/St. Peter's was in the context of an Invisible College workshop, and I joined a group of others in visiting the derelict seminary structures and exploring the grounds via an "audio drift" (Gallagher 2015). During the visit, I was struck by a powerful sense of the challenges inherent in trying to have it both ways – ruin and reuse, process and preservation. It was also clear that philosophy was already coming up against practicality, even before any plans had been implemented. It had been determined that the obligation to make the seminary structures compliant with health and safety legislation would require the removal of the asbestos folded through the fabric of the structure, forcing an unstitching not by rain and time but by crowbar and power saw. The *Rhododendron ponticum*, which had spread unchecked across the site in the century and a half since its introduction as an ornamental, was now classed as a noxious invasive weed, ostensibly threatening the survival of vulnerable indigenous species (Despard and Gallagher 2018). Meeting basic levels of ecological and environmental sanitation would, it seemed, require radical intervention at the site.

Before I returned for my second visit a couple of years later, I reviewed the master plan for the site, which had been prepared by landscape architecture firm erz Limited in 2011. I was particularly interested to tease out a sense of how specific aspects of the site's ruination would be addressed and to understand what fate was imagined for the plants that had occupied the site in the interval since abandonment. The plan was unequivocal about the need for some form of structured action: "All of the traces of human intervention, from medieval to modern, currently exist in a ruined state. These cultural fragments, alongside the unmanaged landscape, overtaken by invasive species, are in an advanced stage of decline" (erz Limited 2011, 6). The specter of "invasive species" was invoked from the outset, even within an expansive vision for reoccupation to proceed through a "generative", "metabolising" process of "incremental placemaking". "A state of permanent flux is integral to the concept", reads the plan, and the overarching principle for visitor experience is to retain "the sense of surprise and discovery that the site contains at present" (ibid, 6). This all sounds good (and familiar, from the Duisburg model), but as the plan becomes more specific, some of the openness starts to close down. The "conservation philosophy" seeks to "accomplish an effective rescue of the buildings under consideration" through selective "repair, upgrade, intervention" (ibid, 20–21). The plan expresses a desire to keep "the qualities of the current 'interim' state of particular locations" through selective retention of "regenerative" species (ibid, 7). Other species would not be so lucky: "Immediate intervention is required to remove the Rhododendron and other problem species [Japanese knotweed, Himalayan balsam] and to create a viable baseline position for new planting and management of the woodland and remnants of the designed landscape" (ibid, 33). Most species subject to "vegetation management" remained unnamed. In the immediate setting of the seminary buildings, "Rank vegetation and debris [are] to be cleared, [and] original surfaces reinstated" (ibid, 23). Apparently entropy was more attractive in the abstract than in its expression through the agency of individual plants.

When I visited Kilmahew again in May 2014, plans had moved on, and while no major work had yet taken place, both rhododendron and asbestos removal were scheduled to begin within the next few months. I visited the site in the company of three people who all knew it much better than I, and I'd decided before going that I'd give myself a discrete task to focus my attention. With the help of a 1967 edition of The Observer's Book of Wildflowers (published the year after the seminary opened), I would attempt to identify the plants that were growing on and around the ruined structures. As at Duisburg, I wanted to understand the role that specific ecological agents were playing in the making of the ruin, and to know, or try to know, the ecology as an antidote to seeing only the obvious aesthetic framing of the ruin as spectacle.

An "Inventory of Trees, Shrubs & Plants in Kilmahew/St. Peter's Estate" had been carried out in 2013, but the seminary complex had not been included in the survey.

We entered the site on an unmarked path that dropped us into the woodland at the top of the estate and made our way down to the seminary along the course of one of the two burns that splice the estate, past thickets of rhododendron, around a small pond. At the seminary, we came up against a barricade: a high fence of steel slats, each sharpened to a threatening point. The fence had accreted layers of expired warning signs, most now illegible. The newest addition was a graphic of a stylized person whose lungs had been invaded by splotchy black spots: "Danger: Asbestos Dust Hazard. Cancer & lung disease hazard. Authorised personnel only". A second sign below read simply, "Dangerous building" with a yellow warning triangle. Across this someone had spray-painted, "LIES!!!" Inside, a sign posted by Reigart Contracts Ltd. ("Demolition. Recycling. Asbestos Removal"), read, "Danger: do not play in this area".

Our plant inventory took care not to disturb any of the building's lethal fabric. We began with the area around the cracked granite altar. The floor was boggy with rain that had fallen into the roofless structure, the puddles laced with a sour stew of human and leaf litter. Rank vegetation and debris rooted in the rotting concrete and in cracks in the walls. We started with a catalog of the ordinary army of first flank pioneers: dandelion, dock, bramble, nettle, plantain, willowherb, hawkweed, sow thistle, hart's tongue fern. We noted the odd specimen of maidenhair fern, forget-me-not, tutsan, suspected saxifrage. Several trees had taken root in the structure as well, their seeds cast on a strong winter wind, perhaps, or carried in the fold of a raver's cuff: ash, willow, birch, Scots pine, hazel, Lawson cypress. One cypress specimen had its roots tangled

FIGURE 16.5 *A high fence of steel slats, each sharpened to a threatening point*

FIGURE 16.6 *We began with the area around the cracked granite altar*

in a snarl of red and white warning tape, but the tree's flat, feathery branches looked healthy and vigorous. The Lawson cypress is native to Oregon and California; it would have arrived on the estate at the behest of one of the nineteenth-century gardeners, who also introduced giant sequoia, California redwood, Douglas fir and Himalayan cedar. Most of the trees in the ruined

FIGURE 16.7 *One cypress specimen had its roots tangled in a snarl of red and white warning tape*

seminary were still saplings, less than head height, but the original specimens are now mature. One giant sequoia near the walled garden has swollen to a girth of almost 7 meters.

We continued our botanizing outside the main seminary building, on the foundation of the original Kilmahew House, which was destroyed by fire in the early 1990s and subsequently demolished. Within the footprint of the house, a dense thicket of birch trees has self-seeded, hiding the rusty hulk of an abandoned car. This is one of the only areas of spontaneous regeneration that will be retained, according to the master plan:

> The strategy for the spaces immediately associated with the seminary buildings is broadly focused on uncovering, repairing and reinstating the original finishes (hard and soft). … The counterpoint to this conventional conservation approach is the retention and management of the regenerative birch trees within the footprint of the former Kilmahew House. The retention of these trees acts as a marker of the interim derelict state of the site.
>
> *(erz Limited 2011, 18)*

The birch trees will be asked to function as a symbolic remnant of a discrete interval in the biography of the site, an interval that will be effectively closed by the restoration of the rest of the complex to its origins. The grove will function as a safe space for the evocation of entropy, if not its actual expression. "The dense block of birch trees echoes the original spatial arrangement, over time filling the physical void of the demolished house. Within this block of trees a discrete space is proposed for quiet, personal discovery" (ibid). Other patches of self-seeded birch will be retained in the tennis court and the walled garden.

The rest of it will go. The odd assemblage of volunteer plants in the seminary structure will give way to a restored "original finish", and their stories will become untellable. For a moment, though, through our act of inventory, they came into focus not as a generic collective but as individuals – the spindly red stalks of the herb robert, the familiar sleeve of the dock leaf, the scrappy saplings of ornamental conifers. They were not only a vegetative backdrop providing a screen of green to soften the gray concrete but also a community of volunteers – hopeful, patient, unaware of the landscape vision that would (eventually, probably) exterminate them. It is difficult to say whether there may have been an alternative to this removal and exorcism. Has anyone tried to interpret the cultural history of asbestos in situ in a contaminated building or through its removal? Such an attempt could be framed, in Simmel's terms, as a "realization of a tendency inherent in the deepest layer of existence of the destroyed" (Simmel 1959 [1911], 263). How will the wholesale removal of the rhododendron alter the experience of this place, and how long will it take for the landscape to heal over the scars? In the unmaking of this site, the role of people as ecological accomplices, enabling the free play of entropic process, will be uneasily paired with a more interventionist role, attending to the excision of offending substances and species.

The potential for Kilmahew/St. Peter's lies, perhaps, in the willingness of those involved in determining its future to expose, rather than obscure, the compromises and contradictions that will trouble any reoccupation of this site. The ruin will be ruined, but exactly how it will be ruined is still an open question. At the 2012 Invisible College event, one of the speakers commented, "You can't tame a ruin. … The moment you make a perceptible constructive intervention … you have destroyed a large part of what gave the ruin its value" (Calder 2012). Given the inevitability of the destruction of the ruin in its current state, the project turns to a framing of negotiation and debate as an end itself, with a continual looping back through the physical structure of the site to make incremental changes and craft provisional solutions. But can you have a master plan and still leave a site unmastered, open to different appropriations and inscriptions? At Duisburg, this seems to have been achieved, if precariously. The pressures at Kilmahew/St. Peter's are different, though, and the stakes are higher. In the end, the hard truth is that the vision for this site relies on securing support from funders and partners who

FIGURE 16.8 *The ruin will be ruined*

may not be content with reassurances about openness, inclusivity and improvisation. The future for this site will involve an opening up, most likely, but this will also mean a closing down of some of the possibilities that the site now holds, fraught with failure and collapse, scavenged and scarred, but still occupied, and appreciated.[1]

Closing

At the close of this chapter, I'm left feeling slightly dissatisfied with my conclusions. Is all that we can hope for honesty about our ambivalence and the contradictions that riddle any attempt to collaborate with ecological and chemical process? I keep thinking about an example from much closer to home. I live in a terraced house in a small Cornish village, and our narrow backyard is hemmed in by neighbors' gardens on three sides. We have lived in the house for four years, and this summer, the benign neglect of our borders came to a gentle sort of crisis, with the neighbors on all three sides requesting that we attend to the overgrown vegetation rooted on our side of the boundary line. We pulled stubborn willowherb from a high stone wall, hacked back sprawling skeins of bramble and uprooted a buddleia that had tunneled under a lattice fence and was lifting the whole fence alarmingly in gusts of high wind. We brought our neighborly relations back into some kind of balance by checking the growth of the plants that we shared our space with. The ethic was pragmatic rather than collaborative. So to suggest that somehow it could be otherwise in places that are much larger and more complex, places that have a responsibility to a wider public, seems somehow disingenuous. But it is still something that I think is worth contemplating, if only because experiments like those taking place at Duisburg and at Kilmahew/

St. Peter's allow us to imagine a different trajectory in relation to remnant ecologies and residual architectures, in which the words "restoration" and "preservation" sit awkwardly. Both of these words privilege recovery of an imagined original state over the discovery of possible future states – which are still rooted in the substance and stories carried forward from the past. And while attempts to reverse the needle may always involve a wavering commitment, they are worth exploring even in their wobble. Perhaps that is the promise of a positive passivity: the willingness to live with the wobble.

Credit

An earlier version of this chapter appeared in *Curated Decay: Heritage Beyond Saving* by Caitlin DeSilvey (University of Minnesota Press, 2017), with the title "A Positive Passivity: Entropic Gardens". Copyright 2017 by the Regents of the University of Minnesota. Used by permission of the University of Minnesota Press.

Acknowledgments

The fieldwork discussed in this chapter was carried out independently, with support from the University of Exeter College of Life and Environmental Sciences. Immediately after my visit to Duisburg Nord in July 2012, I traveled to Bayreuth University, where I presented the work to Matthew Hannah and his students. I benefited greatly from their critical comments on my initial attempts to make sense of my experience. I first visited Kilmahew/St. Peter's when I was invited by Hayden Lorimer and Mike Gallagher to participate in an Invisible College event in September 2012. Many of my observations in this chapter draw on conversations and presentations associated with this initial exposure. I returned to the site on an informal visit with Mike, Hayden and Erin Despard in May 2014.

Note

1 My narration of activity at Kilmahew/St Peter's ends in 2016, when this chapter was originally written. In late 2017, NVA decided to withdraw from the site and suspend capital investment plans, citing "increasing risks, both financial and physical". In June 2018, after the (unrelated) withdrawal of their core funding from Creative Scotland, the organization ceased operations entirely. "St Peter's Seminary rescue arts group to close", 5 June 2018, available at: www.bbc.co.uk/news/uk-scotland-44373883. In May 2019, Historic Environment Scotland published a report for the Scottish Government recommending a 'curated decay' approach for the site. (https://www.historicenvironment.scot/archives-and-research/publications/publication/).

References

Atchison, J. and Head, L. 2013. "Eradicating Bodies in Invasive Plant Management." *Environment and Planning D: Society and Space* 31: 951–968.

Bangstad, T. 2014. *Defamiliarization, Conflict, and Authenticity: Industrial Heritage and the Problem of Representation.* PhD Thesis. NTNU-Trondheim.

Barndt, K. 2010a. "'Memory Traces of an Abandoned Set of Futures': Industrial Ruins in the Postindustrial Landscapes of Germany." In *Ruins of Modernity*, edited by J. Hell and A. Schönle, 270–293. Durham, NC: Duke University Press.

Barndt, K. 2010b. "Layers of Time: Industrial Ruins and Exhibitionary Temporalities." *PMLA* 125(1): 131–141.

Calder, B. 2012. *The Invisible College Public Talk*. The Lighthouse, Glasgow, 8 September.
DeSilvey, C. 2017. *Curated Decay: Heritage Beyond Saving*. Minneapolis, MN: University of Minnesota Press.
Despard, E. and Gallagher, M. 2018. "Media Ecologies of Plant Invasion." *Environmental Humanities* 10(2): 370–396.
ERZ Limited. 2011. *Kilmahew/St. Peter's Masterplan Summary Document*. Accessed 13 November 2011. http://nva.org.uk/wp-content/uploads/2014/11/Kilmahew_StPeter_MASTERPLAN_smallersize.pdf.
Gallagher, M. 2015. "Sounding Ruins: Reflections on the Production of an 'Audio Drift'." *Cultural Geographies* 22(3): 467–485.
Head, L. and Atchison, J. 2009. "Cultural Ecology: Emerging Plant–Human Geographies." *Progress in Human Geography* 33(2): 236–245.
Hetherington, K. 2010. "The Ruin Revisited." In *Trash Culture: Objects and Obsolescence in Cultural Perspective*, edited by G. Pyc, 15–37. Oxford: Peter Lang.
Hitchings, R. 2006. "Expertise and Inability: Cultured Materials and the Reason for Some Retreating Lawns in London." *Journal of Material Culture* 11(3): 364–381.
Hobbs, R. J., Higgs, E. R. and Hall, C. M. 2013. *Novel Ecosystems: Intervening in the New Ecological World Order*. Chichester: Wiley-Blackwell.
Hollis, E. 2011. "Anxious Care and Unsightly Aids." In *To Have and to Hold*, edited by G. van Noord, 54–57. Edinburgh: Luath Press.
The Invisible College. n.d. Accessed 13 November 2018. www.theinvisiblecollege.org.uk/locale.
Keulartz, J. 2013. "The Different Faces of History in Postindustrial Landscapes." *Paper presented at the Society for Ecological Restoration Fifth World Congress*. Madison, Wisconsin, 9 October.
Krupar, S. 2013. *Hot Spotter's Report: Military Fables of Toxic Waste*. Minneapolis, MN: University of Minnesota Press.
Lange, D., ed. 2004. *The Ruhrgebiet Industrial Heritage Atlas*. Essen: Regionalverband Ruhr.
Latz, P. 2001. "Landscape Park Duisburg Nord: The Metamorphosis of an Industrial Site." In *Manufactured Sites: Rethinking the Post-Industrial Landscape*, edited by N. Kirkwood, 150–161. Abingdon: Taylor & Francis.
Latz, T. 2011. "Once Upon a Time …" In *To Have and to Hold*, edited by G. van Noord, 62–67. Edinburgh: Luath Press.
Löfgren, O. 2005. "Wear and Tear." *Ethnologia Europea* 35(1–2): 53.
Lorimer, H. and Murray, S. 2015. "The Ruin in Question." *Performance Research* 3: 58–66.
Lorimer, J. and Driessen, C. 2014. "Wild Experiments at the Oostvaardersplassen: Rethinking Environmentalism in the Anthropocene." *Transactions of the Institute of British Geographers* 39(2): 170.
Marris, E. 2011. *Rambunctious Garden: Saving Nature in a Post-Wild World*. New York: Bloomsbury.
NVA. 2011 "Statement of Intent." In *To Have and to Hold*, edited by G. van Noord, 22–27. Edinburgh: Luath Press.
NVA. n.d. *Kilmahew/St. Peter's*. Video available at NVA Public Art. Accessed 13 November 2018. http://nva.org.uk/artwork/kilmahew-st-peters/.
Pearce, F. 2015. *The New Wild: Why Invasive Species Will Be Nature's Salvation*. Boston, MA: Beacon Press.
Piper, J. 1948. *Buildings and Prospects*. London: Architectural Press.
Pitt, H. 2014. "On Showing and Being Shown Plants: A Guide to Methods for More-than-Human Geography." *Area* 47(1): 48–55.
Pollan, M. 1991. *Second Nature: A Gardener's Education*. New York: Dell.
Rosenberg, E. 2009. "Gardens, Landscape, Nature: Duisburg Nord, Germany." In *The Hand and the Soul: Aesthetics and Ethics in Architecture and Art*, edited by S. Iliescu, 209–230. Charlottesville, VA: University of Virginia Press.
Simmel, G. 1959 [1911]. "The Ruin". In *Georg Simmel, 1858–1918: A Collection of Essays, With Translations and a Bibliography*, edited by K. H. Wolff, 259–266. Columbus, OH: Ohio State University Press.
Storm, A. 2014. *Post-industrial Landscape Scars*. New York: Palgrave Macmillan.
Swanton, D. 2012. "Afterimages of Steel: Dortmund." *Space and Culture* 15(4): 276.

17

HERITAGE ECOLOGIES AS WORLDING PRACTICES

Christina Fredengren

Inherited connectivities

We have entered an era where human-induced climate change and environmental degradation occur at an accelerating speed (Steffen et al. 2015). As Chakrabarty (2009), among many others, has noted, in the Anthropocene, the human has become a major geological agent that moves more materials than the planet itself. While geologists, as keepers of extended timelines, discuss if this is a chrono-stratigraphic recognizable epoch (Zalasiewicz 2016), these times go by many names such as the Anthropocene, the Chthulucene or the Capitalocene (see Haraway 2016). Whereas not all humans are equally exposed to climate dangers, anthropogenic-materializing actions have generated difficult connectivities. In this context, Chakrabarty (2009, 220) has remarked that species thinking, that is, thinking the 'anthropos' in climate change, is a mere placeholder for something that cannot be fully experienced or understood in its entirety. Furthermore, Chakrabarty reasons, the historical narratives at hand do not prepare us for the challenging futures to come, nor can human history continue to be set apart from the deep registers of natural history.

This reasoning presents a problem for both archaeology and heritage studies. First of all, it is problematic to separate natural heritage from cultural heritage, as they are deeply entangled, and the anthropocentrism of the heritage sector may need to be unpacked and to some degrees also challenged (Fredengren 2015; Harrison 2015). Second, there is a need for altered ways of engaging with deep time and what it means to be human at many levels. Many of us have pointed out that climate change, species extinction and environmental degradation are significant human-induced inheritances affecting present and future generations, though still not captured as heritage. Accordingly, it is important to explore how heritage studies, and also archaeological writing, can meet these challenges and be transformed and contribute to fields such as the environmental humanities, critical posthumanism and new materialism, that have dealt with these matters at length (Fredengren 2015, 2018a; Fredengren and Åsberg 2020).

The particular focus of this chapter is how heritage mattering weaves ecologies of relations between and through places, materialities and generations in sometimes unexpected ways. In a case study, we will acquaint ourselves with Husarviken, close to Stockholm University Campus, on the boundary of Sweden's only National City Park. The park is a naturalcultural heritage area where different heritage legislations coincide and influence developments. A primary focus within this park will be the heritage ecologies that gather around Lyell's Oak at Norra Djurgården, in Stockholm. The Oak was described by Sernander (1926) as a "naturminne" – a "nature-monument". The argument for preservation of such heritage was, for Sernander (1926, 226–227), that their natural and national values would lay the foundation for the homeland's proud future. Instead of following the traditional process of investigating the role of heritage in discourses on national identity making (itself a valid but well-trodden route), we will trace other cartographies, temporalities and relations that coincide here. The urban housing development of Norra Djurgårdstaden, located by Husarviken, has won prices for its ecological approach. Simultaneously, critics argue that it communicates that sustainable development can be reached primarily through environmentally friendly technology while glossing over the conflict between the necessity to do something and the resistance to changing lifestyles (Wangel 2013, 17). However, matters are more complex than that.

This chapter will fold out a range of connectivities and explore heritage through several seemingly non-related phenomena that gather in and diffract from this place, as they have consequences for the connections between several generations. It explores some of the emerging relationships between Lyell's Oak, the melting ices of the Arctic, a piece of art and the housing estate, as well as a number of shadow places that trickle into or out of such inheritances – and connect to yet other locales and times – in what may be called messy and queerly temporalized ecologies. It deals with how critical post-humanist feminist thinking can be enrolled to approach situated processes of heritagization. Thereby, it explores heritage places, stories and imaginaries in ways that acknowledge how matter "matters", in complex worlding practices in the Anthropocene and beyond. The aim is to trace how material, immaterial knots between past, present, futures can be observed as situated phenomena and ask questions around how such knots may need to be retied for crafting more habitable worlds – to find other ways for temporal relatings through heritage materialities with different speeds, accelerations and entanglements. It will be argued that heritage phenomena, such as these, have a number of evolving connectivities, as they are parts of (in)humanly induced worldings, anthropogenic bendings and terraformings, with effects for several more-than-human generations to come. And hence, that engagement with these phenomena must stretch beyond conventional ways of dealing with heritage, into notions of heritage ecologies that work on numerous unfolding time-scales.

Knowledge genealogies

Heritage, as captured through discourse analysis, has given rise to the concept "authorised heritage discourses" (AHD), which refers to how social identities and collective memories are formed through attributions of meaning and value that serve present-day political ends (Smith 2006). As such, heritage has largely been framed in anthropocentric ways, understood as generated through human gazes and meaning-making processes, and significant primarily for human use. This social constructivist approach has influenced both research and heritage services for several years and continues to do so through inscription in practice and policy documents (Fredengren 2015). This is aptly observed and captured in Pétursdóttir's (2013) writings as the signs of a new "intangible heritage discourse". In this dominant approach, questions around the politics of heritage labelling and who gets represented are centralized, while material aspects are pushed to the side, together with a number of other actors (Fredengren 2015) that make up heritage ecologies.

However, if we, as is prevalent in this approach to heritage, assume the world exists only through the human mind, we lose sight of any relationship between non-humans (Witmore 2015, 48; Farstadvoll 2019, 27; see also Farstadvoll, chapter 18). Disregard for the impact of material workings can be one of the factors that have shadowed over the material advances of several unfavourable inheritances collating together to induce climate and environmental challenges. Surely, climate change and environmental degradation are at work, despite being unacknowledged by heritage business. However, as material phenomena, they gather forces, demand to be acknowledged and even push themselves into the heritage category and may thereby work performatively to diffract and change it. As Farstadvoll (2019, 18) notes, environmental pollution and extinction are not signs of an *anthropos* in control but rather evidence of peoples' limited foresight into, and authority over, material trajectories. Farstadvoll (2019, 11) builds this reasoning on Olsen's (2012, 83) and Pétursdóttir's (2013) queries regarding what happens when things lose their human privileging as "things-for-us" and start causing trouble and behaving in an unruly fashion. Their reasoning highlights how a range of different involuntary inheritances are at play in the Anthropocene (Olsen and Pétursdóttir 2016) as things continue their work after having passed their roles as useful for humans.

During the Anthropocene, a range of material and immaterial forces have been set in motion. Forces that will bring extraordinary changes to life on earth and have effects for several generations to come. This urges us to refocus, looking beyond the ways heritage forms work discursively for humans (as enrolled in identity creation or memory work), to consider what heritage does and how it works in more-than-representational (to use Lorimer's 2005 term) and material ways. Here, the letting in of a variety of materializing forces is long overdue, not only in academic writing, but also in heritage services and museums (where I still [2019] listen to lectures where archaeology/heritage is described as a particular human gaze on materials from the past). Then others, such as Pétursdóttir (2017), argue instead that archaeology is the study of things. I would like to further this argument by adding that archaeology and heritage studies are involved in the exploration of how matter "matters" in relation to more-than-human worlding practices.

In 2012, Linköping University and the *Posthumanities hub* hosted the conference *Entanglements of New Materialisms*. At this event, I explored how feminism and new materialism could alter heritage studies and the practice of archaeology (Fredengren 2012). I was curious about how critical heritage studies could involve material processes to a greater extent while giving a new impetus towards both critical and affirmative thinking/practices and hence move away from years of engagement with analyses of heritage discourses, delivering "critique only" (Fredengren 2015). I was looking for theories that engaged with the materializing world and that could be used to query the anthropocentrism evident in heritage studies in ways that maintained a critical eye on heritage processes. In a trial piece (Fredengren 2015, 120), I made use of key texts in this field, such as Barad (2007, 2012), Braidotti (2013), Haraway (2008) and Tsing (2012), to approach heritage differently. I proposed a critical, feminist posthumanist take, whereby heritage processes could be understood and analysed as "Baradian phenomena" – in short – as complex material-discursive entanglements of nature-cultures in becoming. In his thesis about the derelict landscape garden of Retiro in Molde, Norway, Farstadvoll (2019) used this definition, favourably expanded upon it and, in places, coupled it with theories of object-oriented ontologies to highlight the "impurities" of heritage in a situated place.

This chapter further unfolds some of the thinking from my previous work, applying these ideas to the case study around Husarviken. This, to explore how the theories mentioned previously could be useful and to open up new ways of engaging with heritage matters. In recent writings within critical heritages studies, object-oriented ontologies (OOOs) are trending (e.g. Pétursdóttir 2017; Harrison 2018; Farstadvoll 2019). The current chapter also shows how critical feminist posthumanist thinking can add to such perspectives, for example, with regard to power-analysis. The chapter deals with how heritages can be unfolded if analysed as ecologies. It is furthered that heritage ecologies, or for that sake museum ecologies, can be explored methodologically in ways similar to how media ecologies are captured (e.g. Fuller 2005; Parikka 2016). This also in ways that free the use of "ecologies" from mainly relating to what commonly has been captured as "nature" and open up analyses of a broader field of agentialities and interlinkages that make up heritage as phenomena. In such ecologies, the media is deeply entangled with the message, and they can be traced by looking at emerging material-semiotic hybridities (where semiotics may not only be reserved for human use). This means to trace out a variety of knotted relations and extensions that make up heritage ecologies, to embrace a wider range of impurities than often compassed by green heritage (as in the natural heritage of parks, gardens etc.), that give rise to situated heritage places and what relations diffract from there and interfinger in other places and times. As demonstrated in the case study of the chapter, such heritage ecologies can be expanded upon by folding out the complex materializing temporal relations between geological processes, thinking, people, animals, melting ice, clay, fungi, trees and gasworks walls and fishing cabins, which all come into play in situated worlding practices that cannot easily be contained within a spatially defined heritage designated area.

Heritage as performative phenomena

The previous reasoning around heritage ecologies is underpinned by theories that allow working with heritage as performative phenomena (Fredengren 2015, 120). Drawing on Karen Barad (2007, 2012), heritage may be understood as "doings", or practices (not necessarily involving human beings as agents) that enframe how matter comes to matter as inheritance. Heritage as a performative phenomenon differs from heritage as a social construct and can be analyzed as coming into being through the intra-action between concepts, ethics and materializing processes. The term intra-action is used instead of inter-action to mark that there are no predetermined, physically bound entities or things from where agency is exercised (Barad 2007, 141). Agentiality is instead coming about through a congealing of a variety of sources and forces.

The term *phenomena* can be explained lexically as referring to particular situations that on the one hand are observed to exist or happen, whereas on the other hand their causes may be unclear. In Barad's process ontology (see for example 2012, 30–33), phenomena are understood more specifically, as they are the smallest analytical unit. They rise through the ongoing differentiation processes of an interconnected world that change through the mutually transformative interplay between different agentialities. This as a part of the open-ended, ongoing and unfolding worlding of the world. However, by interfering in various worlding processes, humans and more-than-humans make agential cuts. This is what Barad refers to as "cutting together-apart" (2012, 46), into this interconnected and unfolding world, where local enactments create new phenomena. Such enactments can appear through a variety of actions, thinking technologies or scientific apparatuses, and such cuts have material and ethical effects. If applied to heritage, it is possible to observe a number of such agential cuts or boundary making processes at play (Fredengren 2015, 120). For example, selection processes, labelling exercises, curatorial or conservation practices but also a variety of materialization processes.

However, with a performative "bracketing off", as is often seen in heritage making, these processes do not necessarily create a true or an all-encompassing representation of heritage (or the story it is supposed to reflect). Instead, heritage labelling can be understood as a lens that may cause unexpected diffractions (compare Barad 2007, 72), where the ways in which heritage is captured and observed also cause it to change in material-discursive ways. For example, heritage labelling may be a shaping apparatus, but as such it may simultaneously set in motion a range of materializing processes that are partly out of human control. The heritage label does not carry the full excess of a phenomenon. Furthermore, many phenomena, such as climate change, work as heritage despite being missed out by heritage labelling. Anthropocene climate change is largely human made, and this inheritance can gather up, entangle and make co-jointed *deep time interventions* with other phenomena, in ways that stretch both beyond human intentions and imaginations. An intriguing example of this is the collective of nuclear reactors that, some 2 billion years ago, summoned their powers in the area now known as Oklo in Gabon (Ball 2004). Here, a nuclear fission reactor "invented" itself, so to speak, in its multiplicity, and created a heritage that still signals a material presence in the landscape with emissions equivalent to that of a nuclear plant. One may ask how these presences throw themselves into and disturb the heritage category, despite not having been formally invited in.

Using Barad's approach, heritage can be considered as evolving phenomena, in states of becoming, where a range of agentialities crossing the nature/culture divide are brought into being. Similarly, as pointed out by Smith (2006), heritage is a "doing", and both approaches thus focus on processes of heritage in-the-making rather than heritage as steady-state phenomena. Going even further, new materialism and critical posthumanist theory trouble the idea that human beings alone are the ultimate producers or the ultimate users of such heritages. Instead it works in queer ways through time to form unexpected relations as gatherings of a variety of materializing sources and forces. Hence, in the format of marked-off sites, museum collections, exhibitions and so on, heritage is the result of intra-active and differentiating processes, where a range of more-than-human others, such as agencies of geology, humans, animals, water, concepts and ethics, co-work to make it develop. Heritage can be shaped by a series of boundary-making processes, where human actions, as one type of matter-bending, influence developments. For example, divisions into natural and cultural heritage and the shaping of localities by bureaucratic apparatuses make agential cuts into an interconnected world and hence interfere in various ecologies. However, much of this goes beyond human knowledge and capture. Whereas a particularly defined heritage may be argued to represent a particular group and their identity, there is always more-than-this at work. Heritage as phenomena acknowledges that there are a range of more-than-representational workings of these matterings. The heritage lens may have formative and structuring powers, but these powers are not all within human control; for example, labelling may have diffractive effects, and what is framed may go off on unexpected tangents. Here, the title of this chapter refers to heritage that can be ecologized by trying to examine and unfold the relations that contribute to its continued becoming. It explores how heritage ecologies work tentacularly as parts of worlding practices (using Haraway's 2016 terms) and, furthermore, work diffractively once in place – forming unexpected alliances – beyond human-induced conceptual capture and making. Haraway (2016, 130) confronts us with the importance "of training the mind and imagination to go visiting", to be curious, to welcome the unexpected, to make conversations with and to take up "the unasked-for obligations of having met". That is, to cultivate the ability to respond – to take up "response-ability", no matter if we meet non-natal kin. The tracing of such visits and worldmakings can be coupled with questions that I ask through my case studies together with Rose (2011), Haraway (2016) and Barad (2012):

- What cuts are made and what phenomena come into being through heritage making?; what collaborative practices are set in motion, what connectivities and mutualities emerge, and which of these become undone by such activities?
- What processes and relations are influenced and temporally stabilized (or otherwise affected) by heritage making – and how are situated heritage phenomena leaking into other phenomena?
- How do such heritage phenomena diffract, direct and entangle living processes in situated places?

The Arctic and melting ice

Here, I would like us to call on and make acquaintance with the thermohaline cycle that transports water and heat around the earth. The thermohaline circulatory flow can be likened to a very slow conveyor belt of water currents that lead from the Arctic, connecting all four oceans. Whilst thermohaline circulation is strong, climate change and faster ice-melt might lessen the salinization that takes place in the Arctic, disturbing and changing the flow with adverse climate effects (Rahmstorf 2006). Climate change is thought to cause deep time interventions that may shift this slow process of water circulation, and the melting ice will lead to water level rise in the oceans. Whereas human-induced climate change, the ice-melt and the beginning of change in the thermohaline cycle are the result of a number of different factors and agencies (among these hyper-consumption and life led in petroleum cultures [see LeMenager 2014]), this is also an example of how a variety of materialization processes take on lives of their own once a chain of events is set in motion. Also, these changes, which will alter living conditions for humans and animals alike, can be considered nature-culture heritages that create bonds between past and future generations in rather unexpected ways.

There are connections between ice, water, time and even the Arctic that are built into the nature-culture landscape in, for example, the middle of Stockholm, and their workings "lense" themselves onto this locality and diffract from here to other places, other times. At Norra Djurgårdstaden, new dwellings are expanding over an industrial area. This work enables better living conditions for middle-class academics, students and others, for whom top-notch apartments and amenities are being designed. Furthermore, this living is conditioned on a variety of green "technofixes", such as easier ways of sorting garbage or a street architecture that allows for the safe passage of animals in the streetscape. In-built cavities and dams in Djurgårdsstaden might take care of excessive waters temporarily, and preparations have been made for at least some of the futurities of extreme weathers predicted by climate scientists.

Here, also, old gasworks walls have been dismantled and industrial grounds configured to make room for new inhabitants. The refurbishment was undertaken despite the objections of heritage organizations such as The International Committee for the Conservation of the Industrial Heritage (TICCIH 2014), which stated that the gasworks were "the most impressive preserved examples of the urban gas supply technology in all of Europe". However, there is much more to the heritage ecologies here than that. In this place, people laboured and their bodies were exposed to hard work and enmeshed with hazardous substances. Coal and petrol were used to produce gas. Not only bodies but also soils were contaminated, and in the ongoing refurbishment of this heritage site, the interrelated, less wanted, toxicities of soil had to be removed. One could draw attention to the cultural histories of class and gentrification in process here, observing that the people living in the new development are those who have the social and economic capacity to afford ecological living. Furthermore, the gasworks and the housing estate also contribute to a worlding with materializing effects connected to the rise and acceleration of the Anthropocene. Thereby, as heritage places, they forward into and onto climate change, global warming, that diffracts back both here and elsewhere, forms of the melting of the ice, rising sea levels and changes to the thermohaline circulation.

Lyell's Oak

Several oaks grow at Husarviken, some of great renown, and the oldest of these trees are approximately 500–800 years in age (Nationalstadsparken 2019). The oaks in this park, together with the grounds, have been under royal influence, more or less, since medieval times and hence spared from development pressures. Oaks are trees that have attracted human attention for a long time; their lives have been managed and appropriated by the upper echelons of society, while it was illegal for commoners to cut them down. Notably, oaks were the privilege of the crown, treasured for the acorns used as feed for farm animals, and, later on, in high demand as ship-building materials for the navy (Eliasson 2003, 50–51, 70). Though not described historically as heritage-making practices, several decisions to care for the oaks could be seen as materializing cuts, privileging the growth of some life-forms above others. The cutting together-apart of oaks and their support ecologies have turned into heritage making, as these trees have been the joint concerns linking up several generations with different and changing motifs. In some cases, such cuts were made with an eye on supplying future extraction economies with raw materials, but they also crafted local ecologies in both intended and unintended ways. This is only one example of how human action has influenced the workings of various ecologies where extraction economies, that is, production systems based on the taking and transformation of non- or only slowly renewable resources, make long-lasting material impacts into an interconnected world. However, such interventions and cuts are not only human-induced, as plants, trees and land contribute their own agential gelling to the materializing temporalities of situated places. Here at Husarviken, a variety of material bendings have let the oaks thrive as a privileged species for hundreds of years.

The park website informs visitors that this is an unusually large population of oaks, one of the largest in northern Europe. Many of them have attracted their own attachments and become knotted together with the lives of others. For example, they are lifeworlds of several birds and insects (Jansson and Antonsson 1994/1995), but trees also work as traps for greenhouse gases. Furthermore, here they contribute to the recreational landscape for people and their pets walking in these un-natural but more-than-cultural forests. Tsing (2012, 141) writes about mushroom picking as an activity that leads people to the unruly areas at the "seams of imperial space, where we cannot ignore the interspecies interdependencies that give us life on earth". The unruly, for Tsing, is a term to indicate ways of paying closer attention to the workings of human-animal-environment relations, which might even invite inter-species thinking and working. Oaks seem to have a particular relationship with fungi such as *Cantharellus pallens*, which in turn draw people out in forested areas during the summer season and lense them onto more wayward cartographies – which in turn might produce other ways of getting close to, knowing and entangling with the world.

Besides, oaks and their connectivities may weave people into various emotional relations. Eliasson (2003, 381–382) highlights that oaks in the late 1700–1800 came to stand for state interference in farmers' affairs and that an oak-hatred consequently developed. Contrary to this, many today cherish and show their attachment to these trees – and the oaks are also mourned when cut down. One example is the relations attested to and emotions raised by people in activist groups, such as Alternativ Stad (Alternative City), a group that mobilized when the so called "TV-oak" (located outside the offices of the Swedish Broadcasting agency) was cut down (SvD 2019). This example furthers and highlights the human-oak entanglements in this area that provide thinking-ground for processes of mourning and loss when phenomena captured as certainties and stabilities, such as the oaks, fall apart or get disassembled.

As I have discussed elsewhere (Fredengren 2016), encounters with deep-time materialities can be powerfully affective as they interfere unexpectedly with everyday temporalities and the long-term liveliness and connectivities that continue beyond the lifespan of temporally bound human bodies. Likewise, thinking through the material lives of trees or the thermohaline makes time and life-processes appear different, which may inspire long-term thinking and perhaps also action, besides being entangled in emotional processes. Morton (2010, 130–131), followed by Holtorf and Högberg (2014), has argued that certain materialities, such as plutonium and nuclear power, are good to think with, as they expand the mind to apprehend long-term cause and effect. That trees and forestry work in this way may also disturb and cause temporal problems to politicians and planners. This has also been highlighted by Nordblad in a paper on the politicization of the long-term in the development of forestry policy (2017). While obvious on one level, it often remains underarticulated that the heritage sector also engages in the politicization of the long term and works in future-bending ways. Furthermore, materialization processes of the long durée of heritage (see Fredengren 2015, 118), as exemplified by oaks or the thermohaline, may provide enticement for *the mind to go visiting* places, times and consequences outside the range of ordinary human life-time. The visiting mind can strike conversation with and open up for posing questions about time and management of resources and also about issues concerning justice and care, in between past-present-future generations. These questions need to be met with a response – although no such response-ability may ever have been asked for.

One rather famous tree is the oak that in the span of more than 300 years has grown from a small acorn shoot to become intimately joined with king Karl XI's (1655–1697) fish cabin – Fiskartorpet. Here, the cabin and the tree are physically entangled (Figure 17.1). They have evolved together and are part of each other's bodies and development, with the oak eating itself into the house bit by bit, roots underneath it, the tree crown over it. Presently, the oak and house are joined by a specially made metal plate to keep rain and rot away, but at the same time, they grow, change and decay together, emitting greenhouse gases. However, their co-becoming is presently managed by human care of both tree and house. They are truly entangled with each other and their humans through stabilizing conservation technologies. Trees, however, feed on decaying wood, whilst the house, so to say, rests on the body of the trees, and they may also develop unexpected relations behind their caretakers' backs. Furthermore, Fiskartorpet's Oak diffracts into geological and science history, with distinct implications for the conception and perhaps also misconception of the Holocene period.

FIGURE 17.1 Fiskartorpet and the oak, Norra Djurgården, Stockholm

Photo: C. Fredengren.

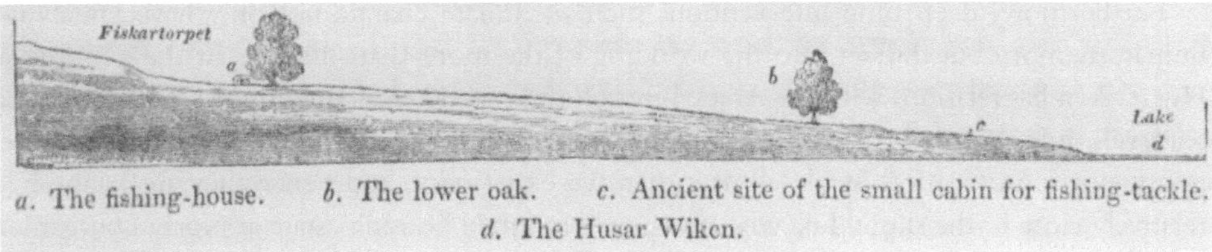

FIGURE 17.2 Lyell's Oak, cross-section, Djurgården, Stockholm (after Lyell 1835, 13)

A bit further down on the same slope, there is a younger oak, Lyell's Oak (Figure 17.2), that together with Fiskartorpet's Oak has been pleated into the history of science. They played a great role in how the Scottish geologist Charles Lyell (1797–1875) began to understand the pace and process with which the waters of the Baltic Sea seemed to withdraw out into the ocean; this was also called *vattuminskningen* (Lyell 1835; Sernander 1926, 42–43). Later research has instead explained the phenomenon as the slow rising of land as it recovers from the last ice age. In this part of the world, isostatic uplift and tilt cause shoreline displacement (see, for example, Hedenström and Risberg 1999). This can be experienced as if the waterline recedes consequently. With basis in the analysis of these two oaks, Lyell argued that the withdrawal of the water occurred at a much slower pace than had been previously observed by others. The later calculations by Sernander (1926, 46) showed that the Fiskartorpet's Oak in Lyell's time was c. 6.53 m above the mean sea level and Lyell's Oak 2.67 m – and an argument was made of a land rise of some 10 cm per 100 years. With these forces, Lyell's Oak and that of Fiskartorpet, which co-conditioned the scientific findings, have slowly, due to the geological forces, moved upwards in the topography, withdrawing and leaving their former ally, the water, behind. Their relationship and the fishing cabin's connection to water have slowly become undone. Other trees have moved into the space recovered from the sea and shoot at the shoreline.

Sernander (1926, 42) described Lyell's and Fiskartorpet's Oaks in heritage terms as "*natur- och kulturminnen*". Whereas the category naturminne (Naturvårdsverket 2019) still exists as a heritage category and covers remarkable trees, rocks and similar, the oaks at Djurgården no longer fall under that classification. It is worth noting that the category naturminne can imply a focus on singularities and exposes what are remarkable things to remember for humans. However, Nationalstadsparken has a general protection through the environmental code. Later heritage categories such as nature- and culture reserves, while still formed anthropocentrically, tend to apprehend heritage more holistically. This can involve to trace them out as green heritage in ecological relations with, for example, animals and vegetation. The Swedish national heritage board captures trees as living archives that can be used to tell stories for humans about landscape development, farming practices and old harvesting methods (Lennartson 2013; Riksantikvarieämbetet 2019) – but there are many other stories that fold out of the encounters with particular trees and their relations that also question anthropocentrism. For example, it is not only humans that remember (see Olsen 2003; Haraway 2016) or only humans that work through stories. Vegetation also remembers the past; the material bodies of the trees keep such memories in their tissue. Their bodies have been altered by connectivities with, for example, farming animals or encounters with landscape conservationists that keep the land open for light to come in, or through scars and trauma (see Storm, chapter 3) – with material effects well beyond human storytelling and heritage labelling.

Furthermore, deep time interventions such as climate change not only have effects for human memory but diffract into the worlding of the more-than-human. At the same time, Husarviken has relations with the Arctic through the sea and consequently also with the rising sea levels. It is clear that the oaks, with lifespans over several hundred years, have led their lives in anticipation of a more stable climate than we experience, and hence they are established relatively close to the shore. Likewise, the newly founded housing estate at Norra Djurgården with its inhabitants has also favoured a shoreline location. Hence, all these actors have a view towards a different future than the one approaching with the meltwaters. Norra Djurgårdstaden has been instructed to plan for sea level rise of more than 2 m above the present in the next 80 years (Miljöbarometern 2015). This takes us to a time when the human children of today will be old and when the waters might move up closer towards Lyell's Oak. Depending on the strength of climate action, the waters will continue to rise over considerable time (see ISPCC 2019; Oppenheimer and Alley 2016). Hence, sea-level rise is linked to climate change as a major future heritage phenomenon. I am not sure if one can have memories of the future, but if so and if plants and other materials have memories, this place has a built-in future-memory, where the oaks and the buildings remember a future that is only gradually changing; a future that looks very much like the stable Holocene and where the waters are not excessively rising.

Shadow places

FIGURE 17.3 Installation of the artwork *The Flesh of a Female Goddess* at Husarviken, Stockholm

Photo: C. Fredengren.

Oblivious to the projected sea level rise, but aware of other environmental questions, a new artwork by Sara Söderberg was established on the shores of Husarviken in 2019 (Figure 17.3). It is an installation piece of handmade brick in pastel colours of yellow, pink and brown – perhaps skin colours. It forms rectangular shapes on the shore. One part can be used as a barbeque, providing residents with a place for burning and consuming, for example, meat. It is named "Köttet av en kvinnlig gud" – the flesh of a female god (Söderberg 2019) – in itself a gender troubled name. The artwork's webpage shows films of how a digger and a shovel make cuts into clay. This depicts the extraction of clay used for the making of bricks. There is supposed to be a text on the artist's webpage that goes with the film – but there is no document under the text-heading (the link is broken) – which opens up for speculation and further reading.

Possibly, the artwork and its digital extension echo ideas from the book *The Death of Nature: Women, Ecology, and the Scientific Revolution* written in 1980. Based on a historical study of documents from c. 1500–1700 AD, Merchant (1980, 193) here argued that whereas nature had earlier been comprehended as an organism, it was from then increasingly seen as being "a system of dead, inert particles". The value system that came with such a mechanical understanding of nature was "fully compatible with the directions taken by commercial capitalism" (Merchant 1980, 193). Hence, in this way, "nature" died, and this perception was also transplanted and connected to similar and contemporary oppression of female bodies. Perhaps instead, the artwork could refer to Bruno Latour's (2017) writings on the necessity for us to face Gaia not primarily as a deity but as a range of methods proposed for exploring how the interaction between living organisms and the non-living fashions the planet. Or the artwork could refer to the terraforming and tentacular Cthulhu of Haraway (2016, 51–57). Maybe it appeals to Stengers' (2015) much more volatile, intrusive Gaia as an assemblage of more-than-human forces that do not care about human reasoning but where humans need to gather forces "against the coming of barbarism" in the wake of climate change (see also Haraway 2016, 61, 175 note 11)?

Regardless of whether the artwork makes a secular or post-secular religious reference, it speaks about the needs for new thoughtful and creative apparatuses to capture, compose and expose the knottings at this place with the extraction economies of the clays that are used in house construction. It could be seen as an intervention that draws attention to the relationship between the site and its shadow places (Plumwood 2008), that is to say, with those places outside this location that are implicated and used to facilitate the housing development. It highlights how a range of different materials had to be extracted in order to make these buildings and how this has caused material effects elsewhere. Hence, the artwork throws light on one of the many hidden relations that have made this urban development possible at this location, revealing the anthropos, as that geological agent, at work. The development at Djurgårdsstaden is conditioned by its relationship to the earth and clays elsewhere, to wounds that are made to the texture of the planet in other locations. "Köttet av en kvinnlig gud" also has a certain amount of bodily "icky-ness" about it – as it meshes together material and worldly references to the fabric of a female(?) body with barbequing. It is also a critique of the human-animal relations maintained during food consumption and points towards a cannibalizing of the earth. Talking to one of the residents reveals that it does not function as a barbecue – there is not enough draft to sustain the fire – and it is not particularly liked as a feature. This artwork is threading yet another line into the materializing stories of this place, stories that underscore the relations with other species, other places and other times. However, it is an artwork that risks launching into negativity about climate and environmental change, in that it "lenses" the burden of climate change onto some specific people's actions and consumer choices. In fact, the matterings of climate change are much more complex than this, where not only capital movement but also a range of other sources and forces are at work. While the development at Norra Djurgårdsstaden can be discussed as a gentrified technofix, the architecture and infrastructure still focus on material enticement towards more sustainable action. The material structures emphasize recycling, re-using and reducing as important activities to lessen the strain on planetary resources. As one resident states – *I am well aware of both the use of technofixes and gentrification in this place; that living here is not for the masses* – but this is one way of acting differently in these challenged times. While it has been a long time since this area, the oaks and their surrounds were outside the "imperial space" (Tsing 2012), maybe unruliness and the provision of locations of refuge can still be found inside such thoroughly managed places and hopefully make room and provide hospitality for a lot of othered others. Hence, this urban-ecologized place is one version of alter-worlding. It may not stretch far enough, and only include a select few, but it nevertheless reflects hope and also insight into the limited range of options situated human persons can have in these times.

As an apparatus, the artwork lenses us onto and temporarily stabilizes some earthly workings for observation but hides others. It is located on the shoreline that may become inundated by sea-level rise through climate change and the melting of Arctic ice. If not balanced by land-rise, these newly built houses, the brick artwork (although rumour says it is a temporary installation) and many shore-bound lifeworlds are at stake for being engulfed by waters. They will eventually turn into submerged archaeologies – and this is a relation, and potential knotting and unknotting, of connections that has gone unanticipated by trees, artists and developers alike, as they remember different futures.

Situating the troubles – Lyell's gradual approach

As stated by Rose (2013, 1) "(t)ime and agency are troubled, relationality is troubled, situatedness is troubled. We are tangled up in trouble". Lyell was deeply entangled in time-making and a major keeper of continuous geological deep time. Carrying on the work of James Hutton, Lyell suggested to name the rather stable climate of the last 11,000 years or so *the recent period*, which later on was named the Holocene epoch – meaning "entirely recent" or "new" – by Paul Gervais (1816–1879), a French geologist who pushed Lyell's idea internationally (Lewis and Maslin 2015, 172–173). As Ghosh (2016, 19–22) has noted, Lyell forwarded a gradual approach to planetary change, an approach that promoted geological explanations around the smooth forces of nature but which also played into the same temporal registers as used elsewhere in society at that time in literature and art. According to Ghosh, it worked to turn a blind eye to the disrupting powers of environmental catastrophes and the fracturing of time prevalent in earlier events, scholarship and storytelling. Many researchers, such as Chakrabarty (2009), Parikka (2016) and Rose (2011), urge unconventional history writing – they advocate a perspective that engages in histories of deep time but which also acknowledges the significant sources and forces of non-human temporal authorship, which stretches across both natural and humanistic science. There is also, as Bastian (2012) points out, a need for new ways of timekeeping and for adjusting to ecological relations that need to be cared about. There is a need to set priorities right in these troubled times and engage in worlding processes differently – to develop new sensitivities for re-lating. There is also a need to acknowledge ruptures of catastrophic times (see also Stengers 2015) both in the past and potentially in the future at this place. Hence, also Lyell's type of time keeping needs to be re-considered. The gradual approach as well as the conventional linear history writing may have lulled us into the troubled times.

Most often, archaeological materials are arranged in linear chronotopes (in narratives, exhibition, typologies etc.). It is more uncommon to analyze archaeology as temporally queer materialization processes that project into each other out of bound of neatly periodized slots (see Fredengren 2018a). Not even the contemporary is a safe space. In the locality around Husarviken and Lyell's Oak, processes of different temporal rhythms interfere, interfinger and burst open the contemporary as the temporalizing mess it is. Here the range of non-linear, oddly entangled materializing processes work at different paces. These processes with their mix of deep and shallow time relatings are more-than-contemporary. Hence, temporal vocabularies need to be expanded to cater to both constancies and change, with an openness for queerly temporalizing becomings, meetings, unexpectancies, latencies and lingerings and lastings, as well as for differing speeds of change, calamities or unexpected eruptions. Here at Husarviken planetary time, the long life cycles of trees, developers' time, anticipations, long-lasting commitments, the times of gentrification, appropriation, outmodedness and innovation coincide and fold together in queerly timed materializing phenomena. They can be understood as multi-agential clock-sites that draw together a range of different materializing temporal ecological relations. This speaks to Norra Djurgårdsstaden, as the place is tuned into a possible slow rise of the water levels – the water might not strike at a gradual pace but might operate through uneven, rupturing and unexpected moves. Rob Nixon (2011) has with the concept of "slow violence" captured how oil spills, toxic wastes from industrial accidents such as that of Bhopal or the nuclear disaster of Chernobyl, have had formerly ignored effects that spread out, seep into and influence the lives of humans, animals and crops over long periods of time. Moreover, with this notion of slow violence, Nixon also underscores the critical question of the "we" of the Anthropocene; "we" as people who are differently situated and differently exposed to its workings, which is particularly significant as the death count in the global south rises and material relations urgently need attention and care. The environmental problems move across conceptual, spatial and temporal scales (Parikka 2016), and in effect such slow violence stretches in between, affects and connects not only human but also multispecies generations in uneven ways for times to come. Against that background, we may ask: Where are the toxic burdens from the old industrial grounds deposited – what relations are they merging into and affecting? What other shadows, bodies and places are produced through their disposal? How do they continue to work in other ecologies, in the Arctic and other places? And how can one develop practices of *slow care*, thinking through and engaging with such heritage materialities?

Storying more-than-human heritage ecologies

Gonzáles-Ruibal (2018, 12) has raised concerns that post-anthropocentric approaches in heritage studies disregard power analyses of, for example, gender and that "the very necessary vindication of things has meant that all-too human relations of power and conflict have virtually disappeared from our accounts" (18). In the same paper, Gonzáles-Ruibal curiously mentions Braidotti, who is an outspoken feminist scholar. There are many versions of post-humanism where Braidotti (2013) and Barad (2007) are situated quite differently compared to some of the OOOs, especially when it comes to analysis of power. They provide tools to analyse the world's becoming through relational and material-discursive changes, and in their analyses, power-relations are of huge significance. Importantly, they underline the fact that discursivities have materializing effects when they deal with gender and species differentiations and matterings. Furthermore, they alert us to the observation that the discursive field is not a privilege for humans only – material-discursive workings are also carried out by the more-than-human. Discursive practices are "(re)configurings" (Barad 2007, 148) that emerge also through intra-actions, where subjects, entities and elements ongoingly, differentially and materially articulate the world – that is, they contribute to worlding outside of, but sometimes conditioned by, anthropogenic speech acts. A prime example of this is the climate change and environmental heritage that slipped into the category next to unnoticed by humans. I previously wrote (2015, 13) that the social constructivist versions of heritagization work to "alienate a range of naturalized others from the analysis and thereby play right into the hands of misogynistic and exploitative forces" by rendering these others mute and un-agentic; in this way, heritageization carries out a materializing power play.

These are theoretical approaches that could be useful for moving critical heritage studies beyond critiques of representations only into a different kind of tracing of heritage ecologies. They can also be useful for addressing some of the problems with social constructivist and culture historical approaches to heritage. As pertinently put by Pétursdóttir (2017, 185), "rather than limiting archaeology's significance to its culture-historical approaches, a wider understanding of the archaeological project is required". Farstadvoll (2019, 40–41) describes such approaches as either overmining or undermining heritage, making use of Harman's (2013) terminology. Such contextually bound or science-grounded analyses of heritage do not exhaust or capture the excesses of a particular heritage place. As has been outlined through the case studies here, there is more to these places than is accounted for by cultural-historical human storytelling or through natural sciences alone, as such methods would not trace the myriad diffractive relations that run through these localities. For example, telling the story of a prime European gaswork technology provides one limited, cultural-historical slice of this place that, when argued as a memory of human industrial achievement, also mis-traces how it is tangled together with a range of different material workings and how it shadows over into the Anthropocene as an unacknowledged part of heritage phenomena. Telling the story of Lyell's Oak as an example of science history misses out on how a situated gradual approach to time-keeping and change and on how that story may have had a lulling effect on policy and building practices. Pétursdóttir (2017, 178) mentions that a culture-historical approach risks a covering over of a range of more-than-human legacies. I agree with this perspective, and the industrial heritage has a lot of disturbing legacies. But I would like to add to this argument, drawing on Barad: both underminings and overminings of heritage and archaeology (such as storytelling, heritage labelling and other political speech acts) have materializing effects. Heritage storytelling can attract conservation activities that stabilize buildings – the maintenance of communication routes within parks – or support the selection of artefacts to remain in museum collections. In addition to this, as dealt with by Iovino and Opperman (2014), Ghosh (2016) and Fredengren (2018b), such human storying is rarely ever human alone but a practice infiltrated by a variety of more-than-human materializing sources and forces. Taken together, this reasoning does not mean that it is futile to study the more-than-cultural histories of, for example, the Arctic, thermohaline, Lyell's Oak and isostatic uplift, gradualism, potentialities and catastrophes, Norra Djurgårdsstaden or the gasworks "tentacularly", to use

Haraway's (2016) term. Such work might launch us into materially bound histories of how energy-cultures are made, which might also provide information on how such cultures work and are folded into other ecologizing and materializing stories converging in this place but also diffracting from it. Not only does this show how such knottings are made but might also give indications of how they both materially and immaterially can be re-tied elsewise – how heritage ecologies may be used to find other ways to re-late and tie knots elsewise.

Here, the relatings between the thermohaline, Lyell's Oak, the artwork and the housing development speak and work at many different temporal registers. And, as we can see, there are myriad cuts, stabilizations, connections and conditions that conjoin and that allow this geography and other related geographies to come into being; yet this place also speaks into the tentacular workings of the Anthropocene. As rightly pointed out by Pétursdóttir (2017, 178), archaeology is a discipline deeply engaged in "stuff that remains". The Anthropocene is full of both resilient, enduring and volatile things and slowly materializing tanglements and processes. The trees come into being with the ongoing materializing temporalities of the isostatic uplift and human care practices, and they play a role in the livelihoods of creatures of many kinds and can be unfolded in their ecological relations. Their livelihood was shaped, together with many other oaks, by the ban on cutting them down "as a species" in the royal park at Djurgården. They got heritageized, and their material and temporal and materializing trajectories were bent through intra-action with humans but also in cooperation with other species a long time ago. Furthermore, they have contributed to particular approaches to geology and time-making, such as that of Lyell, that were of the gradualist school, a school that turned a blind eye to the volatile workings of geological materials.

There are a range of materializing co-workings that alter the trees and the artwork, which also transforms this place around Husarviken into heritage ecologies where anthropogenic action is one of several co-producers. To trace such ecologies, as has been done in this chapter, also points to the notion that heritage definitions are always "leaky" (as there cannot be singularities such as natur- or kulturminnen) – buildings, places, artworks come into being through the gathering of relationalities outside of the bounds of a humanly defined place. This is made apparent by "Köttet av en kvinnlig Gud", which works to draw attention to the shadow places and bodies implied in the building materials. In effect, it also points towards how humans work as geological actors in ways that underpin the working of the new urban development and how this may be conditioned around the gender politics of othering. As Farstadvoll (2019, 7) writes, material heritage has an ecological legacy and as such is "inherited by non-humans". In the current case study, many of the inheritances have an ecological legacy in a range of different places that are implicated by the resource extraction that was necessary for urban development in the past, with material implications for a range of othered others. As argued by Fredengren and Åsberg (2020), we need to be concerned about what inheritances are passed onto future generations and how we work these out not only as humans but also as multi-species ancestors. Analysing heritage in this way also alerts us to the need to tend to the wounds of the world in the many shadow places that are stitched into the materializing temporalities of the planet.

Heritage ecologies and worlding practices

As suggested by the previous case studies, heritage is ecologically/materially leaky when it comes to effected temporal trajectories. Materialization processes gather and collect relations that stretch through geological times via, for example, isostatic uplift, the thermohaline, the long materializing lifecycle of trees, the lives of humans and other critters. However, while heritage storytelling may provide an alternative for connecting natural and cultural history, it is not an orderly storying, as could be portrayed in a temporal sequence of ordinary heritage narratives that focus on singularities. To analyse heritage ecologically in this way could mean to unfold a variety of connectivities and relations that swarm together in the becoming of heritage phenomena and which move in between these and their shadows. It also means to read through the materialities; to relate to how they have "agencement" in a variety of processes; to be aware of cascading effects, wild earth and water movements; to learn to live in ways that expect the unexpected; to re-late; and to understand how a range of temporally excessive processes – processes that outlive the temporal span of human lives – are in motion.

As outlined in Fredengren and Åsberg (2020) and inspired by Rose (2011), further questions can be put to the situated and emerging worldings encountered here. Such questions could be about which and to what extent synchronized connectivities are sustained in this place and if relations between self and others are flourishing. Are the possibilities for life enhanced through heritage making – and if so, in what ways? Which and to what extent are mutualities sustained in a particular heritage ecology (or what type of relations are in place and developing)? Are the relations mutually beneficial and/or asymmetrically *co-evolving?* (see Fredengren and Åsberg 2020). Do they, as the feral biologies discussed by Tsing and picked up by Haraway (2016, 100), provide places of refuge in times when cheap nature is over and when hospitality needs to be practiced? Are there some particular temporal relations that need more time to be cared for?

These questionings might provide an alternative to the critical heritage studies in the AHD tradition that has often delivered critique only. Braidotti (2013, 54) underlines the importance to move beyond the critical mode but to creatively craft places and stories in more affirmative ways, in order, as Haraway (2016, 168) puts it: "to seek and release the energies of the past, present, and future ... for a more habitable, flourishing world". This means to carry out critical work but also to gain strength to provide visions and new figurations but which also allow for feelings and desires that can be incorporated for crafting something different. This approach is also important for the heritage sector and at the very least may serve to lift it out of years of negativity to start engaging in reparative work in broken places. Hence, while it is necessary to keep an eye on the politics of inclusion and exclusion in heritage studies, to be "staying with the troubles" as Haraway advises us to do, this may involve finding more affirmative ways of working (see Fredengren 2015). To tease out the various ecological relations and to trace the relatedness of different subjectivities is one way of dealing with what may be dystopian negativity and inspire us to try to figure out how to live on in challenged worlds. This method could be a way to think through materialities and to start imagining how other ways of worlding the world are possible in different localities.

Haraway (2008, 41) discusses this as the need for finding attachment sites "for participating in the search for more livable 'other-worlds' (autres-mondialisations) inside earthly complexity" where, for example, the inhabitants of Norra Djurgårdsstaden are out searching for such "other worlds" in their own ways and where the oaks could constitute such attachment places – for queerly temporalized *relatings* across generations. However, many of the inhabitants of this place work across temporal boundaries and make deep time interventions. Also, at Husarviken, heritageization processes sometimes interfere with variously situated lives in this place and ask to strike up conversation and ask their minds to go visiting. This way of relating to heritages is a way to explore a place through curiosity and can be taken for many places. Not only does an analysis of heritage worlding ecologies open the eyes to what is in the process of being made – it also anchors climate change in place and in between places. To approach heritage ecologically is to open oneself to querying the excessive amount of relational inheritances that slip outside heritage labelling but which still are passed onto later generations. These inheritances include climate change and environmental degradation, and their making ties together knots of past and present more-than-human generations. Some relations are tightly knotted together in this locality, whereas others are in the process of becoming looser and, ultimately, undone.

If we accept that heritage (as material interventions that make time in queer ways) has a role in creating inference patterns in a variety of ecologies, we should consider that it needs to be examined relationally to see what links alter in heritage processes and how they make a difference, in time, through time and perhaps also how they disturb the making of time and the making of relations. Rose (2011, 10–12) remarks that the conditions of life, set out as relationships and interdependencies between living and non-living, are always work in progress and it is important to pay attention to which "interspecies collaborative projects" and connectivities come into being in situated places. If these observations are used to analyse heritage as connectors in wider ecologies, they could be used to trace how these connectivities operate through exchange of information, energy or support for living things. Furthermore, following Rose (2011), they could be addressed in stories or art, as exchange pathways. By influencing such connectivities, by tying the world together, heritage processes contribute to and influence situated worldmaking, where human actions mark out and acknowledge particular inheritances. However, humans do not make the world alone, and many intra-generational binds (see Fredengren and Åsberg 2020, for a discussion of this term) go on un-noticed. Also, as aptly pointed out by Bastian and van Dooren (2017, 5), "the Anthropocene is about foldings and pleatings, about simultaneous and contradictory temporalities, about the breakdown and (re)formation of new multitemporal relations". As Bastian (2012) argues, time is not given but something that is made – in relation. This is of relevance for analysing which temporal connections come into focus and which relations get prioritized when time is made in the encounter with archaeology and heritage materialities.

As many note, there is a need for new imaginaries, inspiration and approaches for understanding other ways of world-making in these challenging times. Here, Haraway (2008, 2016) works with the concept of "worlding" but contends that she does this differently compared with, for example, Heidegger, who also uses the term. Where Heidegger saw only humans as truly alive, Haraway includes a range of excluded others as having agency in worldmaking (see Rose 2011, 9, 22; Haraway 2016, 11; see also Morton 2019, 9). To track how heritage becomes a part of worlding/worldmaking may imply an investigation of what situated worlds and possibilities for life and death are brought about in situated places: what knots and relations between living and non-living are made and un-made, what type of work-in-progress can be made out in a particular geography and what type of re-latings are implied in the entangled co-becoming of generations. The alterworlding – *autremondialisation* – or *other worlding* of Haraway (2008, 3, 41) captures the work involved in how to learn to retie the knots of mundane multispecies life on earth. In this context, both involvement with queerly working Deep Time materialization processes as well as with practical materialized traditions and knowledge could be useful for future making. Haraway is stretching for "another-globalisation" captured as a more socially just globalization that points to other ways of becoming across human-animal-nature boundaries (2016, 13). I have made a speculative commentary on this (Fredengren 2015, 2016), asking how meetings with queerly timed archaeological materialities could give rise to intensive material encounters that in turn could entice people to pay closer attention to the environment and its worldly materializing processes. This approach opens up for creative, affirmative and experimental work with heritage ecologies as worlding practices. As expressed by Haraway (2016, 12) "it matters what knots knot knots, what thoughts think thoughts, what descriptions describe descriptions, what ties tie ties. It matters what stories make worlds, what worlds make stories".

References

Ball, P. 2004. "Ancient Nuclear Power Controlled by Water." *Nature*, 1 November.

Barad, K. 2007. *Meeting the Universe Halfway: Quantum Physics and the Entanglement of Matter and Meaning*. Durham, NC: Duke University Press.

Barad, K. 2012. "Quantum Entanglements and Hauntological Relations of Inheritance: Dis/continuities, Space Time Enfoldings, and Justice-to-Come." *Derrida Today* 3(2): 240–268.

Bastian, M. 2012. "Fatally Confused: Telling the Time in the Midst of Ecological Crises." *Environmental Philosophy* 9(1): 23–48.

Bastian, M. and van Dooren, T. 2017. "Editorial Preface: The New Immortals: Immortality and Infinitude in the Anthropocene." *Journal of Environmental Philosophy* 14(1): 1–9.

Braidotti, R. 2013. *The Posthuman*. Cambridge: Polity Press.

Chakrabarty, D. 2009. "The Climate of History: Four Theses." *Critical Inquiry* 35(2), Winter: 197–222.

Eliasson, P. 2003. *Skog, makt och människor: En miljöhistoria om svensk skog 1800–1875*. PhD Dissertation, Skogs och lantbrukshistoriska meddelanden nr 25. Stockholm: Kungliga Skogs-och Lantbruksakademin.

Farstadvoll, S. 2019. *A Speculative Archaeology of Excess: Exploring the Afterlife of a Derelict Landscape Garden*. PhD Dissertation. Tromsø: UiT – The Arctic University of Norway.

Fredengren, C. 2012. "Kulturarvets värde för en hållbar samhällsutveckling." In *I valet och kvalet: Grundläggande frågor kring värdering och urval av kulturarvet*, edited by C. Fredengren, O. W. Jensen and Å. Wall, 189–224. Stockholm: Riksantikvarieämbetet.

Fredengren, C. 2015. "Nature: Cultures. Heritage, Sustainability and Feminist Posthumanism." *Current Swedish Archaeology* 23: 109–130.

Fredengren, C. 2016. "Unexpected Encounters with Deep Time Enchantment. Bog Bodies, Crannogs and 'Otherworldly' Sites. The Materializing Powers of Disjunctures in Time." *World Archaeology* 48(4): 482–499.

Fredengren, C. 2018a. "Re-Wilding the Environmental Humanities. A Deep Time Comment." *Current Swedish Archaeology* 26: 50–60.

Fredengren, C. 2018b. "Personhood of Water. Depositions of Bodies and Things in Water Contexts as a Way of Observing Agential Relationships." *Current Swedish Archaeology* 26: 219–245.

Fredengren, C. and Åsberg, C. 2020. "Checking in with Deep Time. Intragenerational Care in Registers of Feminist Posthumanities, the Case of Gärstadsverken." In *Deterritorializing the Future: Heritage in, of and after the Anthropocene*, edited by R. Harrison and C. Sterling, 56–95. London: Open Humanities Press.

Fuller, M. 2005. *Media Ecologies: Materialist Energies in Art and Technoculture*. Cambridge, MA: MIT Press.

Ghosh, A. 2016. *The Great Derangement: Climate Change and the Unthinkable*. Chicago, IL: The University of Chicago Press.

Gonzáles-Ruibal, A. 2018. "Beyond the Anthropocene: Defining the Age of Destruction." *Norwegian Archaeological Review* 51(1–2): 10–21.

Haraway, D. 2008. *When Species Meet*. Minneapolis, MN: University of Minnesota Press.

Haraway, D. 2016. *Staying with the Trouble: Making Kin in the Chthulucene*. Durham, NC: Duke University Press.

Harman, G. 2013. *Bells and Whistles: More Speculative Realism*. Winchester: Zero Books.

Harrison, R. 2015. "Beyond 'Natural' and 'Cultural' Heritage: Towards and Ontological Politics of Heritage in the Age of Anthropocene." *Heritage and Society* 8(1): 24–42.

Harrison, R. 2018. "On Heritage Ontologies: Rethinking the Material Worlds of Heritage." *Anthropological Quarterly* 91(4): 1365–1384.

Hedenström, A. and Risberg, J. 1999. "Early Holocene Shore-Displacement in Southern Central Sweden as Recorded in Elevated Isolated Basins." *Boreas – An International Journal of Quaternary Research* 28(4): 490–504.

Holtorf, C. and Högberg, A. 2014. "Communicating with Future Generations: What Are the Benefits of Preserving for Future Generations? Nuclear Power and Beyond." *The European Journal of Post-Classical Archaeologies* 4: 315–330.

Iovino, S. and Opperman, S., eds. 2014. *Material Ecocriticism*. Bloomington, IN: Indiana University Press.

ISPCC. 2019. "Chapter 4: Sea Level Rise and Implications for Low Lying Islands, Coasts and Communities." In *IPCC Special Report on the Ocean and Cryosphere in a Changing Climate*, edited by H.-O. Pörtner et al., 4:1–169. Accessed 13 August 2019. https://report.ipcc.ch/srocc/pdf/SROCC_Final-Draft_Chapter4.pdf.

Jansson, N. and Antonsson, K. 1994/1995. *Eklandskapet som miljöövervakningsobjekt*. Länsstyrelsen. Accessed 13 August 2019. www.lansstyrelsen.se/download/18.2887c5dd16488fe880d66d5d/1538562087040/Eklandskapet%20som%20miljöövervakningsobjekt.pdf.

Latour, B. 2017. *Facing Gaia: Eight Lectures on the New Climatic Regime*. London: Polity Press.

LeMenager, S. 2014. *Living Oil: Petroleum Culture in the American Century*. Oxford: Oxford University Press.

Lennartson, T. 2013. "Träd och buskar: Månghundraåriga historieberättare." In *Vårda Väl*. Stockholm: Riksantikvarieämbetet.

Lewis, S. L. and Maslin, M. A. 2015. "Defining the Anthropocene. Perspectives." *Nature* 519(7542): 171–180.

Lorimer, H. 2005. "Cultural Geography: The Busyness of Being 'More-than-representational'." *Progress in Human Geography* 29(1): 83–94.

Lyell, C. 1835. "Philosophical Transactions. On the Proofs of a Gradual Rising of the Land Certain Parts of Sweden." *Philosophical Transactions of the Royal Society of London* 121: 1–38.

Merchant, C. 1980. *The Death of Nature: Women, Ecology, and the Scientific Revolution*. San Francisco, CA: Harper & Row.

Miljöbarometern. 2015. *PM – riktlinjer för dimensionerande havsvattennivåer i Norra Djurgårdstaden*. Accessed 15 July 2019. http://miljobarometern.stockholm.se/content/docs/tema/klimat/havsnivaer/PM-riktlinjer-dimensionerande-havsnivåer-Norra-Djurgårdsstaden.pdf.

Morton, T. 2010. *The Ecological Thought*. Cambridge, MA: Harvard University Press.

Morton, T. 2019. *Humankind. Solidarity with Nonhuman People*. London: Verso.

Nationalstadsparken. 2019. "Jaktparken." *Länsstyrelsen i Stockholms län*. Accessed 13 August 2019. www.nationalstadsparken.se/default.aspx?id=2255&ptid=0.

Naturvårdsverket. 2019. *Naturminnen*. Accessed 15 July 2019. www.naturvardsverket.se/Var-natur/Skyddad-natur/Naturminne/.

Nixon, R. 2011. *Slow Violence and the Environmentalism of the Poor*. Harvard, MA: Harvard University Press.

Nordblad, J. 2017. "Time for Politics: How a Conceptual History of Forests Can Help Us Politicize the Long Term." *European Journal of Social Theory* 20(1): 164–182.

Olsen, B. 2003. "Material Culture after Text: Re-Membering Things." *Norwegian Archaeological Review* 36(2): 87–104.

Olsen, B. 2012. "The Return of Things and the Savagery of the Archaeological Object." In *Savage Objects*, edited by G. Pereira, 71–83. Lisbon: Imprensa Nacional Casa da Moeda.

Olsen, B. and Pétursdóttir, Þ. 2016. "Unruly Heritage: Tracing Legacies in the Anthropocene." *Arkæologisk Forum* 35: 38–45.

Oppenheimer, M. and Alley, R. B. 2016. "How High Will the Seas Rise?" *Science* 354(6318): 1375–1377.

Parikka, J. 2016. "Deep Times of Planetary Trouble." *Cultural Politics* 12(3): 279–292.

Pétursdóttir, Þ. 2013. "Concrete Matters: Ruins of Modernity and the Things Called Heritage." *Journal of Social Archaeology* 13(1): 31–53.

Pétursdóttir, Þ. 2017. "Climate Change? Archaeology and Anthropocene." *Archaeological Dialogues* 24(2): 175–205.

Plumwood, V. 2008. "Shadow Places and the Politics of Dwelling." *Australian Humanities Review* 44: 139–150.

Rahmstorf, S. 2006. "Thermohaline Ocean Circulation." In *Encyclopedia of Quaternary Sciences*, edited by S. A. Elias. Amsterdam: Elsevier.

Riksantikvarieämbetet. 2019. *Träd som biologiskt kulturarv*. Accessed 15 July 2019. www.raa.se/kulturarv/landskap/biologiskt-kulturarv/trad-som-biologiskt-kulturarv/.

Rose, D. B. 2011. *Wild Dog Dreaming: Love and Extinction*. Charlottesville, VA: University of Virginia Press.

Rose, D. B. 2013. "Anthropocene Noir." In *The People and the Planet 2013: Transforming the Future, Conference Proceedings*, edited by P. James, C. Hudson, S. Carol-Bell and A. Taing. Melbourne: RMIT University.

Sernander, R. 1926. *Stockholms Natur*. Stockholm: Almqvist & Wiksells Förlag.

Smith, L. 2006. *Uses of Heritage*. New York: Routledge.

Söderberg, S. 2019. *Köttet av en kvinnlig gud*. Accessed 20 July 2019. www.sarasoderberg.com/nds-kttet-av-en-kvinnlig-gud.

Steffen, W., Broadgate, W., Deutsch, L., Gaffney, O. and Ludwig, C. 2015. "The Trajectory of the Anthropocene: The Great Acceleration." *The Anthropocene Review* 2(1): 81–98.

Stengers, I. 2015. *In Catastrophic Times: Resisting the Coming Barbarism*. Translated by Andrew Goffey. Ann Arbor, MI: Open Humanities Press.

Svenska Dagbladet. 2019. *Alla hade en relation till tv-eken*. Accessed 30 June 2019. www.svd.se/alla-hade-en-relation-till-tv-eken.

TICCIH. 2014. *Letter from TICCIH the International Committee for the Conservation of the Industrial Heritage*. Accessed 30 July 2019. www.sim.se/upl/files/134753.pdf.

Tsing, A. 2012. "Unruly Edges: Mushrooms as Companion Species." *Environmental Humanities* 1: 141–154.

Wangel, J. 2013. "Hur hållbara är Hammarby sjöstad och Norra Djurgårdsstaden?" In *Hållbarhetens villkor*, edited by H. Teleman, C. Caldenby, E. Ullstad and F. von Platen, 86–103. Malmö: Arena.

Witmore, C. 2015. "Archaeology and the Second Empiricism." In *Debating Archaeological Empiricism: The Ambiguity of Material Evidence*, edited by C. Hillerdal and J. Siapkas, 37–61. New York: Routledge.

Zalasiewicz, J. 2016. "A History in Layers: What Mark Will We Leave on the Planet?" *Scientific American* 315: 31–37.

18

MOLD, WEEDS AND PLASTIC LANTERNS

Ecological aftermath in a derelict garden

Stein Farstadvoll

FIGURE 18.1 Wildlife in the thicket

Photo: S. Farstadvoll.

Introduction

Envisage an old landscape garden with impenetrable silver fir (*Abies alba*) thickets and creeping carpets of periwinkle (*Vinca minor*) draping crumbling stone terraces underneath a dense canopy of deciduous and coniferous trees. The anthropogenic presence in the lush landscape is revealed by a cluster of derelict buildings, a dilapidated Swiss chalet villa, a gardener's residence, and a garage adorned with a rusty corrugated iron roof. This is Retiro, a derelict nineteenth-century landscape garden and summer estate located in Molde, a small town on the north-western coast of Norway. The Spanish noun *retiro*, which translates to English as "retirement" or "withdrawal", was the name given to the property by the Danish consul and industrialist Christian Johnsen. The name may indeed sound familiar; the garden is named after the famous Buen Retiro Park in Madrid. In his youth, Christian was denied entrance to this wondrous park, which made him swear that he would himself create a garden that was open to everyone (Eikrem 2015, 50). The eight-and-a-half-hectare estate and landscape garden were finished in 1874, nestled in picturesque rural surroundings on the temperate and fertile south facing shore of the Fannefjord. True to Johnsen's promise, the garden was open to all, from the common farmer to the German Kaiser. Today however, the jointly entropic and verdant landscape exhibits another form of radical openness in its apparent dereliction that includes beings and objects that never were included on the original guest list.

This chapter will explore three different facets of Retiro that constitute and continue to shape the place today, namely fungi, invasive organisms, and "feral" artifacts. These examples were chosen because they are rather unconventional in the traditional frame of cultural heritage but no less importantly because they quite literally brought attention to themselves during several seasons of field survey in the garden. These organisms and things are constituent parts of the material "unconscious" (cf. Olivier 2011) that maintains, reshapes, and inhabits Retiro. They exemplify an archaeological record that exist in parallel with common anthropocentric descriptions and historical heritage-characterizations but which vitally contributes to the complex ecology that makes the place.

Framing heritage in this a way invites reflection on how it also affects and is affected by non-humans such as plants, animals, minerals, and fungi. A descriptive engagement with such entities can foster novel insights into a heritage site as a particular ecosystem. The radical "openness" of ruinous places is an underlying theme in the field of contemporary archaeology (i.e. Olsen 2010, 166–173; Farstadvoll 2016), and though focused on the nature of a feral landscape, I hope this chapter can expand this topic with concrete examples. Concern for this openness has, moreover, mostly been focused on relationships between humans and derelict places (Edensor et al. 2011; Moshenska 2014), on acts of remembering and disclosing hidden and overlooked pasts (González-Ruibal 2008), or on how abandonment opens and affords the wondrous and mysterious nature of things (Pétursdóttir 2014). Here I want to draw attention to interactions between anthropogenic objects and other non-humans and to how these also constitute a part of this ecological openness. Based on fieldwork in Retiro – a heritage site bordering the natural and cultural – my aim is thus to weave an intricate picture of how such places are constituted, maintained, and experienced.

The matter of heritage and ecology

For its northerly latitude, Fannefjorden has an unusually mild climate, owing to the large hills protecting the area from cold northerly winds and the towering Sunnmøre Alps to the south

FIGURE 18.2 A muddy garden path

Photo: S. Farstadvoll.

that intermittently spawn warm föhn winds (Jakobsen 1953, 435; Vestad 1961, 11). These climatic circumstances make this an ideal place for growing a wide variety of plants, from exotic perennials like the castor bean (*Ricinus communis*) to large trees like the Lawson cypress (*Chamaecyparis lawsoniana*) (Vestad 1961). The garden, thus, had both native and non-native plants that created an eclectic landscape exhibiting both a local and foreign character. This hybridity was further emphasized in its adornment, with statues of beings from the Greek pantheon set amidst a timber hunting-cabin folly and ponds emulating their natural counterparts found in the surrounding mountains.

No longer routinely raked and refreshed with new coats of gravel, the serpentine paths of Retiro are today pockmarked and muddy. Some are completely grown over after being blocked off by large fallen trees. Yet there are clues of people still tracing the old passages, some of which have been patched with wooden boards in the muddiest sections. Long gone, however, are the gardener and his neat topiary shaping of the boxwood (*Buxus sempervirens*) and yew (*Taxus baccata*) in the flower garden. The garden's dereliction started in the interwar period when the maintenance became too costly (Vestad 1961, 16; Parelius 1982[1953], 58). During the war, Wehrmacht forces took control over the property. The garden was used to hide vehicles, which damaged it to such an extent that a local historian described it as a ruin in 1951 (Jakobsen 1953, 437–439; Parelius 1982[1953], 58). While the northern half of the property has been owned for several decades by the municipality, the remaining section was sold by a descendant of Johnsen in 2005 to a real estate development company. Today the windows in the villa are boarded up with plywood, and the same goes for the large gardener's residence.

As an unexploited seaside property, plans to redomesticate the garden have of course emerged. In a recent assessment, the municipality asserts that "The old Retiro garden is accessible, but not in a state that invites active use" (Molde kommune 2014, 13–14, my translation). One of the proposals to rectify this problem involves "rebuilding" the garden and renovating the villa to use it as a kindergarten (e.g. Molde kommune 2014; Molde Eiendom 2014). A part of the property is also currently threatened by plans for building a new roundabout and road extension. Despite these plans, and though not assigned special status by heritage authorities, both private and public interest groups see Retiro as a valuable heritage site that needs to be rescued from its current circumstance (Farstadvoll 2016). Interestingly, however, this bureaucratic and economic inertia that has for long foreclosed the garden in a state of limbo has, as we shall see, also made it possible for a novel and alternative heritage landscape to develop.

Seeing a grove that once was an open field now covered in birch trees (*Betula pubescens*) stunted by the fierce competition for sunlight, the concerns raised about the future of the property might seem warranted. The sentiment of the municipal planners resonates with the assessment given by biologists that surveyed the area in the early 1990s:

> The efforts to clean up after ravages of the hurricane and nutrient release from the root systems of the trees have meant that the vegetation and appearance of the park now are neither particularly aesthetically inviting nor suitable for moving about in.
>
> *(Jordal and Gaarder 1995, 128, my translation)*

The 1992 New Year's Day Storm is regarded as the most damaging cyclone to hit Norway in modern times. Many of the garden's coniferous trees had grown tall in the absence of pruning, making their towering trunks vulnerable to the strong gusts of wind. Once the colossal organisms fell, the spaces left open in their wake, together with the accumulated biomass, triggered an ecological succession (cf. Franklin, Shugart and Harmon 1987) – new things started to grow, pioneering species such as raspberry (*Rubus idaeus*), fireweed (*Chamaenerion angustifolium*), and birch. This demonstrates that regardless of the biologists' judgment, Retiro's present situation might not be as uninviting, aesthetically or otherwise. Some sections of Retiro can be hard for people to move through, but the local badgers (*Meles meles*) and roe deer (*Capreolus capreolus*) traverse the dense underbrush without any problems. The nature of Retiro, thus, raises awareness of a world that extends beyond the usual conceptual frames of cultural heritage use or statutory duties of the municipality.

Retiro is a place that exposes its heritage as an emergent presence that stubbornly continues to exist despite, and because of, the lack of deliberate maintenance. A motivation for engaging with such a place is to disclose the fact that these places are also constituted by non-humans, which often are overlooked (cf. Harrison 2015; Olsen and Pétursdóttir 2016; Kobiałka 2018), and how each one of them, in different ways, is involved in the unceasing formation of heritage sites. The definition of heritage this chapter operates with thus includes things that are generally thought not to belong under this term, such as modern trash, spores, and disposable plastic spoons. I regard heritage not as something we inevitably have the chance to define and choose but as something inherited, whether we like it or not (cf. Edensor 2016; Farstadvoll 2016; Olsen and Pétursdóttir 2016). This may represent an alternative way of understanding heritage. Nevertheless, thinking heritage beyond the anthropocentric is not farfetched.

Tangible heritage is ecological because it cannot be treated apart from the "discursive and material conditions and practices" that ensure its maintenance as heritage sites and objects over time (Rubio 2016, 60). In Retiro, fungi, plants, and other non-human entities are part of the material conditions and practices that aggregate the park in its nascent feral state. To think ecologically, one must acknowledge the interplay between things and beings (cf. Bryant and Joy 2014, xi), but ecology also deals with the "transformation and flux of energy and matter" that these relationships and interactions create, maintain, dissolve, and become affected by (Begon, Howarth and Townsend 2014, 5). This chapter also aims to demonstrate that heritage is not a homogenous phenomenon (cf. Fredengren 2015) but rather something that emerges through a wide array of organisms and things. Even though objects should not be reduced to their momentary relations (cf. Harman 2011), ecology shows us that it is important to see how things and beings relate and interact, or withdraw from each other, in order to begin to understand a place. Ecology thus forms an argument against the perception of purity and separation that often exist in contemporary heritage management concerning both the "natural" and "cultural" environment (cf. Latour 1993; Lowenthal 2005; Harrison 2015; Shotwell 2016). Retiro can be classified as an example of a "fourth nature", "novel ecosystem", or "eco-fusion" (cf. Kowarik 2005; Morse et al. 2014; Rotherham 2017), as a place where the anthropogenic heritage is inseparable from what is usually regarded as the natural environment, and vice versa. This "feral" character reflects global developments, such as pollution, habitat destruction, global warming, mass extinction, and so on. However, instead of trying to eliminate presumed historical and organic imperfections, places such as Retiro can invite an opportunity to speculate about what lies ahead. In the following, I will trace a few of the feral stories I've come across through my work in the garden and how these manifest Retiro's intricate heritage.

FIGURE 18.3 Jelly fungus on a painted stick

Photo: S. Farstadvoll.

Fungal legacy

In the center of the garden's north-eastern quadrant is a small clearing with an understory crowded with wood ferns (*Dryopteris filix-mas*). The forest floor consists mostly of decomposing leaf litter interlaced with green moss and fallen twigs, which each year become dimmer under an expanding canopy of invasive sycamore (*Acer pseudoplatanus*) and silver fir. In May 2015, a gnarly and truncated branch could be seen propped up against one of the small sycamores growing along the perimeter, its bark stained with splotches of blue, green, red, and yellow paint. A truly odd and apparently nonsensical object, but what made it even weirder was a gelatinous growth protruding from several places along its surface. The substance was reminiscent of the old folk belief in star-jelly or the "rot of the stars", which is a slimy matter found on the ground after shooting stars had been seen striking through the sky (Hughes 1910; Belcher and Swale 1984). Today it is understood that this substance is most likely the fruiting body of some kind of jelly fungus (Nieves-Rivera and White 2006, 24). Based on its relatively large and lobed body, the specimen inhabiting the painted branch might be the yellow brain fungus (*Tremella mesenterica*). The yellow brain fungus is not itself digesting the branch but is rather a parasite on other fungi that metabolize the dead wood (Zugmaier, Bauer and Oberwinkler 1994). Its pale-orange, milky white and translucent fruiting body mingles with the topography of the branch, lending its own aesthetic presence to the weird chromatic collage. The astronomer Martin Beech commented that the connection between fungi and meteors in folk belief is not so farfetched even if the existence of gelatinous meteors belongs in myth, because "[a]s metaphors of decay meteors and fungi share a common image" (Beech 1989).

In October the same year, the branch was still there, but the gelatinous growths were gone and replaced with gaping scars exposing moldering heartwood. The holes were the only trace left by the fungi, apart from the rotting wood beneath the bark. The pungent smell of humus, decomposing plant matter, infuses the air in Retiro during the summer and fall. *Geosmin* is the scientific name of the chemical that makes this odor, a name derived from the Greek words for "earth smell". It is commonly a metabolic by-product of fungi and other microorganisms living in the soil (Paterson, Venâncio and Lima 2007). The same chemical is also responsible for the evocative *petrichor* smell following rainfall. Strangely enough, human noses are particularly sensitive to this fungal odor (cf. Polak and Provasi 1992), which can be described as earthy and moldy (Rousseaux et al. 2014, 115). The reason for our sensitivity may be that it is a sign of spoiling food, but it is not always a smell we avoid, because geosmin is also what gives red beets their earthy taste (cf. Lu et al. 2003). Like an invisible cloud, the odor permeates every corner of the garden but can also drift with the wind and extend an olfactory presence beyond the borders of the property. Fungi are everywhere in the garden and most other places across the world.

Retiro has an overabundance of cellulose in form of living and dead wood but also in form of anthropogenic structures such as the villa and the gardeners' residence. The buildings are themselves massive biotic objects, with an inner structure composed of large, horizontally interlocked timber logs. Where there is cellulose, you will find fungi. Under the exfoliating white paint on the walls of the villa, one can glean the remnants of ochre paint, but there are also other odd transgressions marring the weather-beaten façade. The white paint unintentionally acts as a contrasting agent to highlight the presence of algae and the rather insidious and all-around infamous genus of the fungi kingdom – the black mold (*Stachybotrys*). The same effect can also be located in one of the dilapidated bedrooms in the gardeners' residence, where the ceiling is covered in thousands of black spots giving away the presence of sporangiums brimming with black spores. The mold, specifically *S. chartarum*, which is often associated with buildings (Miller 2011, 184), can create health problems for both animals and people by producing airborne mycotoxins (cf. Vesper et al. 2000). The fungus thrives in humid cellulose-rich environments, and the buildings offer an irresistible habitat (cf. Miller, Rand and Jarvis 2009). Without being perfectly balanced by keeping humidity in check through heating and maintaining a waterproof layer of paint, old wooden houses such as the ones in Retiro become fungal gardens brimming with readily available nutrients. Humans, thus, are inadvertently talented at creating environments perfect for black mold to colonize.

In the basement of the villa, which consists of meter-thick walls of mortar and stone, one of the boarded-up windows and its frame have at some point been ripped straight out of the wall. This was undoubtedly the work of humans, but the burglars were most likely aided by fungal collaborators. Looking closely at the wooden frame, one can observe that the wood is brittle and crumbly, facilitating the forced entry by making the wooden frame more persuadable. This is the blocky mosaic of "brown rot", caused by fungi metabolizing carbohydrates and cellulose in the wood, leaving behind modified lignin (Schmidt 2006, 135–138). Fungi can transform things with their metabolic abilities, making them more malleable in such a way that it literally opens up new paths. Even in a highly fragmented state, wood that has been affected by brown rot lasts longer in humus because of its modified lignin (Schmidt 2006, 137). Such lasting fragments and detritus are some of the *nachleben* (cf. Tamm 2015) or "afterlives" initiated by a heterogenic collective of fungi. The derelict garden that we encounter today is just as much the afterlife of fungi as of human influence in the past and represents distinct but fused human and non-human legacies (cf. Farstadvoll 2019). An example of this can be encountered on the exterior wall of the privy connected to the gardener's residence. The wall is coated in a biological collage of microorganisms, creating a patchwork of colors consisting mostly of green and black among the remaining patches of pale-pink paint. Human eyes, however, are attuned and attracted to the vibrant blue-green fungal impressions dotting the biotic tessellation. These striking traces probably consist of the pigment *xylindein* produced by fungi of the genus *Chlorociboria* (cf. Edwards and Kale 1965). This "green rot" is a secondary metabolite produced within the fungi's hyphae, which diffuses into and stains the substrate (cf. Schmidt 2006, 120; Robinson and Laks 2010). This kind of "green" wood has been employed in several instances in artistic work, such as in woodcraft (Rolfe and Rolfe 1925, 161) and as a pigment in paint (Blanchette, Wilmering and Baumeister 1992, 230). At Retiro, this pigment recalls a fungal aftermath, a reminder that non-humans also create redundant pasts affecting the present. Why fungi have different colors is still largely a mystery, which alludes to an ecology that, in our perception, extends beyond simple cause and effect (Karlsen 2014).

Having focused on the "aftermath", it is interesting to note that the fungi inhabiting Retiro can as well be part of an ecological heritage, or mycelium, extending to a time *before* the garden was constructed. A pastureland dotted with meadows, copses, and farmland, still remembered by remnants of old stone fences and clearance cairns poking through the carpet of moss and accumulated vegetation. Before the garden's construction, the property was described as unkempt, hilly, and the least pretty place in the area (Jakobsen 1953, 436), which resonates with the biologists' assessment of present-day Retiro quoted in the introduction. Parts of this autochthonous landscape were integrated into the garden's architecture,

FIGURE 18.4 Looking up at a moldy roof in the gardener's residence

Photo: S. Farstadvoll.

interweaving the new organisms with old, instigating ecological upheaval and succession. The early stage of the garden was in a sense defined by an anthropogenically initiated "throwntogetherness" (cf. Massey 2005). In the golden days when it was tenderly cared for by the hands of attentive gardeners, the garden was perhaps more thrown together, ecologically speaking, than in its presently derelict character. By this, I mean that a previously unmanaged ecology was replaced with a precarious and artificial ecology that hinged on a balance provided by the constant upkeep by human hands and tools.

Fungi are part of Retiro's nutrient-scape (cf. Farstadvoll 2019), a mycological topography of metabolizable tissue and dangerous mycotoxins. Even the gelatinous yellow brain fungus is edible for humans, despite its unappetizing appearance. Encountering a place through its nutritional properties and taste is fundamentally ecological, and one might even claim that ecology is the oldest science, because humans have always been driven to understand and adapt to their non-human environment (Begon, Howarth and Townsend 2014, 5). In turn, it is equally important to note that non-humans also adapt and react to anthropogenic influences. Fungi have almost all the same senses as humans – they can "taste" chemicals and sense surfaces (Bahn et al. 2007). When walking through Retiro, plants, fungi, and other organisms sense the soles of our boots, our limbs, and respiration, which comes into contact with the environment. The realization that it is not only we who "sense" a place but that other beings can also sense our presence and the environment ought to open up possible reflections on what we think heritage and heritage experience is.

In heritage conservation, fungi is seen as one of the major agents leading to degradation of material culture (e.g. Caneva, Nugari and Salvadori 2008; Sterflinger 2010). In the eyes of ecologists, however, fungi are major actors in creating and maintaining biodiversity and general ecological processes such as soil formation (cf. Peay, Kennedy and Bruns 2008; Wallander 2014). Fungi do not discriminate between tree stumps in the forest or venerable heritage architecture. Heritage is nutrition and habitats for fungi, while mycophagy is part of the culinary heritage of many human cultures (cf. Pieroni et al. 2005). In some interpretations of Leviticus, fungi in buildings is even referred to as "leprosy on houses" (Kauserud et al. 2007), recalling its presence also in ancient contexts. As they literally permeate every environment on earth, fungi can be unruly and impossible to control (cf. Money 2007; Tsing 2015). As such, it is an inseparable and inherent part of the aesthetics and ecology of Retiro, facilitating patterns and interactions that necessitate and create the presence of the garden and which playfully overflow and percolate into the neighboring environs. Without fungi, Retiro would not be, or at least it would be something else. When structures such as Retiro are assembled, with their biotic landscape architecture and massive wooden buildings, ecological relationships and struggles are inevitably established. Cohabitating with black mold is not necessarily preferable, but it is an unavoidable part of every entanglement with wooden buildings. As such, fungi are emphatic reminders that heritage can't be excised and placed alone on its own pedestal, elevated above and beyond the very stuff it consists of.

Abducted aliens

The fungi growing in Retiro are generally endemic to the local environment or cosmopolitan species found around the globe. In contrast, a substantial amount of the plants found here are species translocated from foreign biotopes. After the gardener and the dutiful "weeding-women" ceased their work, many of these non-native plants succumbed to the alien climate

FIGURE 18.5 Crawling under the Japanese knotweed

Photo: S. Farstadvoll.

and were outcompeted by other plants. However, some of them survived and even flourished (e.g. Vestad 1961, 16). After being left to its own device for half a decade, Retiro has therefore become an unintended "laboratory" for studying the spread of introduced species (cf. Jordal and Gaarder 1995, 62, 128). In a recent survey of the biodiversity on the property, the biologists recorded several alien species such as silver fir, sycamore, and lesser periwinkle (Gaarder and Vatne 2013, 8). In conclusion, the report recommended exterminating the non-native species that exhibit high risk for the local environment. Many of the original plants that survive thus have the double characteristic of being both heritage and pests.

One of the most notorious inhabitants found in the garden is the Japanese knotweed (*Fallopia japonica*): a hardy perennial that annually grows smooth stems that form near-impenetrable thickets. As the name implies, the plant was originally native to East Asian countries such as Japan (Conolly 1977). It is sometimes described as a plant that can outlive the gardener and the garden itself (Bailey and Conolly 2000, 93). The most substantial part of the knotweed is its network of rhizomes, a large web of roots located underground that enables a knotweed colony to spread out and infiltrate its surroundings. Its rhizomes can extend as far as 7 meters away from the stand and reach as far down into the soil as 2 meters (Child and Wade 2000). It has an amazing ability to adapt to different soil and climate types (e.g. Rouifed et al. 2012). In an old catalogue from the mid-nineteenth century, knotweed is said to have many positive properties, such as being both ornamental and medicinal and not least inextirpable (Bailey and Conolly 2000, 94). In non-native habitats, it usually propagates and spreads asexually through vegetative fragments from its rhizomes and other parts of the plant (cf. Conolly 1977; Hollingsworth and Bailey 2000). This ability to grow clonal colonies through vegetative fragments is one of the reasons the plant is so hard to exterminate and is, thus, viewed as a terrible pest, listed among *100 of the World's Worst Invasive Alien Species* by the International Union for Conservation of Nature (Lowe et al. 2000). In Norway, it is recorded on the so-called "black list" as a highly invasive and foreign species and placed in the "severe" impact category (Gederaas et al. 2012).

The Japanese knotweed was probably introduced to Norway in the nineteenth century (Schübeler 1886, 592; Fremstad and Elven 1997, 7), making it an enduring addition to the biodiversity of the region, even if it at the same time threatens it (e.g. Goodenough 2010; Schlaepfer, Sax and Olden 2011; Richardson and Ricciardi 2013). It is not known when the knotweed was first introduced at Retiro, but its presence was first recorded in the early 1960s in the flower garden by the villa and in the landscape garden west of the carriage entrance (Vestad 1961). The knotweed inhabiting Retiro today is not one hegemonic blob but consists of several separate stands. The colonies are mostly located in the southern part of the garden, which has open areas with a lower density of trees. The places where the knotweed grows mark the general areas where they were planted and function as lingering memories of an aesthetic order of the past. It is quite rare to see Japanese knotweed bloom in Norway, and for some reason it is becoming an increasingly scarce phenomenon (cf. Fremstad 2012). But Retiro remains one of the places one can see the small white flowers in full bloom in late autumn.

Recently a large number of the knotweed colonies in Retiro started to wither and disappear. It turned out the colonies had been sprayed with herbicide. Not many days after the colonies were sprayed, however, the brown patches of wasteland quickly became botanical battlegrounds of pioneering species fighting to inherit precious square meters of sunlight. In a twist of irony, one of the pioneer species was the herbaceous perennial creeping jenny (*Lysimachia nummularia*), which is listed in the "high impact" category, just below the knotweed (cf. Gederaas et al. 2012). Moreover, several smaller colonies of knotweed have survived the attack, hiding amongst the dense vegetation around the gardener's residence. Small patches have also moved beyond the property border, temporarily escaping the poison. Being an example of heritage that fights backs with a rhizomatic persistence, there might still be a possibility to see the knotweed bloom in the garden. Thus, because of their autopoietic persistence, plants such as knotweed can be regarded unforeseen futures, memories of consequences that overlooked or underestimated their ecology.

FIGURE 18.6 Spanish slug (*Arion vulgaris*)

Photo: S. Farstadvoll.

The Spanish slug (*Arion vulgaris*) is another sticky alien that has made the garden its home. In northern Europe, the brown slug is frequently found in gardens, wastelands, cemeteries, ditches, and small grasslands (cf. Kozłowski and Kozłowski 2011; Slotsbo 2014), which are usually anthropogenic landscapes. The slug eats a wide range of plants and animals it can catch (Hatteland et al. 2013, 14). In Norway, this rather insidious slug has become one of the most reviled icons of invasive species, probably because of massive media coverage and the animal's recognizable appearance (cf. Qvenild, Setten and Skår 2014). The narrative regularly repeats stories about the slug as a pest in gardens and the attempts and methods to exterminate the gastropods.

The slug is a rather recent inclusion to Retiro. Being first recorded in Norway in 1988 (Hatteland et al. 2013), the town of Molde was one of the three places the species was first found (see von Proschwitz and Winge 1994). Partly similar to the knotweed, slugs of the *Arion* genus can self-fertilize and start colonies from one individual (Hatteland et al. 2015, 317). The Spanish slug took advantage of an inviting opening in the local ecology but was not alone in its actions. The eggs of the slug are usually found in soil and can consequently be dispersed through dirt sticking to shoes or as stowaways in potted plants (cf. Slotsbo 2014), which makes their distribution ecologically intertwined with our movements and artifacts. The slug is certainly not unique to this particular garden, but it is nevertheless one small piece in the material composition, ecologically rewiring the place through its sluggish rhythms and materiality. Whatever definition one may give tangible heritage – as places – it will always and inevitably involve spaces that are shared.

In the instance of Retiro, the knotweed did not invade a pristine ecology but was rather "abducted" and grafted into a different environment. Retiro has never been "pure", not even before the garden was constructed. These alien species were translocated to a place that already was in some way different from an imagined pristine environment, untouched by humans and their non-human entourage (e.g. Tassin and Kull 2015). If the foreign plants and animals didn't belong here, then Retiro can be categorized as an invasive place. The knotweed and all the other non-native organisms that make Retiro can represent a sort of virulent or perhaps "pathogenic" aesthetics. If you are not learned in botany, exotic and aesthetically alluring species such as the knotweed or goat's beard (*Aruncus dioicus*) might not seem that out of place. However, the Anthropocene and current global ecological crises can in a paradoxical way appear disquietingly "attractive" – from enormous algal blooms draining lakes of oxygen triggered by agricultural runoff to beautiful red sunsets created by anthropogenic aerosol suspended in the atmosphere. Just as we cannot always control what we inherit (cf. Olsen and Pétursdóttir 2016; Pétursdóttir 2017), it also follows that we do not always know who or what is the "heir" of a place. Consequently, memories are ecological objects if they are given room to be intertwined with a material environment (e.g. Olsen 2010; Olivier 2011).

Feral artifacts

The first things that come to mind when delving into the ecology of a garden are "natural" objects, such as trees, flowers, and soil. However, Retiro is also inhabited by a diverse range of anthropogenic artifacts and structures, such as the ruins of a Victorian hothouse, stacked stone terraces, man-made ponds, drainage channels, clearance cairns, and broken chain-link fences. These objects all have their own respective places in the local ecology: for example, the fences dictate the movement of large mammals, and the efflorescent concrete of the hothouse

FIGURE 18.7 A plastic lantern and tree litter

Photo: S. Farstadvoll.

slowly leaches mineral nutrients into the surrounding soil. At the dawn of the Anthropocene, the presence and interaction of manmade objects with ecosystems has come into focus. An interesting manifestation of this is that display cases in the zoological exhibit at the natural history museum in Oslo have been infiltrated with modern trash. One of the displays shows a taxidermied male mallard (*Anas platyrhynchos*) swimming in a pond accompanied by a plastic bottle. Similar to the duck, the bottle has been given a placard with taxonomic information, such as species, family, order, offspring, size, and distribution. The placard for the "species" *plastic utrem* is certainly more illustrative and metaphoric than a serious attempt to make an artifact taxonomy. Nevertheless, this placard is a subversion of a nature/culture dichotomy by treating an anthropogenic artifact in the same way as living organisms.

When the last swan (*Cygnus olor*) that graced the ponds in Retiro died and ended up as stuffed animal on a wall in the villa (Parelius 1982[1953], 58), new organisms as well as things moved in. In the course of surveying Retiro, the aforementioned *plastic utrem* had a ubiquitous presence, not to mention its close relative *alum utrem* and *vitrum utrem*. While plastic and aluminum bottles are relative newcomers, the glass bottles have run wild in the garden since its construction. Their presence is vaguer than the newcomers', because most of them have had the time to sink into the soil thanks to the tilling of earthworms, not least because they have become substrates for a range of different organisms. When seen as part of an ecosystem, regardless of their potentially damaging qualities, objects such as plastic bottles can be understood both as invasive and pioneering objects. Ecosystems take time to incorporate and react to new objects. A good example is when windows constitute a collision hazard for birds because they have not had the time to adapt their senses to the materiality of glass (e.g. Martin 2011).

Within a grove of spruce trees (*Picea abies*) in the south-western quadrant of the garden, I stumble over the remnants of plastic lanterns, a recent example of invasive objects. A large cemetery borders Retiro in the west, which is a likely origin for these lanterns, but the question is how they ended up under a spruce tree. The most likely explanation is that birds such as crows, magpies, and seagulls have discovered that the stearin in lanterns and candles is edible (e.g. Higuchi 2003), even if it is not digestible like tallow. The lanterns found at cemeteries are usually made out of brittle plastic that birds have learned can be smashed open by dropping them from heights. This example of animal vandalism has been observed at the local cemetery (cf. Heen 2012). Birds are important for the dispersal of seeds for a large range of plant species (e.g. Howe and Smallwood 1982), but as seen in Retiro, they can also disperse man-made objects. Though not seeds, the lanterns are objects that affect the environment in different ways, such as by dispersing involuntary memories, absorbing and leaching chemicals, and sheltering small ponds of water, forming habitats for microorganisms. Birds, modern mortuary practices, and dense vegetation create a strange ecological interplay with weird mixtures of anthropogenic and organic components.

As a contrast to the recent material inclusions drifting into Retiro, there are older remnants persisting in the evolving landscape. These objects are often well established, meaning they have had the time to become integrated with the local ecology. One notable example is an assemblage of creamy-white statue fragments inhabiting the undulating depression that once was the "Atlantic Ocean" pond, now a mire covered with grass and small saplings drawing nutrients from the old sediments. The busiest paths in the garden congregate towards the pond, where the brim forms a natural roundabout. The creamy-white fragments were originally part of a statue depicting Triton riding a hippocamp while blowing into a conch. There

FIGURE 18.8 An ephemeral pond in the curl of the hippocamp's tail

Photo: S. Farstadvoll.

is evidence that these vestiges draw the attention of passersby, for example, that the more ornamental fragments have been poached while the unwieldy and indistinct pieces have been left alone. This is one example of environmental selection and sorting found at Retiro, where sometimes it is the least "attractive" things that persist. After observing the haphazard collection of fragments over four years, it is evident that they are moved, mostly by humans. Some have probably been posed to create picturesque scenes for photography, and other times they have evidently been played with by children. The remaining fragments have, however, not yet strayed far from the original pedestal.

It is not only humans that have been drawn towards the ceramic shards, as apparent by the layer of biofilm covering their surfaces. These microscopically thin layers consist of algae, cyanobacteria, lichen, and other microorganisms, which act as a living and dynamic patina. On specific types of substrates, some types of biofilms have been documented to have bioprotective properties (e.g. Cutler et al. 2013); in other words, they can help preserve the object they grow on. Inadvertently, these microorganisms care for the substrate they extract nutrients from and live on and are in turn preyed upon and generate a habitat for creatures such as the Spanish slug. In the center of the curly tail of the hippocampus, a small pond regularly forms from rainwater. In this artificial cavity, the fake marble is stained green from the algae and bacteria taking advantage of the humidity. On closer inspection, the surface of the ceramic material is porous and rough, inadvertently creating a topography where organisms such as lichen, algae, bacteria, and moss can literally get a hold. Anthropogenic materials regularly interface with other non-humans, as for example through microscopic pores on the ceramic surface of the statue fragments or the depression of the "Atlantic Ocean", collecting organic detritus that decomposes and becomes a source of nutrients for plants. Retiro has a reciprocal and inherent relationship with a "heritage biota" (cf. Viles and Cutler 2012), which makes it the place it is today. In that sense, autopoietic organisms and allopoietic objects, such as disposable plastic lanterns, are both equally important in their own unique ways for the being and becoming of a place.

Conclusion

Regardless of what the future may bring, Retiro is neither retiring nor retreating but rather developing and advancing, notwithstanding the inevitable ecological upheaval that looms on the horizon in form of urban "revitalization". This chapter has engaged with Retiro as it is encountered today in its derelict state. The presence of fungi demonstrates that the heritage of Retiro in many ways depends on the very matter it is made of – organisms that usually go unmentioned and are even regarded as damaging. The introduced plants and Spanish slug highlight the tension and dichotomy between foreign and native and demonstrate that heritage may exist as open places leading to unforeseen "heirs". The many fragments of distant and recent pasts that constitute the landscape allude to the complex afterlife of anthropogenic objects, where they gain unintended roles in the local ecosystem.

These examples demonstrate a small piece of the complex non-human undercurrents, originating both in distant and recent pasts, which make up the specific ecology of a place like Retiro. The artifacts that contribute to the feral landscape can infiltrate consciousness as well as bodies, like black mold spores clinging to lungs. The human presence at Retiro is ephemeral, activated and partly dictated by Retiro's ecological character. Despite their disproportional impact on the environment, there are mostly no people present within its boundaries, and they make up an unsubstantial part of the biomass. As heritage, the coagulating collage of human and non-human presences reflect an ecology in transition, heading in unknown directions. Heritage definitions that also include non-humans force us to acknowledge that the intricate ecology of a site also informs the human experience but without reducing a place to human conclusion – there is always something more just behind what we currently experience. This thought also correlates with an idea that science and other endeavors seeking to understand things are often more about discovering and broadening the number of known objects than reducing them to exactitudes (cf. Harman 2013, 97–99).

This begs the question of how places of heritage can be explored as inseparable from a diverse non-human realm, crisscrossed with the strange *umwelts* of black mold, knotweed, and fragments of ceramic statues. Retiro is an inviting place but not only for the reasons that most likely become highlighted in the context of history, archaeology, or municipal planning. To grasp the full extent of heritage places and their ecologies, it is, I would argue, necessary to establish connections also to things that might usually be seen as inconsequential or even irrelevant. This ecological practice is one of joining futures, of holding different loyalties and severing bonds, of acknowledging the particular presence of fungi and polystyrene packaging. The derelict garden, as observed by the biologists surveying it, is, I would claim, an unintended experiment for the dispersal of plants and other introduced lifeforms. But it is also a heritage experiment, absorbing some of the material aftershocks of the modern and prosperous society that envelops it. Like a mire that can absorb torrential rain and avert floods, or a mangrove forest that protects the land from devastating storms, Retiro may from an ecological perspective also hold an absorbent and filtering character, inviting things in while rejecting others. In one way, Retiro is reconstructing, or perhaps more succinctly, *growing* dynamically without any clear plan for an imagined future. The appreciation some ecologists have started to show for the potential beneficial aspects of novel ecologies on the preservation of some species (e.g. Maclagan, Coates and Ritchie 2018) can serve as an inspiration to look closer at the ecological complexity of heritage environments that sometimes get left behind in the hunt for a pristine past. This is not a heritage that asserts comforting histories about who we were or are but rather one that makes visible the inheritance of an uncertain future.

References

Bahn, Y.-S., Xue, C., Idnurm, A., Rutherford, J. C., Heitman, J. and Cardenas, M. E. 2007. "Sensing the Environment: Lessons from Fungi." *Nature Reviews Microbiology* 5: 57.

Bailey, J. P. and Conolly, A. P. 2000. "Prize-Winners to Pariahs – A History of Japanese Knotweed *s.l.* (Polygonaceae) in the British Isles." *Watsonia* 23: 93–110.

Beech, M. 1989. "Shooting Stars and Gelatinous Fungi." *Mycologist* 3(2): 144.

Begon, M., Howarth, R. W. and Townsend, C. R. 2014. *Essentials of Ecology* (4th edn). Hoboken, NJ: Wiley.

Belcher, H. and Swale, E. 1984. "Catch a Falling Star." *Folklore* 95(2): 210–220.

Blanchette, R. A., Wilmering, A. M. and Baumeister, M. 1992. "The Use of Green-Stained Wood Caused by the Fungus Chlorociboria in Intarsia Masterpieces from the 15th Century." *Holzforschung* 46(3): 225–232.

Bryant, L. and Joy, E. A. 2014. "Preface: Object/Ecology." *O-Zone: A Journal of Object-Oriented Studies* 1(1): i–xiv.

Caneva, G., Nugari, M. P. and Salvadori, O., eds. 2008. *Plant Biology for Cultural Heritage: Biodeterioration and Conservation*. Los Angeles, CA: Getty Publications.

Child, L. E. and Wade, M. 2000. *The Japanese Knotweed Manual: The Management and Control of an Invasive Alien Weed*. Chichester: Packard Publishing.

Conolly, A. P. 1977. "The Distribution and History of Some Alien Species of *Polygonum* and *Reynoutria*." *Watsonia* 11: 291–311.

Cutler, N. A., Viles, H. A., Ahmad, S., McCabe, S. and Smith, B. J. 2013. "Algal 'Greening' and the Conservation of Stone Heritage Structures." *Science of The Total Environment* 442: 152–164.

Edensor, T. 2016. "Incipient Ruination: Materiality, Destructive Agencies and Repair." In *Elements of Architecture: Assembling Archaeology, Atmosphere and the Performance of Building Spaces*, edited by M. Bille and T. F. Sørensen, 348–364. London: Routledge.

Edensor, T., Evans, B., Holloway, J., Millington, S. and Binnie, J. 2011. "Playing in Industrial Ruins: Interrogating Teleological Understandings of Play in Spaces of Material Alterity and Low Surveillance." In *Urban Wildscapes*, edited by A. Jorgensen and R. Keenan, 65–79. London: Routledge.

Edwards, R. L. and Kale, N. 1965. "The Structure of Xylindein." *Tetrahedron* 21(8): 2095–2107.

Eikrem, T. H. 2015. "Villa Retiro – ein arena for nettverksbygging og allmenn rekreasjon." *Romsdalsmuseet årbok* 2015: 45–57.

Farstadvoll, S. 2016. "Blant Restene Av Retiro: Forfall Og Kulturminner." *Primitive tider* 18: 161–181.

Farstadvoll, S. 2019. "Growing Concerns: Plants and Their Roots in the Past." *Journal of Contemporary Archaeology* 5(2): 174–193.

Franklin, J. F., Shugart, H. H. and Harmon, M. E. 1987. "Tree Death as an Ecological Process." *BioScience* 37(8): 550–556.

Fredengren, C. 2015. "Nature: Cultures. Heritage, Sustainability and Feminist Posthumanism." *Current Swedish Archaeology* 23: 109–130.

Fremstad, E. 2012. "Parkslirekne Reynoutria japonica (tidl. *Fallopia japonica*)." *Artsdatabankens faktaark*, No. 246.

Fremstad, E. and Elven, R. 1997. "Fremmede planter i Norge. De store Fallopia-artene." *Blyttia* 55: 3–14.

Gaarder, G. and Vatne, S. 2013. *Retiroparken i Molde. Grunnlagskartlegging av naturmangfold*. Miljøfaglig Utredning, notes 2013–2043.

Gederaas, L., Moen, T. L., Skjelseth, S. and Larsen, L.-K., eds. 2012. *Alien Species in Norway – With the Norwegian Black List 2012*. Trondheim: The Norwegian Biodiversity Information Centre.

González-Ruibal, A. 2008. "Time to Destroy: An Archaeology of Supermodernity." *Current Anthropology* 49(2): 247–263.

Goodenough, A. 2010. "Are the Ecological Impacts of Alien Species Misrepresented? A Review of the 'Native Good, Alien Bad' Philosophy." *Community Ecology* 11(1): 13–21.

Harman, G. 2011. *The Quadruple Object*. Winchester: Zero Books.

Harman, G. 2013. *Bells and Whistles: More Speculative Realism*. Winchester: Zero Books.

Harrison, R. 2015. "Beyond 'Natural' and 'Cultural' Heritage: Toward an Ontological Politics of Heritage in the Age of Anthropocene." *Heritage & Society* 8(1): 24–42.

Hatteland, B. A., Roth, S., Andersen, A., Kaasa, K., Støa, B. and Solhøy, T. 2013. "Distribution and Spread of the Invasive Slug *Arion vulgaris* Moquin-Tandon in Norway." *Fauna norvegica* 32: 13–26.

Hatteland, B. A., Solhøy, T., Schander, C., Skage, M., von Proschwitz, T. and Noble, L. R. 2015. "Introgression and Differentiation of the Invasive Slug *Arion vulgaris* from Native *A. ater*." *Malacologia* 58(1–2): 303–321.

Heen, E. 2012. "Trist at noen tar påskeliljene. Blomstene forsvinner fra gravgården." *Romsdal Budstikke*, 24 March: 8.

Higuchi, H. 2003. "Crows Causing Fire." *Global Environmental Research* 2: 165–167.

Hollingsworth, M. L. and Bailey, J. P. 2000. "Evidence for Massive Clonal Growth in the Invasive Weed *Fallopia japonica* (Japanese Knotweed)." *Botanical Journal of the Linnean Society* 133: 463–472.

Howe, H. E. and Smallwood, J. 1982. "Ecology of Seed Dispersal." *Annual Review of Ecology and Systematics* 13: 201–228.

Hughes, T. M. 1910. "'Pwdre Ser' (The Rot of the Stars)." *Nature* 83: 492–494.

Jakobsen, J. C. 1953. "Litt om hagedyrkingen i Bolsøy før og no." In *Bolsøyboka: En natur-og samfundshistorisk skildring av Bolsøy prestegjeld og herred. Tredje del bind 1*, edited by J. Olafsen-Holm, 424–441. Molde: Mindor Bolsø & Co.s Trykkeri.

Jordal, J. B. and Gaarder, G. 1995. *Biologisk mangfold i Molde. Del 1. Hovedrapport.* Molde: Molde kommune.

Karlsen, P. 2014. "Hvorfor har soppene farger?" *Sopp og nyttevekster* 10(4): 23–25.

Kauserud, H., Svegården, I. B., Sætre, G. P., Knudsen, H., Stensrud, Ø., Schmidt, O., Doi, S., Sugiyama, T. and Högberg, N. 2007. "Asian Origin and Rapid Global Spread of the Destructive Dry Rot Fungus *Serpula lacrymans*." *Molecular Ecology* 16(16): 3350–3360.

Kobiałka, D. 2018. "Living Monuments of the Second World War: Terrestrial Laser Scanning and Trees with Carvings." *International Journal of Historical Archaeology* 23(1): 129–152.

Kowarik, I. 2005. "Wild Urban Woodlands: Towards a Conceptual Framework." In *Wild Urban Woodlands New Perspectives for Urban Forestry*, edited by I. Kowarik and S. Körner, 1–23. Berlin: Springer.

Kozłowski, J. and Kozłowski, R. J. 2011. "Expansion of the Invasive Slug Species *Arion lusitanicus* Mabille, 1868 (Gastropoda: Pulmonata: Stylommatophora) and Dangers to Garden Crops – A Literature Review with Some New Data." *Folia Malacologica* 19(4): 249–258.

Latour, B. 1993. *We Have Never Been Modern.* Cambridge, MA: Harvard University Press.

Lowe, S., Browne, M., Boudjelas, S. and De Poorter, M. 2000 [2004]. *100 of the World's Worst Invasive Alien Species: A Selection from the Global Invasive Species Database.* The Invasive Species Specialist Group (ISSG) a Specialist Group of the Species Survival Commission (SSC) of the World Conservation Union (IUCN).

Lowenthal, D. 2005. "Natural and Cultural Heritage." *International Journal of Heritage Studies* 11(1): 81–92.

Lu, G., Edwards, C. G., Fellman, J. K., Mattinson, D. S. and Navazio, J. 2003. "Biosynthetic Origin of Geosmin in Red Beets (*Beta vulgaris* L.)." *Journal of Agricultural and Food Chemistry* 51(4): 1026–1029.

Maclagan, S. J., Coates, T. and Ritchie, E. G. 2018. "Don't Judge Habitat on Its Novelty: Assessing the Value of Novel Habitats for an Endangered Mammal in a Peri-Urban Landscape." *Biological Conservation* 223(July): 11–18.

Martin, G. R. 2011. "Understanding Bird Collisions with Man-Made Objects: A Sensory Ecology Approach." *Ibis* 153(2): 239–254.

Massey, D. 2005. *For Space.* London: Sage.

Miller, J. D. 2011. "Health Effects from Mold and Dampness in Housing in Western Societies: Early Epidemiology Studies and Barriers to Further Progress." In *Fundamentals of Mould Growth in Indoor Environments and Strategies for Healthy Living*, edited by O. C. G. Adan and R. A. Samson, 188–210. Wageningen: Wageningen Academic Publishers.

Miller, J. D., Rand, T. G. and Jarvis, B. B. 2009. "Stachybotrys Chartarum: Cause of Human Disease or Media Darling?" *Medical Mycology* 41(4): 271–291.

Molde Eiendom. 2014. *Barnehage på Retiro-området? En vurdering av realisme, utfordringer, økonomi og gjennomførbarhet.* Molde: Molde kommune.

Molde kommune. 2014. *Detaljregulering for Retiro.* Plan No. 200611.

Money, N. P. 2007. *The Triumph of the Fungi: A Rotten History.* Oxford: Oxford University Press.

Morse, N. B., Pellissier, P. A., Cianciola, E. N., Brereton, R. L., Sullivan, M. M., Shonka, N. K., Wheeler, T. B. and McDowell, W. H. 2014. "Novel Ecosystems in the Anthropocene: A Revision of the Novel Ecosystem Concept for Pragmatic Applications." *Ecology and Society* 19(2): article 12.

Moshenska, G. 2014. "Children in Ruins: Bombsites as Playgrounds in Second World War Britain." In *Ruin Memories: Materialities, Aesthetics and the Archaeology of the Recent Past*, edited by B. Olsen and Þ. Pétursdóttir, 230–249. London: Routledge.

Nieves-Rivera, Á. M. and White, D. A. 2006. "Ethnomycological Notes. II. Meteorites and Fungus Lore." *Mycologist* 20(1): 22–25.

Olivier, L. 2011. *The Dark Abyss of Time: Archaeology and Memory.* Lanham, MD: AltaMira Press.

Olsen, B. 2010. *In Defense of Things: Archaeology and the Ontology of Objects.* Lanham, MD: AltaMira Press.

Olsen, B. and Pétursdóttir, Þ. 2016. "Unruly Heritage: Tracing Legacies in the Anthropocene." *Arkæologisk Forum* 35: 38–45.

Parelius, N. 1982 [1953]. "Retiro på Fannestranden." In *Hedersskrift til Nils Parelius på 70-årsdagen den 24. mai 1982*, edited by B. Austigard, O. R. Grüner and J. Julnes, 55–63. Molde: Romsdal Sogelag.

Paterson, R. R. M., Venâncio, A. and Lima, N. 2007. "Why Do Food and Drink Smell Like Earth?" In *Communicating Current Research and Educational Topics and Trends in Applied Microbiology*, edited by A. Méndez-Vilas, 120–128. Badajoz: Formatex.

Peay, K. G., Kennedy, P. G. and Bruns, T. D. 2008. "Fungal Community Ecology: A Hybrid Beast with a Molecular Master." *BioScience* 58(9): 799–810.

Pétursdóttir, Þ. 2014. "Things Out-of-Hand: The Aesthetics of Abandonment." In *Ruin Memories: Materialities, Aesthetics and the Archaeology of the Recent Past*, edited by B. Olsen and Þ. Pétursdóttir, 335–364. London: Routledge.

Pétursdóttir, Þ. 2017. "Climate Change? Archaeology and Anthropocene." *Archaeological Dialogues* 24(2): 175–205.

Pieroni, A., Nebel, S., Santoro, R. F. and Heinrich, M. 2005. "Food for Two Seasons: Culinary Uses of Non-Cultivated Local Vegetables and Mushrooms in a South Italian Village." *International Journal of Food Sciences and Nutrition* 56(4): 245–272.

Polak, E. H. and Provasi, J. 1992. "Odor Sensitivity to Geosmin Enantiomers." *Chemical Senses* 17(1): 23–26.

Qvenild, M., Setten, G. and Skår, M. 2014. "Politicising Plants: Dwelling and Invasive Alien Species in Domestic Gardens in Norway." *Norwegian Journal of Geography* 68(1): 22–33.

Richardson, D. M. and Ricciardi, A. 2013. "Misleading Criticisms of Invasion Science: A Field Guide." *Diversity and Distributions* 19(12): 1461–1467.

Robinson, S. C. and Laks, P. E. 2010. "Wood Species Affects Laboratory Colonization Rates of Chlorociboria sp." *International Biodeterioration & Biodegradation* 64(4): 305–308.

Rolfe, R. T. and Rolfe, F. W. 1925. *The Romance of the Fungus World: An Account of Fungus Life in Its Numerous Guises, Both Real and Legendary*. London: Chapman & Hall.

Rotherham, I. D. 2017. *Recombinant Ecology – A Hybrid Future?* Cham: Springer.

Rouifed, S., Byczek, C., Laffray, D. and Piola, F. 2012. "Invasive Knotweeds Are Highly Tolerant to Salt Stress." *Environmental Management* 50(6): 1027–1034.

Rousseaux, S., Diguta, C. F., Radoï-Matei, F., Alexandre, H. and Guilloux-Bénatier, M. 2014. "Non-Botrytis Grape-Rotting Fungi Responsible for Earthy and Moldy Off-Flavors and Mycotoxins." *Food Microbiology* 38: 104–121.

Rubio, F. D. 2016. "On the Discrepancy Between Objects and Things: An Ecological Approach." *Journal of Material Culture* 21(1): 59–86.

Schlaepfer, M. A., Sax, D. F. and Olden, J. D. 2011. "The Potential Conservation Value of Non-Native Species." *Conservation Biology* 25(3): 428–437.

Schmidt, O. 2006. *Wood and Tree Fungi: Biology, Damage, Protection, and Use*. Berlin: Springer.

Schübeler, F. C. 1886. *Norges Væxtrige: et Bidrag til Nord-Europas Natur-og Culturhistorie, Volume 2*. Christiania: Aschehoug.

Shotwell, A. 2016. *Against Purity: Living Ethically in Compromised Times*. Minneapolis, MN: Minnesota University Press.

Slotsbo, S. 2014. *NOBANIS – Invasive Alien Species Fact Sheet – Arion lusitanicus*. Accessed 23 March 2018. www.nobanis.org/fact-sheets.

Sterflinger, K. 2010. "Fungi: Their Role in Deterioration of Cultural Heritage." *Fungal Biology Reviews* 24(1): 47–55.

Tamm, M. 2015. "Introduction: Afterlife of Events – Perspectives on Mnemohistory." In *Afterlife of Events: Perspectives on Mnemohistory*, edited by M. Tamm, 1–26. Basingstoke: Palgrave Macmillan.

Tassin, J. and Kull, C. A. 2015. "Facing the Broader Dimensions of Biological Invasions." *Land Use Policy* 42: 165–169.

Tsing, A. L. 2015. *The Mushroom at the End of the World: On the Possibility of Life in Capitalist Ruins*. Princeton, NJ: Princeton University Press.

Vesper, S. J., Dearborn, D. G., Yike, I., Allen, T., Sobolewski, J., Hinkley, S. F., Jarvis, B. B. and Haugland, R. A. 2000. "Evaluation of *Stachybotrys chartarum* in the House of an Infant with Pulmonary Hemorrhage: Quantitative Assessment Before, During, and After Remediation." *Journal of Urban Health* 77(1): 68–85.

Vestad, Ø. 1961. *Eldre hager i Molde og omegn.* Master's Thesis. Norwegian College of Agriculture.

Viles, H. A. and Cutler, N. A. 2012. "Global Environmental Change and the Biology of Heritage Structures." *Global Change Biology* 18(8): 2406–2418.

von Proschwitz, T. and Winge, K. 1994. "Iberiaskogsnegl – en art på spredning i Norge." *Fauna* 47(3): 195–203.

Wallander, H. 2014. *Soil: Reflections on the Basis of Our Existence.* London: Springer.

Zugmaier, W., Bauer, R. and Oberwinkler, F. 1994. "Mycoparasitism of Some Tremella Species." *Mycologia* 86(1): 49–56.

19

HERITAGE AND THE VISUAL ECOLOGY OF THE PLANTATIONOCENE

Rui Gomes Coelho

Introduction

The significance of plantations in the world economy is gaining increasing visibility as monoculture now occupies more land than ever, like soy production in Brazil, the oil palms in Indonesia, and the olive groves in Portugal. These are places where large-scale exploitation of labor and natural resources and big land ownership have a deep history intertwined with the rise of modernity, colonialism, and the capitalist world-system. Born in great houses, the utopias of indefinite economic development encouraged cash crop production, replacing the traditional management of agriculture and forestry while irremediably disrupting biodiversity and the livelihood of rural communities. The heritage of plantations is embodied in our daily cup of coffee, which arrived at the leisurely tables of industrialized societies through the labor of those who toiled in American coffee and sugar plantations between the sixteenth and the nineteenth centuries (Mintz 1985; Moore 2016; See Moore et al. 2019).

Jason Moore (2015, 3) challenges us to think about capitalism as a world-ecology, in which we perceive the "accumulation of capital, the pursuit of power, and the co-production of nature in dialectical unity". The plantation is a localization of this ecological order, insofar as commodity revolution, organized environmental disruption, forced displacement, and racialization of labor were all experimented on its grounds. Donna Haraway and a variety of scholars named this world-ecology "the Plantationocene", encouraging us to reflect on the centrality of the plantation in our lives and how it is inevitably part of our common heritage (See Haraway 2015; Haraway and Tsing 2019; Haraway et al. 2016).

The world-ecology of capitalism in Plantationocene could only be articulated with a new way of "seeing and ordering reality" (Moore 2016, 109). It was only possible because the bodily senses of its agents were gradually tuned, readied to engage the new order, whether through the physical violence of labor or the sweetness of its products. The project of visuality was essential to this process of subjectification (Mirzoeff 2011, 474–475). A social order exists through representation and authoritative claims to look but also as lived experience, so it cannot last without a sensorial configuration in which it appears as given and natural. Archaeology – or the scientific production of knowledge about the past – emerged in the nineteenth century as a way of making the past legible through the lens of the capitalist world-ecology, naturalizing it through a specific visual apparatus that included documentation practices, particularly cartography and photography (See Olsen et al. 2012, 79–101; Shanks and Svabo 2013; Hamilakis and Ifantidis 2015). In other words, archaeology is the product of the plantation complex of visuality (Mirzoeff 2011) and contributed to turn the plantation into a form of heritage. By mediating the plantation ecology in the present by the means of visual documentation and explanation, heritage is also a cornerstone of the Plantationocene.

During my research in 2014, I encountered a stack of old black-and-white photographs in Boa Esperança, a plantation in the state of Minas Gerais, Brazil (Figures 19.1–19.7). Dating back to the late nineteenth to early twentieth century, this stack is animated by the relationships that characterized the institution of slavery and its enduring legacies in the present (cf. Bennett 2010, 20–38). As objects, these images are more than just representations of a certain past; they are active agents of a visual ecology that continues to embody the struggles for emancipation and social justice in a post-slavery society. The photographs from Boa Esperança belong to Mr. Barros, a descendant of the planter and slave owner who founded the plantation. The ontological status of the images is ambiguous insofar as they circulate with unclear permission from its owner. However, by circulating in this manner, they have also become an alternative form of heritage or a diffuse means to mediate the past of struggle – in a present in which violence over racialized subjects continues to be part of everyday life. In Brazil, heritage discourses about slavery have been historically constrained by the myth of "racial democracy" and only recently started to be systematically challenged in the public sphere (Araujo 2012).

In this chapter, I argue that heritage scholars, particularly archaeologists in the field of landscape archaeology, continue to reproduce the planter's imagination. I invite the viewers to think of ways in which we can disrupt the visual-affective traps of the Plantationocene. In order to do so, I will consider the unauthorized reproduction of photographs from Boa Esperança as an act of countervisuality, a means of self-emancipation, or what I will call here *aquilombamento*. I draw inspiration from Beatriz Nascimento's understanding of the *quilombo* as a process, not necessarily bounded to a particular site, through which the subalterns generate alternatives to exploitation and experiment with freedom (Nascimento 1985).

I will contrast them with images produced by myself during a project of landscape archaeology, which I discuss as a means of visual entrapment (Figures 19.8–19.13) (See Gell 1998, 68–72). I will show both sets of images – the stack from the *fazenda* and my own imagery – as negatives and video stills. Considering photography as "a practice and a form of human relations" (Azoulay 2019, 366), the images are displayed in this chapter in the ways in which they have been socialized in order to let us question their ontological status as documentary evidence.

Visiting the plantation

Boa Esperança is an estate by the river Preto in Minas Gerais. The *fazenda* was created in the nineteenth century during the expansion of the coffee monoculture in the Paraíba Valley. I visited the site in November 2014 as part of my research on the sensorial regime of slavery embodied in the productive landscape of this particular valley. Initially, I was also curious to find out the extent to which the experience of slavery was embedded in the social memory of the valley's communities. I realized very soon, however, that most families working in the plantations moved there sometime in the twentieth century, long after the 1888 emancipation. The old planters' class, too, had almost disappeared from the valley by the end of the nineteenth century, after a wave of bankruptcy that ensued the fall of coffee prices in the world market (Stein 1985).

When my friend A. informed me that Boa Esperança was one of the few *fazendas* left in the hands of its founding family, we decided to take a trip to the *fazenda*, which – although only about 40 kilometers from one of the largest municipalities in the state of Minas Gerais – is mostly accessed by rough dirt roads occasionally surrounded by dense woods. The region is still partly covered by Atlantic forest, the predominant ecosystem when coffee planters started moving there in the early nineteenth century. A. was well acquainted with Mr. Barros, the current owner, who welcomed us with cups of coffee.

The date of Boa Esperança's foundation is uncertain. It was created by the baron of São José das Três Ilhas, José Bernardino de Barros, as part of an estate that included three other plantations by the mid-1870s. Boa Esperança held the planter's great house and was the largest of his *fazendas*. In 1876, the baron was one of the richest men in the region; his remarkable wealth included 726,000 coffee bushes of different ages, tended to by 265 enslaved workers (Freire and Saraiva 2016). By comparison, only about 9% of the planters controlled more than 100 enslaved workers in the neighboring region of Vassouras (Salles 2008, 157). The baron was therefore part of the land-owning elite that wielded tremendous power over the free and the unfree social networks in the region. It was this social capital that allowed him to get credit from banking institutions, coffee planters, merchants, and his own brother

FIGURE 19.1 Fazenda Boa Esperança. Domingos Braga, Minas Gerais

Source: Private collection.

even after the plantation system started to show serious weaknesses. The credit was likely used to purchase enslaved people and to build the luxurious residence that I visited at Boa Esperança (Freire and Saraiva 2016). The Barros' family residence was at the center of a vast ecosystem founded on the systematic exploitation of land and people. This ecosystem resulted in extreme violence over enslaved workers and in dramatic environmental changes (Pádua 2002). This ecosystem is still in place despite the shifts that it went through with the formal abolition of slavery.

When the institution of slavery was abolished in 1888, some of the former enslaved workers left the plantations. They joined communities of self-emancipated slaves called *quilombos*, moved into towns, or resorted to seasonal rural work without permanent residence. Some remained in the plantations and continued to work there as wage laborers, tenant farmers, or sharecroppers (Gomes 2004). Boa Esperança was one of these places that continued to employ former slaves after emancipation. Like elsewhere in the Americas, however, freedom was purely nominal; bondage continued through different means. Since cash was scarce, workers were often paid in credit that they then used for purchasing groceries and other items from the landowner. Accelerating their accumulation of debt was a regime of fines that were applied to a wide range of offenses, such as stealing food, disrespecting the overseer, or working elsewhere without the planter's permission (Freire and Saraiva 2016). Former slaves and European immigrants working at the plantation were thus trapped in debt-bondage while performing nominally free labor. The Barros family kept detailed records of these exchanges in several account books.

This transition from an economy based on the labor of enslaved workers to an economy in which people were all nominally free generated an ambiguous interface between the biomes of freedom and unfreedom. New strategies of social control and resistance appeared throughout the nineteenth century, as the ecology of slavery changed, and formal emancipation entered people's lives. The shifting character of those biomes allowed for the reconfiguration of bondage relations as well as resistance strategies. Despite the abolition in 1888, the heritage of slavery continued to flourish in the ambiguity of this "transient ecotone" (Neilson 1993).

Like accounting, photography was another technology that the baron could mobilize to control resources with precision and consolidate his power in the plantation ecology. The history of photography's appearance in Brazil indicates the elites' need to describe and appropriate a vast territory: Swiss immigrant Hercule Florence, who worked as an illustrator in a Russian scientific expedition led by Baron Georg Heinrich von Langsdorff between 1824–1829, developed several techniques to help document the natural world with ease and accuracy. In 1833, some years before Louis Daguerre announced his most famous invention in Paris, Florence had already called one of those techniques "photographie". Remarkably, the first image he produced was that of a jailhouse. Florence's photography was not popular enough to last (Kossoy 2017); however, his story illustrates how scientific and technical developments make "nature legible to capital accumulation" (Moore 2016, 86).

When photography became popularized in Brazil during the 1840s, it was primarily used to make individual and family portraits to circulate among private audiences. Photographers also documented landscapes, plantations, and social types. The documentation of dramatic and wild landscapes was intended to lure foreign audiences with the exoticism of an untamed land, distant from civilization. The same can be said about the photographs of social types, particularly of enslaved people and Amerindians. Landscapes and social types were printed on

FIGURE 19.2 Fazenda Boa Esperança. Domingos Braga, Minas Gerais

Source: Private collection.

cards and circulated among visitors to Brazil as souvenirs (Muaze 2017), reinforcing the role of photography as a technology of representational colonialism (Wallis 1995).

Images of productive landscapes such as plantations had a different social function. According to historian Mariana Muaze (2017), plantation photographs indexed modernization – the development of an orderly productive landscape – and concealed the contingency of modernity to slave labor. This contradiction reflects the tensions that permeated Brazilian society in the nineteenth century. The tremendous expansion of the institution of slavery in the Paraíba Valley occurred in the age of abolitionism and continued well after the formal interdiction of the transatlantic trade in 1850. All the while, Brazil was a liberal, constitutional empire that was presented by its elites as an industrious, modern nation-state (Marquese, Parron and Berbel 2016). Coffee plantations were supposed to represent Brazil as a forerunner in new agricultural management and technologies and as provider of one of the most demanded commodities in the world.

French-born photographer Marc Ferrez was a key figure in the process of bringing photography into the plantation ecology. He refined existing photographic methods and experimented his own techniques in order to get a particular visual narrative about landscapes. In this narrative, shots of plantation nuclei were usually centered in a horizontal frame and taken from far away. This kind of framing suggested a world balanced between the regularity of plantation outlines and the cadenced silhouette of the hills. Pictures of enslaved people

FIGURE 19.3 Fazenda Boa Esperança. Domingos Braga, Minas Gerais

Source: Private collection.

working in rows of coffee bushes or in groups, resting, or getting ready for work completed the narrative of this orderly plantation ecology and "pacified slavery" (Muaze 2017, 18). The violence of slavery and environmental depredation were discreetly expunged from this visual narrative and turned into a visual game of symmetry and regularity, in which enslaved people and resources could be clearly individualized and itemized. The audiences and goals of these photographs were quite straightforward. Ferrez had his work commissioned by planters' organizations, got several awards within and outside Brazil, and exhibited his images in world fairs. In 1885, for example, Ferrez had his plantation images projected on a screen just outside of Paris (Muaze 2017, 9–10). The success of these visual events promoted the acceptance of a contradictory Brazil among popular audiences and, more importantly, helped naturalize a particular way of looking at the world.

While Mr. Barros talked with my travel companions in the kitchen, I walked around the house taking pictures and videos of rooms. As I crossed the main hallway, I spotted a card box of photographs on a table placed in a corner. The box was barely covered, so I was able to see a small pile of glass plate negatives (Figures 19.1–19.7). Once I returned to the kitchen, we discussed the many sorts of historical objects that I had seen around the house, but Mr. Barros did not comment on the photographs.

I thought it was not appropriate to pursue the topic at the time. I was already quite content with the opportunity of visiting the great house at Boa Esperança, which had hardly changed

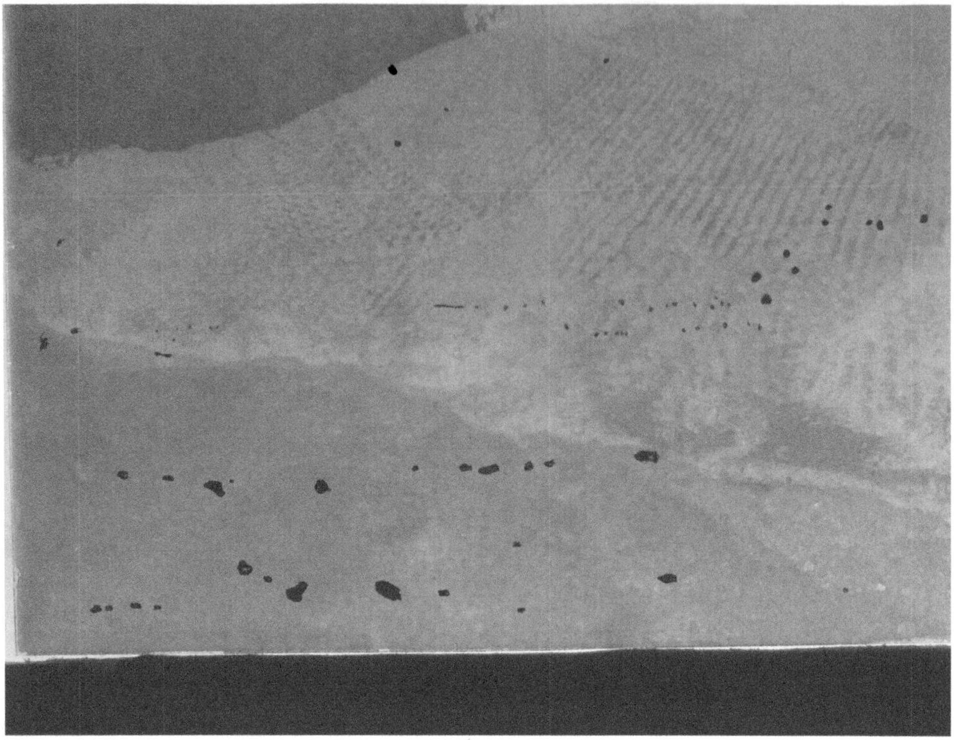

FIGURE 19.4 Fazenda Boa Esperança. Domingos Braga, Minas Gerais

Source: Private collection.

since the time of its construction in the late nineteenth century. With its wallpaper, furniture, bookcases, and paintings on the walls, the house felt like it was abandoned even before its founder moved in. Yet, the suspension of this place in time was deceptive. Mr. Barros moved through each room with great ease, as if he had spent his life there, whereas the last person to actually live in the house was his mother. After her passing, Mr. Barros decided to keep everything as it was.

The photographs in the card box were just a detail for me, while I focused on my own visual engagement with that remarkable space. After all, it seemed quite ordinary that another photographer had visited the plantation over a hundred years before me.

Escaping the plantation

After we left the *fazenda*, A. commented that the stack of photographs that I had seen in the house was probably from the baron's time. My friend never had the chance to look at them himself because there was never a chance to ask Mr. Barros, who was known for being very protective of his family's old possessions. In addition to the objects and collection of images, there was also an important series of account notebooks and other written materials from the late nineteenth century. Mr. Barros appreciated the interest that scholars had expressed over the years regarding the collections. However, unsystematic work on the archive made it hard

FIGURE 19.5 Fazenda Boa Esperança. Domingos Braga, Minas Gerais

Source: Private collection.

for the visitor to make sense of what was already known about it. I was told that a local teacher had the chance to handle the photographs years before while working on an assignment for his education degree. The teacher asked someone in the neighboring town of Porto das Flores to digitize the images, who ended up keeping copies. Somehow, this led to a rumor in which the copies were now circulating within the community without Mr. Barros' permission. Eventually, I managed to get a flash drive with a copy of those photographs from someone in Porto das Flores. I, too, became part of that network of unauthorized viewers.

The photographs in Mr. Barros' archive comprise a large number of family portraits, taken either indoors or outside the great house. They include portraits of the elderly baron, José Bernardino de Barros, alone or with his son, Antônio (Figure 19.6), and grandchildren. There are also a couple of images from a wedding ceremony, likely Antônio Bernardino de Barros', who is the grandfather of the current owner. In addition to the photographs of the family, there is a series of images that document the groups of workers and the plantation's landscape. These include slopes covered by rows of new coffee bushes and other crops, tracts of forest about to be cleared out, animal pens, and various farm buildings.

The photographs were made with the dry plate process, also known as gelatin dry plate negative. This technique was first invented in 1871 by Richard Leach Maddox and improved later in the 1870s. It simplified the photographers' lives by minimizing the number of tasks required to do immediately after the shot and by decreasing the exposure time. For this

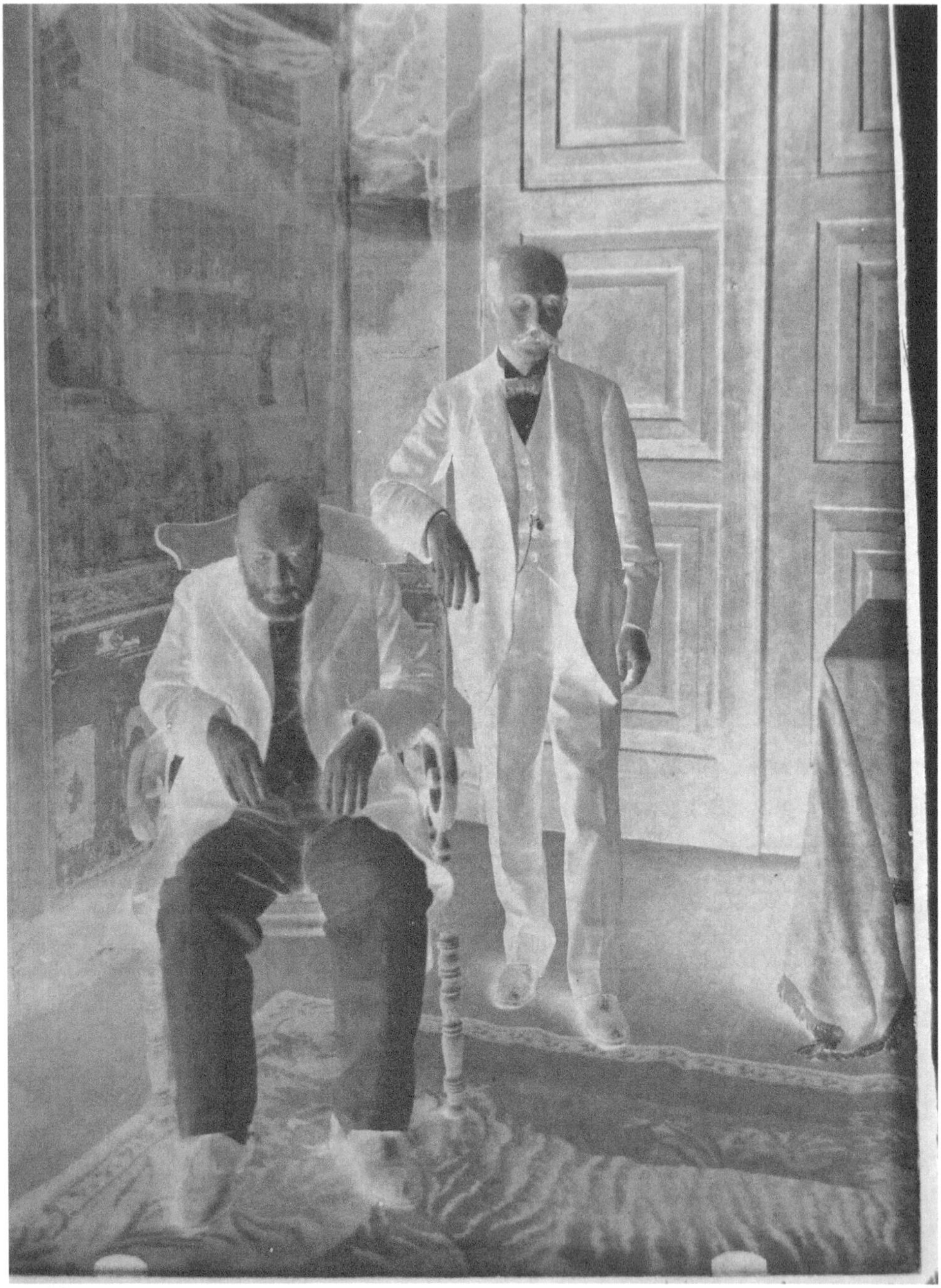

FIGURE 19.6 Fazenda Boa Esperança. Domingos Braga, Minas Gerais

Source: Private collection.

reason, landscape photography became much easier (Kennel 2009, 45). Marc Ferrez is credited for having introduced the dry plate method for the first time in Brazil, which is connected to his dedication to document landscapes (Muaze 2017, 7). The world popularization of this technique occurred around Ferrez's time, with the dissemination of commercially manufactured plates and developers since 1879. These materials were sold in standardized sets and were ready to use by photographers (Kennel 2009, 45).

There are scans of labels from two dry plate boxes in Mr. Barros' archive, one being produced in Rochester, New York, by Eastman Kodak (founded in 1888) and the other by the Belgian trademark Agfa (created in 1897). The dry plate process was gradually replaced by flexible film since 1897 and reached its peak around 1900. According to Mr. Barros, the baron died in 1912, which gives us the timeframe of 1888–1912 as the likely dates for the photographs in his archive. These were the years that followed the abolition of slavery in 1888 and witnessed the ruin of most coffee planters in the Paraíba Valley. It is likely that the images were made by a regional photographer based in Juiz de Fora or perhaps Valença, who regularly traveled across the region's plantations.

Planters and visitors introduced photography technologies in the valley as a medium to generate a certain narrative about the plantation ecology. This plantation world was productive and modern but also contingent to the institution of slavery. Like painting (Marquese 2007, 2010), photography helped slavery become more acceptable among the ruling classes of imperial Brazil by concealing the scars of enslaved workers and turning a landscape of depredation into an orderly one that could be easily recognized by the progressive minds of Europe (Muaze 2017).

However, the introduction of cameras in the valley was also a way of introducing a particular kind of modernity in the plantation ecology. The expansion of coffee production in nineteenth-century Brazil was accompanied by a set of disciplinary practices that added different layers of experience to slavery's mode of production. This new form of slavery, called "Second Slavery" (Tomich 2004; Tomich and Zeuske 2008), was characterized by a new sensorial configuration in which the plantation landscape was devised to naturalize slavery, enforce stricter forms of self-discipline, and experiment with new labor routines (Coelho 2017). Objects such as account books mediated the transformation of slavery into a regime of modern labor management, in which enslaved people became productive units (Rosenthal 2018). Slaves were turned invisible and trapped in numbers and bound stacks of paper. Similarly, photography was a device that introduced a particular order in the plantation. It set the limits of the visible in the plantation ecology by framing a separation between planters' families and groups of workers, by always portraying workers on productive spaces, and by documenting property. More than simply technologies of representation, account books and photography shaped the plantation world by directing peoples' gazes and reducing them to numerals and notes on pieces of paper.

As Nicholas Mirzoeff (2009, 77) once put it, photography turned time into an object just like slavery turned people into objects. Mid-nineteenth-century planters in the United States embraced the possibilities offered by the new technology in the plantation ecology; they had slaves photographed and then the daguerreotypes circulated among potential buyers. Once all bodily senses were mobilized to naturalize the brutality of slavery, photographs and people became interchangeable values (Mirzoeff 2009, 81). The planter's gaze shaped a particular form of visuality, which in turn commodified both images and people.

The photographs at Boa Esperança portray people and places right after the abolition of slavery in Brazil. It was a changing world in which planters clanged to all devices available to them in order to maintain their power. The small family archive of photographs grounded the *fazenda* in the planters' order when the social configurations had started to shift. At that point in time, it might have seemed certain that land was getting exhausted because of monoculture and continuous destruction of the forest. Paintings from the 1870s–1880s already showed the effects of decades of depredation (Marquese 2007). Nevertheless, the dry plate process immobilized the always changing productive units of the plantation: coffee bushes, tracts of forest, standing workers, livestock. Photography chemicals offered a durable chance for the survival of the plantation ecology when it was already showing traces of wear due to the irresolvable contradictions of a plantation society and the inherently self-destructive monoculture system.

Boa Esperança's photographs continue to materialize a form of confinement. They still index the commodification and control of people, nature, and resources in the past, as well as the power structures that enabled so. However, the power of those images goes beyond the effect of representation; they also restrain our gaze to a form of visuality that emerged within the confines of slavery. Early plantation photographers crafted a visual narrative in which symmetry, regularity, and itemization became a proxy for the violence of slavery. The visuality of slavery remained unchallenged despite the abolition of the institution and throughout the post-emancipation process in which former plantation societies reconfigured notions of freedom.

There is an additional layer of captivity in the ontological status of the photographs I encountered in the *fazenda*. Besides the plantation owner and some of his family members, the images display many anonymous individuals about whom we are unlikely to find any reliable information in surviving archives. One of the ultimate effects of human commodification was the reduction of individuals to units of production and numerals. The anonymization carried by circulation of photographs beyond the lives and memory of those depicted corresponds to another form of itemizing people and turning them into bare units, ready to be subjected to the gaze of modern viewers. By remaining in the hands of Mr. Barros, the images acquire a specific "haptic temporality" that provides tangibility to a visual ecology that is somewhere between kin represented by family albums and the enforcement of law and order as seen in portraits of convicts (Campt 2017, 69–100). Thanks to their materiality (See Edwards 2002), those images continue to reproduce the old planter's dream of halting shifts in the status of his property; they are the last stronghold of a bygone era of prosperity.

At Boa Esperança, however, the visual violence of the plantation world is currently being challenged from within. One of the most drastic forms of resistance among the enslaved in the plantation ecology was running away. The self-emancipated – *quilombolas* – posed a double threat to the plantation system by disrupting its economy and for presenting a radical alternative to the many forms of violence that enslaved people were subjected to in the plantation. It means that the *quilombola* threatened the plantation system as an individual, someone who annulled the investment of a planter and interrupted a certain labor routine. More importantly, the fugitive effectively challenged private property, the quintessential foundation of liberal capitalism.

Photography, like account books, presented itself as a means to document and control but also to entrap people who were subjected to the plantation system. Like the enslaved, former slaves who remained in the *fazenda* or newly arrived immigrants were bonded to the planter by annotations and numerals that accounted for credits and debits, materializing coerced social ties that would otherwise remain unpalpable. Photographs rendered those relationships visible by displaying not only people, the plantation subjects, but also the land and buildings that resulted from coerced labor. People continue to be collectively subjected to the visual violence of the plantation as heritage, which remains unchallenged in the ways we represent

FIGURE 19.7 Fazenda Boa Esperança. Domingos Braga, Minas Gerais

Source: Private collection.

it. Yet, unlike the runaways, we have very limited means to break the bonds that tie us to the plantation ecology as viewers, now that we are all nominally free.[1]

Some people in Porto das Flores, neighbors of Mr. Barros whom I came across during my visit to the *fazenda*, found an ingenious way to break the visual cycle of plantation violence embedded in Boa Esperança's photographs. The unauthorized reproduction and circulation of those images amid whispers and unsaid cautions resonates the audacity of *quilombolas*, no matter how politically unintentional the action might be.

The story of Boa Esperança's fugitive photographs resonates with the visual *aquilombamento* of Elizabeth Ramsey, a woman enslaved in Texas who exchanged photographs with her freed daughter, Louisa Picquet. In response to Picquet's delivery, the planter tried to sell her mother and sent back a daguerreotype of Ramsey in her best outfit. Nicholas Mirzoeff (2009, 80) pointed out that despite the attempted transaction of the planter, it is plausible that Ramsey tried to impress her daughter. In fact, we know that the enslaved engaged with visual representations of their condition, changing and subverting them (See Chaney 2008). In a way, the image embodied her own idealized escape.

Trapped in the plantation

In 2014, I witnessed the *aquilombamento* of a photographic archive and took a small part in it by copying and circulating its images. Yet the agency of those dry plates from Boa Esperança went far beyond my own role in the emancipation journey. At the time, I did not think of those photographs as more than documentary evidence of a nineteenth-century plantation landscape; I expected an archive in which I could identify, describe, and analyze a certain past. The documentary processes of landscape archaeology reduce materiality to units of analysis, which are supposed to explain everyday experience and social change through time. This approach is a product of the plantation complex of visuality (Mirzoeff 2011) and shares an epistemological genealogy with the planter's effort to control, quantify, and commodify his property. The works of archaeology and plantation planning result from the emergence of scientific rationalism in the seventeenth century and are part of European efforts to extend its dominion across the world. Both enabled economies of knowledge that continue to extract resources and produce inequality across race, class, state, and academic contexts. As tools of domination, maps are a good example of a materiality shared by planters and archaeologists.

My preoccupation with the documentary character of the photographs from Boa Esperança prevented me from understanding something important: that the dry plates and I parted ways when I tried to study them while they continued – and continue – to circulate without permission. Trapped in the rationalist imagination, I was unable to notice the radical potential of objects that escape their subjection in the plantation world. As an archaeologist, my attitude was akin to the planter's. Regardless of the escapes of enslaved workers, planters continued to foster a future for the plantation world, unable to fully understand the threat posed by the radical attitude of the self-emancipated. To this day, I am not sure if landscape archaeology can escape the plantation insofar as its questions and methods are trapped in the planter's imagination.

In 2019, I produced a short video about my earlier experience in the Paraíba Valley (Figures 19.8–19.13). The piece focused on movement and how labor routines enforced a particular sensorium on the plantation landscape. In this sensorial regime, all bodily senses were convoked by the planters to naturalize the institution of slavery. By directing the gaze and movement of people within their domains, planters expected to create a self-disciplined

FIGURE 19.8 Rui Gomes Coelho, 2019. *The Road to Massambará*. Still from video

FIGURE 19.9 Rui Gomes Coelho, 2019. *The Road to Massambará*. Still from video

population of enslaved workers. This was a non-narrative strategy that aimed at controlling a vast, hilly landscape that could not be otherwise effectively surveilled. The enslaved, on the other hand, responded with their own sensorial strategies. For example, they developed a spiritual landscape to which the planters' class could not have access to and played with sound, music, and ritual in order to establish an aural landscape. These non-narrative processes were nevertheless material and left traces behind that I sought to comprehend in my research (Coelho 2017).

The video is a document of my work, but it is also a self-reflection process that started in the field and is far from being complete. I used to walk several hours a day on the plantation roads. In the nineteenth century, slaves spent most of their time walking back and forth from the fields, along rows of coffee bushes and clearing up the old forest for new crops. As the years passed, coffee monoculture exhausted the land and enslaved workers were forced to walk further away, where the remaining parts of forest were burned with the promise of another coffee harvest (Stein 1985). On my long walks, I engaged with the affective traces of an environmental catastrophe and an oppressive institution upon which the modern world was built. But the cadence of my steps also provided me with the time to reflect on the ways in which people escape the traps of enslavement and its contemporary legacies.

I called the video *The Road to Massambará*, thinking of one of the trails that I decided to walk during my fieldwork. The day I did it, I was overcome with memories from my family's narratives of our ancestral land of Alentejo. I was thousands of miles away, yet the embodied memory of Portuguese latifundia, its labor struggles and years of hunger, manifested itself during that walk on my way to Massambará (Coelho 2018). The video was a result of that affective event, a non-narrative reflection on how different temporalities intersect the archaeologist's personal experience during fieldwork. The piece starts with a scene at dawn, a woman and her children walking away from a *terreiro* in the *quilombo* of São José da Serra. The *terreiro* is the name given to the central plantation plaza where the enslaved worked, but it is also how people

FIGURE 19.10 Rui Gomes Coelho, 2019. *The Road to Massambará*. Still from video

FIGURE 19.11 Rui Gomes Coelho, 2019. *The Road to Massambará*. Still from video

name Afro-Brazilian religious spaces. This scene is followed by myself walking on a plantation trail, and a succession of plantation scenes, mostly still, of spaces, people, and routines, which are interrupted by a series of photographs from Mr. Barros' archive. The video ends with my return to the margins of the plantation. That is where I found people performing *jongo*, praying in the forest, and going about their everyday lives at the *quilombo*. The entire video was juxtaposed by sounds recorded in the Paraíba Valley.

FIGURE 19.12 Rui Gomes Coelho, 2019. *The Road to Massambará*. Still from video

FIGURE 19.13 Rui Gomes Coelho, 2019. *The Road to Massambará*. Still from video

I prepared this video for a session organized by my colleagues Yannis Hamilakis and Eva Mol at the Theoretical Archaeology Group meeting in Syracuse, New York. I was prepared to be questioned about the video's flow, which could be understood as an essentializing approach to the plantation world and the opposition between masters and slaves or even an exoticized representation of the valley. However, most of the comments I received aestheticized the video: It was something "beautiful". I was flattered by these comments but felt uncomfortable by the idea of having produced a beautiful representation of the plantation world.

My colleagues certainly did perceive a tension in the video between the representation of the planter's world and the *quilombo*. Yet how could the signifiers of slavery be taken as beautiful things? This question has been following me ever since, as it became quite clear to me that my photographic sensibility captured images that matched the plantation's visual narrative of symmetry and regularity. Such sensibility is not innocent. My images, like Marc Ferrez's or the pictures taken by the anonymous photographer at Boa Esperança, were not just the result of my biological ability to see and the technological apparatus of my camera. Rather, those images and my own visual subjectivity are all part of a social configuration that is far from overcoming the multiple legacies of colonialism, slavery, and extractivism. The plantation complex of visuality is embedded in my work as an archaeologist. The beauty of the plantation is a trap (see Gell 1998, 68–72).

Conclusion

The Plantationocene is one of the most significant configurations of the contemporary world-ecology. Its social legitimacy is contingent on visuality to the extent that the inequality and violence generated in the plantation world are represented and embodied in visual objects. Heritage scholars have questioned the centrality of vision in the modern sensorium and how it continues to constrain the potential of heritage practices (Hamilakis 2014). This centrality is historical and is irremediably connected to the development of the plantation as site of social experimentation. Thus, as heritage practitioners, we are doubly trapped in the plantation world; we look at the landscape in terms coded by the planters' class inasmuch as we continue to reproduce the planters' imagination with our maps, photographs, and other forms of representation. Heritage production is trapped in the visual ecology of the Plantationocene.

Despite the traps set by the planters' imagination, this visual ecology hosts disruptive agents that question the Plantationocene from within. More than just representations of the past, the dry plates from Boa Esperança are animated objects to the extent that they embody racial and class relationships configured in the context of plantation slavery. The plates affect people, attracting the curiosity of researchers and community members, the neighbors of the *fazenda*, who wish to rescue them from arrest. As negatives, these images make us question the "transparency of the visual" (Campt 2012, 127). Their circulation within the community and beyond distributes a particular kind of agency that emerged in the plantation world – resistance. The *aquilombamento* of Boa Esperança's photographs is an act of countervisuality (See Mirzoeff 2011). It resonates with the audacity of the self-emancipated and is an invitation for the modern viewers to question the heritage of a visual ecology that continues to legitimize inequality and violence.

Acknowledgments

I wish to thank Maurício Barros, Annibal Magalhães, Iran Souza da Conceição, Hande Sarikuzu, Þóra Pétursdóttir, Imani Jackson, Inês Beleza Barreiros, Andrés Mario Zervigón, Virginia McBride, Yannis Hamilakis, Lauren Yapp, Ana San Martín, and anonymous members of Porto das Flores' community for stimulating this conversation and providing critical insights. I am also indebted to my colleagues at Brown University who commented earlier versions of this chapter, particularly in the seminar on "Slavery's New Materialisms" led by Seth Rockman, in Ariella Aïsha Azoulay's "Photographic Archives" research group, and in the Joukowsky Institute's archaeology colloquium.

Note

1 Mr. Barros' dry plates run parallel to a more controversial archive. The Peabody Museum of Archaeology and Ethnology at Harvard University keeps hostage a series of daguerreotypes commissioned by Swiss-born biologist Louis Agassiz in 1850. The images depict several enslaved individuals in South Carolina, including a man named Renty and his daughter Delia. Not unlike the photographs of social types in Brazil, Agassiz photography was a means to document, classify, and commodify individuals. However, Agassiz stepped forward by intending to use photography in order to produce evidence for his scientific racism (Wallis 1995). The images have been rediscovered in 1976 and used ever since as illustrations of *antebellum* America and the violence of slavery. Like at Boa Esperança, the visuality of the plantation world remained unchallenged inasmuch it remained in private hands. In the mid-1990s, artist Carrie Mae Weems used the images in her series *From Here I Saw What Happened and I Cried*, an installation in which she combined image and text in order to render visible stories of enslavement and its emotional violence. At the time, Weems agreed to not publish the images without Harvard's permission. She broke the arrangement, and Harvard threatened her with a lawsuit, ending the conflict by purchasing the artworks and consolidating its power over the images (Arellano and DeBello 2018). The photographs did not remain quiet. In 2019, Tamara Lanier, a descendant of Renty, filed a lawsuit against Harvard over the use of her ancestor's portrait. The lawsuit occurred after several unsuccessful attempts to get Harvard to acknowledge its role in the production of the photographs and the violence it continues to exert among the descendants. In the lawsuit, Lanier argues that Harvard has been benefiting from the possession and circulation of those images and claims to be entitled to the daguerreotypes as a matter of equity (Kinsella 2019; See also Azoulay 2019, 146–148).

References

Araujo, A. L. 2012. "Public Memory of Slavery in Brazil." In *Slavery, Memory and Identity: National Representations and Global Legacies*, edited by D. Hamilton, K. Hodgson and J. Quirk, 115–130. London: Pickering & Chatto.

Arellano, E. and DeBello, H. 2018. "The Zealy Daguerreotypes: Power and Possession." *Harvard Political Review*, online. Accessed 14 September 2019. http://harvardpolitics.com/culture/zealy-daguerreotypes/.

Azoulay, A. A. 2019. *Potential History: Unlearning Imperialism*. New York: Verso.

Bennett, J. 2010. *Vibrant Matter: A Political Ecology of Things*. Durham, NC: Duke University Press.

Campt, T. M. 2012. *Image Matters: Archive, Photography, and the African Diaspora in Europe*. Durham, NC: Duke University Press.

Campt, T. M. 2017. *Listening to Images*. Durham, NC: Duke University Press.

Chaney, M. A. 2008. *Fugitive Vision: Slave Image and Black Identity in Antebellum Narrative*. Bloomington, IN: Indiana University Press.

Coelho, R. G. 2017. *Sensorial Regime of "Second Slavery": Landscape of Enslavement in the Paraíba Valley (Rio de Janeiro, Brazil)*. PhD Dissertation. Binghamton, NY: Binghamton University.

Coelho, R. G. 2018. "Topographies of Freedom." *Anthropology Now* 10(1): 72–92.

Edwards, E. 2002. "Material Beings: Objecthood and Ethnographic Photographs." *Visual Studies* 17(1): 67–75.

Freire, J. and Saraiva, L. F. 2016. "'Pertence a Liberta Firmina, 09 de Maio de 1888': Relações de Crédito na Transição do Trabalho Escravo para o Livre em uma Região Cafeeira – Zona da Mata Mineira." *17º Seminário sobre a Economia Mineira, Diamantina*, online. Accessed 13 September 2019. https://diamantina.cedeplar.ufmg.br/portal/download/diamantina-2016/396-676-1-RV_2016_10_09_00_45_43_876.pdf.

Gell, A. 1998. *Art and Agency: An Anthropological Theory*. Oxford University Press.

Gomes, F. S. 2004. "Slavery, Black Peasants and Post-Emancipation Society in Brazil (Nineteenth Century Rio de Janeiro)." *Social Identities* 10(6): 735–756.

Hamilakis, Y. 2014. *Archaeology and the Senses: Human Experience, Memory, and Affect*. Cambridge: Cambridge University Press.

Hamilakis, Y. and Ifantidis, F. 2015. "The Photographic and the Archaeological: The 'Other Acropolis'." In *Camera Graeca: Photographs, Narratives, Materialities*, edited by P. Carabott, Y. Hamilakis and E. Papargyriou, 133–157. Farnham: Ashgate.

Haraway, D. 2015. "Anthropocene, Capitalocene, Plantationocene, Chthulucene: Making Kin." *Environmental Humanities* 6: 159–165.

Haraway, D., Ishikawa, N., Gilbert, S. F., Olwig, K., Tsing, A. L. and Bubandt, N. 2016. Anthropologists Are Talking - About the Anthropocene. *Ethnos* 81(3): 535–564.

Haraway, D. and Tsing, A. 2019. *Reflections on the Plantationocene: A Conversation with Donna Haraway & Anna Tsing Moderated by Gregg Mitman.* Madison, WI: University of Wisconsin-Madison.

Kennel, S. 2009. *In the Darkroom: An Illustrated Guide to Photographic Processes Before the Digital Age.* Washington, DC: National Gallery of Art.

Kinsella, E. 2019. "'Morally, Harvard Has No Grounds': Inside the Explosive Lawsuit That Accuses the University of Profiting from Images of Slavery." *Artnet News*, online. Accessed 5 September 2019 https://news.artnet.com/art-world/harvard-university-slaves-images-1500412.

Kossoy, B. 2017. *Hercule Florence: El descubrimiento aislado de la fotografía.* Madrid: Cátedra.

Marquese, R. 2007. "A paisagem da cafeicultura na crise da escravidão: as pinturas de Nicolau Facchinetti e Georg Grimm." *Revista do Instituto de Estudos Brasileiros* 44: 55–76.

Marquese, R. 2010. "O Vale do Paraíba cafeeiro e o regime visual da segunda escravidão: o caso da fazenda Resgate." *Anais do Museu Paulista* 18(1): 83–128.

Marquese, R., Parron, T. and Berbel, M. 2016. *Slavery and Politics: Brazil and Cuba, 1790–1850.* Albuquerque, NM: University of New Mexico Press.

Mintz, S. W. 1985. *Sweetness and Power.* New York: Penguin.

Mirzoeff, N. 2009. *An Introduction to Visual Culture.* New York: Routledge.

Mirzoeff, N. 2011. "The Right to Look." *Critical Inquiry* 37(3): 473–496.

Moore, J. W. 2015. *Capitalism in the Web of Life.* London: Verso.

Moore, J. W. 2016. "The Rise of Cheap Nature." In *Anthropocene or Capitalocene? Nature, History, and the Crisis of Capitalism*, edited by J. W. Moore, 78–115. Oakland, CA: PM Press.

Moore, S. S., Allewaert, M., Gómez, P. F. and Mittman, G. 2019. "Plantation Legacies." *Edge Effects*, online. Accessed 12 September 2019. https://edgeeffects.net/plantation-legacies-plantationocene/.

Muaze, M. A. F. 2017. "Violence Appeased: Slavery and Coffee Raising in the Photography of Marc Ferrez (1882–1885)." *Revista Brasileira de História* 37(74): 1–30.

Nascimento, B. 1985. "O conceito de quilombo e a resistência cultural negra." *Afrodiáspora* 6–7: 41–49.

Neilson, R. P. 1993. "Transient Ecotone Response to Climatic Change: Some Conceptual and Modelling Approaches." *Ecological Applications* 3(3): 385–395.

Olsen, B., Shanks, M., Webmoor, T. and Witmore, C. 2012. *Archaeology: The Discipline of Things.* Berkeley, CA: University of California Press.

Pádua, J. A. 2002. *Um Sopro de Destruição: Pensamento Político e Crítica Ambiental no Brasil Escravista (1786–1888).* Rio de Janeiro: Jorge Zahar Editor.

Rosenthal, C. C. 2018. *Accounting for Slavery: Masters and Management.* Cambridge, MA: Harvard University Press.

Salles, R. 2008. *E o Vale era o Escravo. Vassouras, Século XIX. Senhores e Escravos no Coração do Império.* Rio de Janeiro: Civilização Brasileira.

Shanks, M. and Svabo, C. 2013. "Archaeology and Photography: A Pragmatology." In *Reclaiming Archaeology: Beyond the Tropes of Modernity*, edited by A. González-Ruibal, 89–102. New York: Routledge.

Stein, S. J. 1985. *Vassouras: A Brazilian Coffee County, 1850–1900.* Princeton, NJ: Princeton University Press.

Tomich, D. 2004. *Through the Prism of Slavery: Labor, Capital and World Economy.* Lanham, MD: Rowman & Littlefield Publishers.

Tomich, D. and Zeuske, M. 2008. "Introduction, The Second Slavery: Mass-Slavery, World-Economy, and Comparative Microhistories." *Review* 31(2): 91–100.

Wallis, B. 1995. "Black Bodies, White Science: Louis Agassiz's Slave Daguerreotypes." *American Art* 9(2): 38–61.

20

I SHED TEARS, LEFT, AND FORGOT

The common frog, mosquitoes, and Grandmother Pine stayed

Suvi Tuominen and Marko Marila

Preface

The objective of this photo chapter is to portray a selection of Second World War heritage sites located in Finnish Lapland from an ecological and multispecies perspective. By this perspective, we refer to the idea that a heritage site unfolds as an ecology of the totality of entities present, each with their unique take on what a heritage site is and means and what can become of it – and of heritage in general, for that matter. The starting point for this chapter, then, is that all of these agents perform heritage in their own peculiar ways. In other words, we contend that heritage creation is a dispersed material process constituted by a plethora of actualities; that a heritage site is made possible through a form of multispecies performance.

As bodily human beings interested in the heritage ecologies of these sites – as well as the types of knowledge they afford – our aim is also to assess the relevance of sensory attentiveness and intuition as forms of knowledge production. We do this by simultaneously foregrounding the body as a medium for experiencing the "interplay of actualities in nature" (Whitehead 1956, 167) and as a medium for transgressing our own bodily boundaries, habitual gazes, and movements at the sites. Through intuitive movement, the body becomes a perceptual grid that calls for a deeper somatic engagement with the various nonhuman forces operating at the heritage sites (Bender, Hamilton and Tilley 2007; Bennett 2010; Whatley, Brown and Alexander 2015). This perceptual grid could be seen as a surface where various materials and actors intersect and bring new forms of knowledge into being. In this way, we want to underscore the importance of the "multiplicity of feeling" (Debaise 2017), and we contend that movement and somatic engagement with heritage sites can become the very method for sensing *with* things rather than knowing *of* things.

Besides paying special attention to the importance of the body and somatic engagement with heritage sites, we wish to denote the importance of unconstrained and playful experimental intentionality. We believe that by adopting a sense of somatic experimentation, it is possible to develop a specific kind of speculative attentiveness towards things and to immerse more deeply in the ecologies of heritage sites. This approach, we believe, will not only expose new interesting artefactual details but will also engender new perspectives that might be relevant in terms of conceptualising heritage sites as ecologies in their own right. The ways in which insects and other lifeforms, for instance, dwell in and constantly research their material remains calls out for more attention and sensitivity from the human visitor's behalf (cf. Ingold 2000).

By also bringing forth historical events that took place at these very sites during the Second World War, we wish to highlight that various nonhuman agents are entangled with and partake in the formation and interpretation of their material layers of memory. These agents are not simply passive objects floating in poetic realms but rather actors within the social, political, and historical spheres of these heritage sites. Although the main focus of the chapter is to steer the reader's attention to the various nonhuman forces operating at our chosen localities, these agents should not be treated as separate from the social and political histories of the sites. Rather, they should be treated as fragmented gateways towards a past that seems unruly and unreachable from the current bodily perspective.

Instead of giving too much space to ratiocination or verbalisation, this chapter operates primarily within the realm of images and atmospheres. It is our hope that the images and atmospheres, combined with lyrical fieldnotes, succeed in performing and exposing the carnal acts between bodies human and nonhuman, living and nonliving (Bolt 2004, 190). By using photography as a medium for conveying these thoughts, we also wish to emphasise the performative power of the image. The images are not to be treated simply as representational of these heritage sites but rather as indicative of the surfaces of movement that both reveal

FIGURE 20.1 Barbed wire is a ubiquitous material commonly found at Second World War heritage sites. Its suffocating effect is felt by plants and animals alike. Kankiniemi PoW campsite, Inari, Finland

Photo: M. Marila.

FIGURE 20.2 Suvi practicing somatics with Grandmother Pine. Somatic practice and movement was used at the sites as a speculative method for sensing *with* things

Photo: M. Marila.

and express care towards those possible multiplicities – and ecologies – of feeling that easily become excluded from conceptualisations of heritage.

The sites

Hundreds of Second World War ruins can be found immersed in Lapland's landscape, mostly in the wilderness areas. Some prisoner-of-war camps, for example, are remarkably visible in these remote areas, and one can discover well-preserved log houses, remains of barbed wire fences, gates, and rubbish pits. The cultural heritage status of Second World War sites in Finland has remained somewhat undetermined due to lack of official recognition and ownership. These are also vulnerable sites that are occasionally looted (Seitsonen 2018, 69).

The Second World War sites in Lapland are entangled with both the recent past and with the contemporary landscape. The material remains of the Second World War reach out from this landscape with their haunting presence, reminding us of what they once were (e.g. Herva 2014). The sites feel dislocated and disoriented, like visitors in a peripheral landscape (also Seitsonen 2018, 70–71). Locally, this heritage holds social significance and value, and the sites perform themselves as places for multiple meanings and layers of memory (Seitsonen 2018, 144–147). The sites also serve as backdrop for paranormal experiences and ghost sightings (Harjumaa 2008; Herva 2014; Koskinen-Koivisto 2016).

We visited and photographed at two of these sites, the Kankiniemi PoW camp and the Haukkapesäjoki PoW camp, both located near the town of Inari in Finland.

FIGURE 20.3 Remains of a wooden cargo crate at Haukkapesäjoki, Inari, Finland

Photo: M. Marila.

What seems characteristic of these two sites is the way they disguise themselves in the surrounding nature. Mushrooms, grass, flowers, and newgrown trees are pushing through the rotten wooden structures, giving new forms to the sites themselves. The sites are also in constant movement due to the number of species of insects and animals who frequently visit or occupy them. Forms of life form these sites. Forms appearing as green spirals.

> Spirals reaching out from the grey logs.
> Spirals in connection with emergent forms of life, "each with its own experience, enjoying individually, and yet requiring each other" (Whitehead 1956, 70).
> Blink. And the frog jumped.
> The smell of the moist bryophyte.
> Leaf beetles making love.
> Spirals changed their colour, appeared and disappeared.
> A gesture of healing (Stengers 2018, 81).
>
> Blink.
> We walked through the swamp. The magical white reindeer observed.
> They knew.
> I will stay. You will go.
> Grandmother Pine whispered.

FIGURE 20.4 Remains of an unknown structure hide in a small pond next to the ruined Kankiniemi PoW camp

Photo: M. Marila.

FIGURE 20.5 A common frog at the Haukkapesäjoki PoW campsite

Photo: M. Marila.

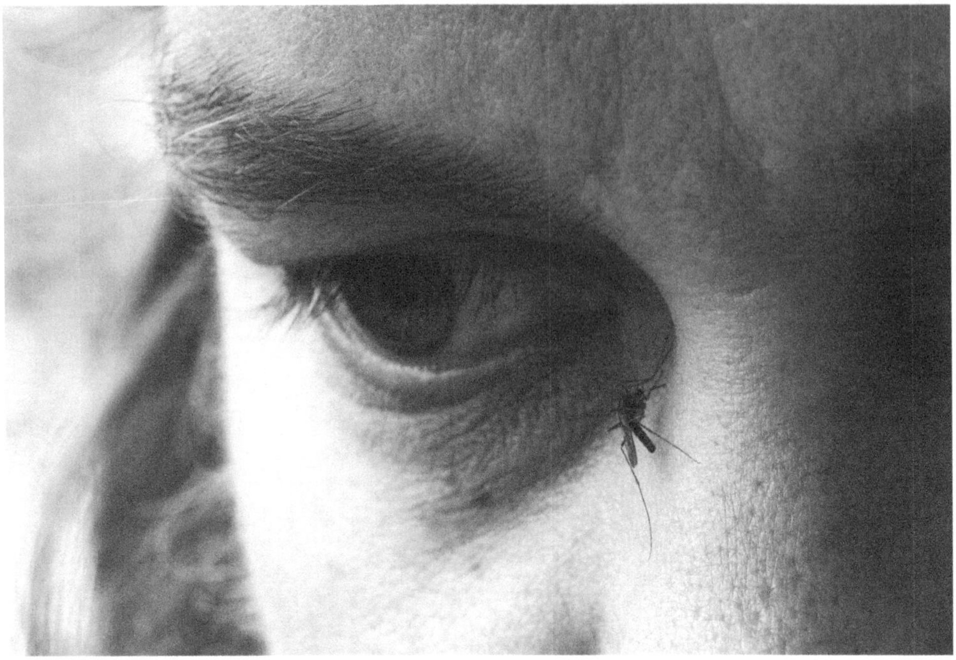

FIGURE 20.6 Lapland is notorious for the huge amount of mosquitoes in the summer. Their presence choreographs the visitors' perception of the sites

Photo: S. Tuominen.

Elsa

In the summer of 1944, mere weeks before the Lapland War between Finland and Germany broke out, Finnish dancer Elsa Puolanne (1906–1996) performed to Finnish, German, and Austrian soldiers in Finnish Lapland and Northern Norway.[1]

Puolanne was one of the first Finnish dancers to adopt a dance style conceptualised as the German *ausdruckstanz* (expressionist dance). In her formative years, Elsa studied with many German teachers, and the dance style she adopted has later been interpreted as entangled with National Socialism (Makkonen 2007). Therefore, a formalist interpretation of dance could connect the form of dance and movement with the political and ideological milieu of the time (see Burt 1998). After the Second World War ended, the contribution of Elsa Puolanne and her contemporaries to the Finnish dance art was considered unprofessional. Dance critic Raoul af Hällström (1946, 13) has stated that "liberated Europe cannot base its artistic credo on mouldy and often completely rotten German museum pieces". In his statement, Hällström referred both to the dancers' performances and to their collaboration with Nazi Germany. In Finland, free dance, having its basis in German expressionist dance, was considered representative of National Socialist ideologies and therefore despised after the war had ended (Makkonen 2007, 123).

FIGURE 20.7 Leaf beetles make love in the midst of the remains of an unruly past

Photo: M. Marila.

FIGURE 20.8 Remains of a wooden cargo crate provide conditions for fungi to thrive. Haukkapesäjoki, Inari, Finland

Photo: M. Marila.

Ideological and political narratives have therefore come to dominate interpretations regarding the dance styles that emerged in Europe in the 1930s. However, it is also possible to explore Elsa's movement as a solely material remain completely disentangled from the social and ideological forces that existed back then. I (Suvi) wanted to dissolve some of the previous interpretations I had read about Elsa's movement and speculate with the direct material experiences she might have felt as a performer in Lapland in 1944. I wanted to explore Elsa's movement not simply as a representation of political ideas but as an archaeological artefact among other material remains of the Second World War in Lapland.

> So I repeated and reconstructed the postures I had seen in pictures.
> Nearby the prisoner-of-war camps.
> I repeated her postures.
> Her posture.
> I repeated her posture.
> The posture became movement. Now. Gone. Now. Gone. Now. Gone.
> *(Hay 2016)*

FIGURE 20.9 Suvi repeating and reconstructing postures known to have belonged to the choreographic repertoire of Elsa Puolanne. Haukkapesäjoki, Inari, Finland

Photo: M. Marila.

Always repeating and disappearing.
Movement from within, with, without. Movement towards the past.
Movement with, within, towards the rocks, the bryophyte, the sky, the grey logs, the rusty cans, the pines, the slowly decomposing camps.
Towards the future.
Blink, and the frog jumped.
I repeated her posture. A "pattern of movement made, remembered and recreated."

(LaMothe 2015, 26)

It became the past.
The frog jumped into the future.
I shed tears. I was becoming other.
Elsa is gone. A Ghost. I shed tears, but my fingers and balance found her.
A Minor Gesture.

(Manning 2016)

My Ghost.
I repeated their movement.
I left and the common frog stayed.

FIGURE 20.10 Suvi repeating and reconstructing postures known to have belonged to the choreographic repertoire of Elsa Puolanne. Haukkapesäjoki, Inari, Finland

Photo: M. Marila.

FIGURE 20.11 A common frog contemplating the ecology of a Second World War heritage site near Haukkapesäjoki, Inari

Photo: M. Marila.

Note

1 Makkonen (2007, appendix 13) reports ten performances in Lapland between 22 February and 8 March 1944, and "many performances" between 14 May and 14 July 1944.

References

Bender, B., Hamilton, S. and Tilley, C. 2007. *Stone Worlds: Narrative and Reflexivity in Landscape Archaeology*. Walnut Creek, CA: Left Coast Press.

Bennett, J. 2010. *Vibrant Matter: A Political Ecology of Things*. Durham, NC: Duke University Press.

Bolt, B. 2004. *Art Beyond Representation: The Performative Power of the Image*. New York: I.B. Tauris.

Burt, R. 1998. *Alien Bodies: Representations of Modernity, Race and Nation in Early Modern Dance*. New York: Routledge.

Debaise, D. 2017. *Nature as Event: The Lure of the Possible*. Durham, NC: Duke University Press.

Hällström, R., af. 1946. "Moderni tanssi – nykyajan tanssi." *Taiteen maailma* 9: 12–13, 21.

Harjumaa, P. 2008. *Aaveriekkoja. Lappilaisia mystisiä tarinoita ja kummituksia*. Rovaniemi: Lapland University Press.

Hay, D. 2016. *Using the Sky*. New York: Routledge.

Herva, V.-P. 2014. "Haunting Heritage in an Enchanted Land: Magic, Materiality and Second World War German Material Heritage in Finnish Lapland." *Journal of Contemporary Archaeology* 1(2): 297–321.

Ingold, T. 2000. *The Perception of The Environment: Essays in Livelihood, Dwelling and Skill*. London: Routledge.

Koskinen-Koivisto, E. 2016. "Reminder of Lapland's Dark Heritage. Experiences of Finnish Cemetery Tourists of Visiting the Norvajärvi German Cemetery." *Thanatos* 5(1): 23–41.

LaMothe, K. 2015. *Why We Dance: A Philosophy of Bodily Becoming*. New York: Columbia University Press.

Makkonen, A. 2007. *One Past, Many Histories. Loitsu (1933) in the Context of Dance Art in Finland*. Doctoral Thesis. University of Surrey.

Manning, E. 2016. *The Minor Gesture*. Durham, NC: Duke University Press.

Seitsonen, O. 2018. *Digging Hitler's Arctic War: Archaeologies and Heritage of the Second World War German Military Presence in Finnish Lapland*. Doctoral Thesis. University of Helsinki.

Stengers, I. 2018. *Another Science Is Possible: A Manifesto for Slow Science*. Cambridge: Polity.

Whatley, S., Brown, N. G. and Alexander, K. 2015. *Attending to Movement: Somatic Perspectives on Living in This World*. Axminster: Triarchy Press.

Whitehead, A. N. 1956. *Modes of Thought*. Cambridge: Cambridge University Press.

PART VI
Epilogues/reflections

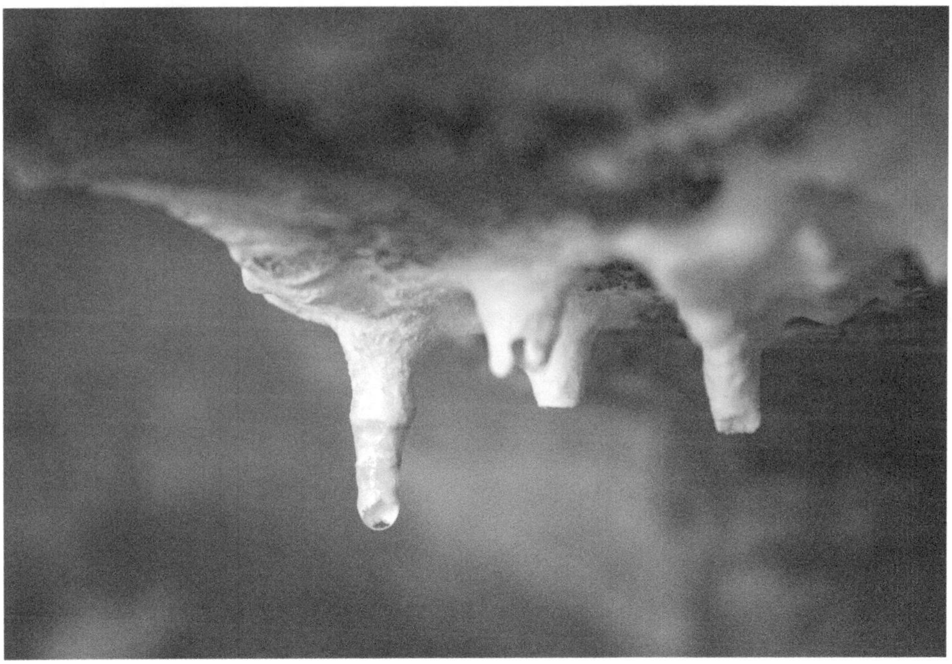

21

INHERITANCE

Timothy Morton

I like to tell my students that the humanities are not candy sprinkles on the cupcake of science. Quite the opposite. Just as the study of medicine requires "pre-med" classes, as they are called in the United States, so the sciences require pre-science, and humanities is pre-science. Why?

A fact is an interpretation of data. If a fact were permanent and solid, there could be no new discoveries. Science with all its incredible progress exists because a fact is not the same thing as the real. What is given, data (Latin, *dare*, to give) is what we have to go on, or as Roger Waters puts it in Pink Floyd's song "Breathe", "All you touch and all you see/Is all your life will ever be".[1] I sometimes think that Waters was channeling David Hume when he said that.

Data is what we have. The real is inaccessible. Hume is the philosopher on whom all modern science (science as such) rests. Hume destroyed the idea that causes and effects can be directly seen – what remains are patterns in data, correlations between one data set and another data set. Causality – links that are stronger than correlations – becomes statistical. It is better to say that 97% of scientists believe that global warming is real and is caused by humans than to insist that it's true and to threaten to torture those who don't agree. Agreement is better than obedience – it's more accurate, let alone more moral.

A fact is an interpretation of data, which means that understanding how interpretation works, and how important it is, is a vital aspect of learning how to be a scientist. And this is why the humanities are beyond important. Science teaches you how to have certain kinds of fact, validated by certain specific institutions such as peer review and scientific consensus. The humanities, however, are all about *how to have a fact at all.* Beyond velocity, spin, temperature, momentum … there are all kinds of fact, contested, socially constructed, and so on – which is quite, quite different from saying that one can say anything at all. Most of theory class is all about how you can't say that this poem can mean anything you want it to mean. Most of theory class teaches you quite the opposite.

A terrible thing has happened. Scientists have skewed interpretations of global warming data to appeal to global warming deniers. They didn't want to put people off. This was a terrible mistake, because it was a game played without the scientists realizing (because they hadn't taken enough humanities classes) that the game space was owned by the deniers. And deniers can always find a way to resist whatever facts are offered to them. For a start, they have the chance to, because the setup is that science offers facts cooked up specially for them, and they have the power in that situation to go, "Oh, this isn't the dish I ordered. Bring me something with more chicken".

This means that global warming data is now far, far more frightening than the scientific consensus about global warming facts. And it also means that politically and ethically, in the widest and the most personal senses, we are totally unprepared to deal with global warming. Without meaning to – meaning quite the opposite, tragically – scientists colluded with global warming deniers.

Scientists get stuck in styles of fact-creation, just as they get stuck in certain sartorial choices. Scientists don't have to wear white sneakers and jeans and pretend not to have a style at all, like people who say they don't have an accent because they are surrounded by people who sound the same. Scientists don't have to be caught in booby-trapped fact factories. They can find some wiggle room in which to generate facts that are more in sync with the data, less coerced by socially powerful reactionary forces. To do this, they need to do things like studying how to read a poem. Yes that's right: I'm a humanist and I'm saying the humanist thing – poetry can save the world – and I'm saying it in the most urgent and deep way possible. In order to figure out how to coexist nonviolently with other lifeforms, scientists should sit around in classrooms for hours, looking at Rorschach blots made out of words. I mean it.

The horrible difference between global warming facts and global warming data indicate something very profound, very salutary. The reason there can be a difference is because there is an even more fundamental crack in the universe, the gap between data and things. Again, since Hume, no one has been able successfully to argue that we have access to things in themselves. What we have are thing data. And there's no good reason measurements are the best or only ways of "realizing" that data in the same way that a conductor "realizes" a piece of music. I peel a banana. I have banana peel. I measure the banana. I have banana measurements. I lick the banana. I have a banana lick. I think about the banana. I have banana thoughts. I write about the banana. I have banana writing. The banana becomes conscious and develops the power of speech. The banana goes on a chat show and talks about itself. "I found myself becoming self-aware during a set of examples by the philosopher Timothy Morton about object withdrawal … it was a very traumatic beginning". What the banana itself would have, like the rest of us, would be banana interview or banana autobiography. Even the banana can't access the banana banana. Even the banana obtains only banana data.

Data is what has been given – the clue is in the word, which is Latin for "what has been given". It is the passive past participle of the Latin verb "to give". Data is the past. A tracery of particles in a diffusion cloud chamber. Patterns of bird migration. The rhymes in verse two. All these are things that have happened already, which is why we can observe them. This has nothing to do with the speed of light in particular – it's much more simple and straightforward than that. A fact is an interpretation of data, which means that a fact is an interpretation of the past. Changing how we interpret data literally means changing the past. The humanities is about changing the past so that the future can be different. So that the future of the future – the very possibility that things can be different at all – remains in effect. The humanities hold open the gate to the future by changing the past.

Data is the past; things are the future

In this volume (chapter 4), Levi Bryant writes the following:

> In this regard, the earth, in its physical reality, has been transformed through thought and the practice that accompanies thought, becoming an embodied reflection of human thought. However, what a strange reflection this is! Unlike a faithful reflection in a mirror or picture in a photograph, the earth is akin to the painting in Oscar Wilde's *The Picture of Dorian Gray*, where it endlessly ages and reflects the ravages of how he has lived his life, while Mr. Gray remains eternally young. The correlate of our thought, the earth, increasingly presents itself as a ruined wasteland transformed by our thought and practice, while humanity still regards itself through the distorted lens of the bloom of innocent youth. In this regard, humanity does not recognize itself in its own painting.

This is exactly the point. It is based on the even deeper point that global warming, as Bryant has argued, shows the Earth at the heart of our being, rather than as a mere correlate of our thought to be known and used. The world – as opposed to the Earth – is at an end, because one can't in good faith treat Earth and everything on it as a horizon or background for human meaning.

I look at my face in the mirror. I see a map of everything that has happened to my face. It's a record, for example, of the terrible acne I had when I was 19. So bad was it that I was one of the first people in the United Kingdom to receive treatment with isotretinoin, or Accutane as it's called in the United States. The scarring that remained wasn't as bad as if I had had smallpox, but it was bad. Every time I see my face, I remember a moment at which I couldn't bring myself to glance in a shop window, like the childhood fear of seeing a monster in the mirror, for fear that the monster would turn out to be actual me.

When you look at the Moon, you're seeing a map of all the meteorites that happened to the Moon. You are looking at the past. This is far less strange than it sounds, in at least one sense – the interesting question is, how come it seems strange at all? The past isn't a point on a line. It's happening "now". That's why history is important, isn't it? Otherwise what happened would not be available at all. No one would ever suffer from it, because no matter how "close" to the "present" the past is, it would always be separated from it by some infinitesimal amount. And if this were in fact the case, nothing could happen, because nothing could be the cause of anything. Because they refer to deep terrestrial time, geology, ecology, and biology give us some potent examples. For example, as you read this, you and what you are reading represent the current state of the Big Bang. The Big Bang is happening "now", and the current Universe is simply the way the wavefront looks for the moment.

And you and what you are reading represent the current state of something we can call the Bacteriacene. Three and a half billion years ago, anaerobic bacteria perished in the toxic environment they had created for themselves, the environment called oxygen – or they didn't, because they were accidentally swallowed by other single-celled organisms. Eventually, according to the logic of endosymbiosis, the anaerobic bacteria, which still had its own DNA, became part of the deep structure of animals (the mitochondria) and plants (the chloroplasts), so deep that one might just as easily describe animals and plants as vessels for bacteria – they are why we move and why plants photosynthesize.

Moreover, you and what you are reading represent the current state of the Anthropocene. Now that we are familiar with this temporal logic, we can identify waves and explosions within the basic explosion – the Anthropocene has at least four "beginnings", and again, although this sounds strange, once you play with this idea for a while, it becomes ever so much easier than thinking time as a series of now-points, because it is not riddled with paradox. The Anthropocene started in 10,000 BCE when humans began to settle and store grain. The Anthropocene started when European colonialism redistributed millions of lifeforms around Earth – including human slaves. The Anthropocene started when the steam engine, motor of industrial capitalism, was patented in 1784 – the uncanny accuracy of the date first used by Thomas Crutzen, who invented the term, lines up perfectly with the uncanny accuracy of the date named by Karl Marx as the beginning of this economic phase. The Anthropocene started when humans began to release radionuclides in 1945, a moment at which Earth systems data are seen to spike alarmingly because of human burning of fossil fuels, the so-called "Golden Spike".

The significant thing is that *all these dates are correct.* Because time is more like a liquid than a point on a line – because in a way, time just is the liquidity of a thing – it makes no sense to locate it at a point, just as locating a wave packet at a point will make it impossible to determine its momentum (the Heisenberg Uncertainty Principle).

But just what is the Moon, anyway? What might it be? How long will it last? Who is this Timothy Morton person who pretends to have written this essay? Exactly what is Earth, what are bacteria, what (to ask a more significantly habitual question) is a human being? Questions like that bring up another quality: the Moon is the future. Of all the entities I have used as examples, the human being is the one we tend to be most curious, or even uniquely curious, about in this respect, as if humans are forever new and the other stuff is just inertly there, as Levi Bryant observes in this volume, in the extraordinary paragraph I have quoted.

So what we are looking at when we look at the Moon or our face is both the past and the future, sliding over each other, irreducibly separate – two waves that don't touch.

What's wrong with automation? An algorithm is actually a very easy and common thing to make and to use, and it has nothing in particular to do with electronic signals embedded in silicon wafers.

What's wrong with automation is that the past is eating the future.

We are already living inside an AI: it's called capitalism. Capitalism is an adaptive AI that learns to extract more and more value from lifeforms, so that there you are on Facebook, when suddenly you realize, all your data has been sold and there is a President Trump.

But data is open. Data isn't facts, not yet. Data means "what is given" in Latin. Some theologians now argue that givenness as such is the appearance of divinity or the sacred in our world. But we don't have to go that far to find it wonderful. What is inherited, beyond automation, beyond the past being built out into a predictable, programmable "future", is what is given.

A sketchy phenomenology of the future future

Ian Bogost's book *Play Everything* is about how to find the future in the past (2016). It is therefore an unintentional, and because of that very, very fine, manual for how to start a revolution. The humanities in general are tasked with changing the future by changing the past. Since interpretations of form (what this chapter has been calling the past is what cultural criticism broadly calls "form") are infinite in the strong sense – they are strictly uncountable – and since these interpretations are strictly "the future" – they exist in an irreducible not-yet – there resides some kind of reserve of futurality in the material, physical, psychic, ideological, poetic, and otherwise cultural (to use some badly formed "record store" type labels) that we have to hand. Precisely because of the un-closeable gap between future and past, the way they criss-cross without touching, new things can happen. If this were not the case, nothing could happen at all. The future would simply be the metastasis of the present, an endless build-out like a wooden platform extended from a tree forever in every direction.

Since a thing does not ever coincide with itself, and since this non-coincidence occurs at any and every point whatsoever (physical, ideational, hypothetical ...), a thing is intrinsically a self-contradictory and ambiguous "toy"-like entity. A toy is defined as a thing that can enter play, and play is self-contradiction. Gregory Bateson (1972) succinctly observes that when his cat gives him a playful nip, the cat is saying "This is a bite and this is not a bite". This playful contradictoriness is again irreducible, or as Jacques Lacan (1981, 48) puts it, "What constitutes pretense is that, in the end, you don't know whether it's pretense or not". Therefore play is a foundational category of revolutionary praxis. When a child skips along the pavement of a shopping mall, trying to avoid the cracks, she is playing with a pre-existing structure, figuring out how to achieve wiggle room in a world created and defined by adults (Bogost 2016).

I may not know exactly how we get there, but after 30 years of thinking about ecological politics, I think I can describe *how it will feel*, which isn't at all a matter of "subjective experience" but is in fact how the contours of ecotopian change will appear to anyone at all, despite who they think they are. I endeavor here to name the phenomenological chemicals, so that when they start to show up, we will have a sense that we're on the right track.

What, for example, is the vehicle of this phenomenology? It would be about treating "objects", entities, things in general, more benevolently than we do now, about not reducing them to reified or commodified (or both) objects-for-us. Are there any examples lying around in the cultural junkyard for us to think about? The Millennium Falcon of *Star Wars* serves as a potent and easy to remember one. The Falcon stands for the pure givenness of a thing. So let's assume art is not just the past – a bunch of decisions made by a bunch of author functions, deliberately or not – art is also the future, because what is this poem anyway, and the number of good readings of this poem are uncountable, a.k.a. infinite, in the nice way.

And let's assume that really corny music and movies (the ones not to talk about at art openings without affecting the urge to vomit tastefully) are talking about what we really want, a.k.a. that future future thing, what we want whether we admit it or not, because look at all those billions of people interested in discovering the superpower of telekinesis. And if you can discover it, it must already be there, and therefore it must be alienated. So the future future is all about how to unalienate things. We are all wanting that stuff, wanting it with our bottoms on the soft seats of cinemas and sofas in living rooms.

So let's assume Blake is right. Let's assume that energy is eternal delight, and let's assume that we get to the future future (the different one) by what he calls "an increase in enjoyment". Let's assume that what is wrong with our contemporary world is not that there is too much pleasure – a common left-wing assumption – but that there is not nearly enough (also a left-wing thought, but less common). Thus consumerism is a terrible photocopy of the kinds of pleasure that a more just society would afford. And the very concept of pleasure within consumerism, the concept derived from utilitarianism, is ugly and flawed. Surely polar bears continuing to live is pleasurable for them. Surely humans could learn to appreciate or otherwise enjoy polar bear pleasure. They have given so much to the Coca Cola corporation to enhance some people's pleasure – we have all learned how to handle those weirdly feminized bottles. Surely we can learn how to enjoy polar bear pleasure.

Let's call this increase Warp Drive. Seems right because when, in *Star Trek*, humans discover warp drive, they almost immediately get rid of money and people spend their lives working on getting better, a.k.a. aesthetic play, like the William Morris vision of the future.

This is where *Star Trek* lets us down a bit. *Star Trek* is a corporate kind of a world and the Enterprise and so on are basically floating open-plan offices. White men kind of got along at the expense of everyone else outside the smoky gentleman's club, then more and more people got enfranchised in more and more ways. So the solution is in part not just about rules and ethics and morals, it's about enjoyment. How to be chummy. Make the planet chummy for the first time. That would be my detourned MAGA hat. Never let academics design hats.

And that means we are talking aesthetics, which means we are talking politics. Should we put a dot of green next to that dot of silver? Since you can't escape the aesthetic dimension, I'm so sorry, Benjamin fans, doing a Keith Moon and blowing up its toilets won't work, and why should the Nazis have all the fun?[2] And we aren't talking about aestheticizing politics here – maybe Benjamin was right, but accidentally. Nazism means gilding the lily, aestheticizing something that was already pretty artistic. And according to the aesthetic dream logic, the Enterprise can easily become a Death Star – another open plan office. Just consider that heavily weaponized one in the second reboot, *Star Trek: Into Darkness*.[3]

So we have to turn to the Falcon, which gives us a much more touchy-feely version of what hyperspace is like: a phenomenology of hyperspace. There is, to be sure, a small amount of it in Star Trek. It usually gives you the chance to have some kind of brown bag lunch. Those are three words that should never have been put together. Also, *Star Wars* was appropriated by Reaganism (the military plan and the idea that the USSR was an "evil empire"), and it's always an amusing tactic to re-appropriate. Moreover, the Falcon is a Millennium one. Nitpicking scholars of radical puritanism in the seventeenth century might prefer it very much if it was the Millenarian Falcon, but that is unnecessarily a bit technical. In brief, I like my utopias without an apocalypse. The way we talk to each other about what's coming next is far too apocalyptic, fatalist, cynical, and nihilist.

The thing about the Falcon is that it is all about how to be in a living room. First you get in out of the street. This might involve firing a blaster at people trying to stop you while walking backwards up a slope from the putative driveway to the front door: we gather that it's a violent neighborhood. In the largest sense, this would be the neighborhood of militarized planetary space where military intelligence infiltrates Extinction Rebellion and encourages the protesters to fly drones over Heathrow just in time for the media backlash.

Once you're safe inside, you have to learn how to sit in a brownish leatherette swivel chair. This is autobiographically pleasant for me because 1977, when we first witnessed the Falcon stuff, I had enjoyed my first brownish leatherette swivel chair experiences thanks to my Atlantic-hopping granddad.[4]

Then, it doesn't matter who you are – you can be L3 or those furry bird beings in Star Wars 8, or Han or Chewbacca. You get to fly the thing. But you're flying while looking out of mullioned widows, which are Lancaster Bomber windows (it is a strange anecdotal detail that my great uncle Barnes Wallis invented them in addition to the bouncing bomb), but they're also the windows of the Victorian shop in *Alice Through the Looking Glass*. When you get to hyperspace by pushing a lever that looks like the main one on a jumbo, you find yourself in hyperspace. And hyperspace is where you realize you're in a living room. You can tinker with things. You can play chess. You can have sexy chit chat.

So I think we know where to find warp drive. It's somewhere in the living room. Where exactly is it? My guess is it's somewhere in those Yuletide simulation pagan fairy lights or somewhere in the Halloween decorations. Disco, that other phenomenon of 1977, and in particular house, which evolved in black gay cultures in Chicago and Detroit in the mid-1980s and is a musical distillation of disco, are "hyper" versions of domestic space where anyone can fly, no matter what their race class gender sexuality and so on. *Star Trek* has the manuals on why this is a good idea, but *Star Wars* lets you feel it as you imagine yourself flying about in the sinthomic falcon, that weirdly inconsistent Irigaray-like sand-dollar-vulva-1970s-Corvette.[5]

The Falcon is a symbol for that quality of a thing that the philosophy of object-oriented ontology tries to get at: its radical contingency. It's garbage, what was cast away abjectly, no longer of use for humans or "as-structured" for them (to use the technical Heideggerian language), waste that that you have to use because the other getaway ship just got blown up. In the seventh episode, Rey yells "The garbage will do!" as she runs from the TIE fighters with Finn. The Falcon is a thing you win in a card game – the prequel *Solo* finally lets us see the game where Han Solo wins it from Lando Calrissian.[6] The Falcon is a thing that happens to you, confusing that active–passive binary that is along with the word survival and the human/nonhuman binary, mixed with the law of noncontradiction, destroying earth nicely.

There's something special about the ontological status of the Falcon. There seem to be four kinds of spacecraft, and three of the four are permutable. The Enterprise can turn into the Death Star. Even the TARDIS, the *machina ex* which Doctor Who is *dea*, can turn into one of those little Honda-like fighter ships that Obi-Wan has to fly in the prequels.[7] And all the spaceships we can imagine can turn into the TARDIS.

But nothing can turn into a Millennium Falcon.

The Falcon is a disco imagined as an introverted space where anyone can coexist with anyone. A house as in house music invented by black gay people in Chicago in the 1980s. Techno or house or EDM or whatever is about how to find the future in the past: about finding freedom, wiggle room quite literally, in a space of maximum repetition constraint, like the predictive pulsations in the club, where Deckard is chasing Zhora in Ridley Scott's film about 2019.[8] Those pulsations were made by Vangelis, who didn't get to be in Yes because of their worry that he might be able to retire all of them.

Given the weight of the past, all that past, Generation Z must be feeling terrible. I recently (in September 2019) interviewed Extinction Rebellion Youth, the political quintessence of Generation Z at this moment. My only question to them was, "How can I be of assistance to you?" Can you imagine what it's like being in Generation Z? As much as I dislike PR advertising labels, I'm fond now of being an X-er. X is a high-scoring letter in Scrabble, for a kickoff. And the X implies something fun and foolish about one's parents, the Boomers: "Who on earth are these people? Oh – they're our children!" We understand a little bit what it's like, in an existential inner sort of a way at least. But Generation Z rightly cries out that everyone else did it to them – broke planet Earth, and now it's on them to fix it? Everyone else, for twelve and a half thousand years. Marx talks about history weighing like a nightmare on the brains of the living (Marx 1996). He had no idea how much that would weigh for Generation Z.

In a way, how dare we older people try to help, given that we don't even understand fully the problem we've created for them and given how we love to shock everyone and make them feel stupid and evil, shocking them with climate and extinction data dumps on page one of the newspaper and paralyzing them with eco guilt on the editorial page? But then again, who else is going to do it? It's like that sentiment of Beckett that Adorno loved – I can't go on, I must go on … fail better (Beckett 2014). The possibility that things might be different – the radical futurality of a thing – still slides across the past, the nightmare form of a thing, not touching, looking like an X shape, a chiasmus, only separated in a third dimension. There is always, in everything, including every language, even the languages of physics, mathematics, and logic, a hidden side, an unconscious format, an unspeakable style that literally is the future. No matter how much logistical churning paves over it, the future is still open.

I hold a book called *Heritage Ecologies* in my hands. I see what I have been given.

Notes

1 Pink Floyd, "Breathe," *The Dark Side of the Moon* (EMI, 1973).
2 I refer of course to Walter Benjamin, "The Work of Art in the Age of Mechanical Reproduction" (1973).
3 J.J. Abrams, dir., *Star Trek: Into Darkness* (Paramount Pictures, Skydance Media, Bad Robot, 2013).
4 George Lucas, dir., *Star Wars* (Twentieth-Century Fox, 1977).
5 The concept of the sinthome, an external symptom, is Lacanian. For Lacan, the sinthome is neither symptom nor fantasy but "the point marking the dimension of 'what is in the subject more than himself' and what he therefore 'loves more than himself'" (Žižek 1992, 132).
6 Ron Howard, dir., *Solo: A Star Wars Story* (Lucasfilm, 2018).
7 For example, in George Lucas, dir., *Star Wars 2: Attack of the Clones* (Twentieth-Century Fox, 2002).
8 Ridley Scott, dir., *Blade Runner* (Warner Bros., 1982).

References

Bateson, G. 1972. "A Theory of Play and Fantasy." In *Steps to an Ecology of Mind*, edited by G. Bateson, 177–193. New York: Ballantine Books.

Beckett, S. 2014. "Worstward Ho." In *Nohow On: Company, Ill Seen Ill Said, and Worstward Ho*, edited by S. E. Gontarski. New York: Grove Press.

Benjamin, W. 1973. "The Work of Art in the Age of Mechanical Reproduction." In *Illuminations*, edited by Hannah Arendt, 217–251. London: Harcourt, Brace and World.

Bogost, I. 2016. *Play Anything: The Pleasure of Limits, the Uses of Boredom, and the Secret of Games*. New York: Basic Books.

Lacan, J. 1981. *Le séminaire, Livre III: Les psychoses*. Paris: Éditions du Seuil.

Marx, K. 1996. "The Eighteenth Brumaire of Louis Bonaparte." In *Later Political Writings*, edited by T. Carver. Cambridge: Cambridge University Press.

Žižek, S. 1992. *Looking Awry: An Introduction to Jacques Lacan through Popular Culture*. Cambridge, MA: MIT Press.

22

ECOTONE

Þóra Pétursdóttir

"We don't walk *on* the earth but *with* the earth."
(Levi Bryant, Chapter 4)

Opening

In *A Land*, Jacquetta Hawkes writes that

> It is only the pathetic shortness of human life that gives each individual a sense of permanence of his [*sic*] background. The land we all walk upon has been under the sea many times, and it will be submerged again.
>
> *(1951, 43)*

When removed from the context of Hawkes' poetic writing about her homeland, where she entangles personal, historical and geological histories and temporalities in an evocative and analytical tale, it is easy to mistake her words for a rather cynical depiction of the human being and its role in the extended legend of land and earth. That, however, was not at all her intention. Weaving together her passion for archaeology, geology, biology and history, her aim was rather to embrace the complex ecology of that legend and the significance of recalling this complexity also when focusing on the minute scale(s) of human heritage. In that sense, Jacquetta Hawkes' writing can be viewed as a prophecy for the current environmental turn in the humanities. And, as stated by the author of her biography, Christine Finn (2020), the core of Hawkes' message is therefore no less timely today and even emblematic for "the future of an endangered planet".

Heritage Ecologies draws on a motivation similar to Hawkes' in that its aim is to envision the complex physical, temporal and mnemonic ecologies that constitute heritage matters, broadly defined. From the perspective of the various places or loci of heritage – be that a museum, archive, landscape, object, archaeological site or structure – a thread running through the chapters in this volume is the wish to think heritage ecologically. To focus on how different stories, chronologies, entities, species and kinds interact in heritage processes and how they come together *in place* to cultivate contingencies while withholding others. The aim is neither a "flattened" image nor to abolish differences between parts but rather to think heritage as a meshwork and thus, as central to Bryant and Joy's definition of ecology (2014, xi), through "relations and interactions between entities" rather than within any single one of them.

Place

Places of heritage may sometimes be envisioned as fortifications set apart from their surroundings – directly so in the case of museums and archives and indirectly in the case of landscapes, archaeological sites and cultural traditions. Whether tangible or intangible, in situ or not, the naming, defining and safe keeping of heritage may reflect a notion of place – resembling Doreen Massey's more general framing – "as closed, coherent, integrated, as authentic, as 'home', a secure retreat" (2005, 6). Such notions of place can also be seen as mirroring the significance given "representativity" as a central criterion in the definition, performance and management of heritage (e.g. Bærenholdt et al. 2004; e.g. Staiff, Watson and Bushell 2013), where the demarcation of place enables the purification and control of the time, memory and meaning mediated.

Many of the contributions to *Heritage Ecologies* have in common that they seek to complicate understandings of heritage places, not merely so by critically deconstructing a black-boxed image of authenticity, purity and order but rather by envisioning how an often fragile entity is being constituted and held together by an ecology of agents, human and non-human. This is not to reject the distinctiveness or particularity of a heritage object but to question an image of its pre-given coherency and to explore the "throwntogetherness" (Massey 2005; Stewart 2008) of place: how the event of place is a dynamic constellation of different elements, times and memories, a sort of gathering or "coming together of trajectories" (Massey 2005, 140–141). This is evident for example in Fredengren's and Farstadvoll's work on urban parks and nature reserves; in LeCain's work on the residues of mining; in Baker's, Bangstad's and Storm's approaches to industrial landscapes; and in Tuominen and Marila's experimental engagement with the heritage of war in Lapland. Moreover, and to nuance a possibly narrow idea of what constitutes a "place", this objective is also very clear in Coelho's and Sterling's work on photographs, which Sterling frames as a "move from a representational view of photography to one that prioritises the ecological emergence of specific images and photographic assemblages" (chapter 7).

These ecological approaches challenge an understanding of place as settled and lucid, as something that becomes disturbed only by *external* forces, and instead bring forth the internal complexity and porousness of its confines (cf. Domínguez Rubio 2014, 2016; Desilvey, chapter 16). As shown in many of the contributions, this further means contesting the "locality" of place through concern for its complex multi-spatiality and temporality – or even its being as a "hyperobject" (cf. Morton 2013). In Morton's definition, hyperobjects may indeed appear as a specific and possibly sinister category of things, but in many ways, it is also a definition that embraces the affirmative characteristics of heritage matters as framed in *Heritage Ecologies*, that is, that they are durable, viscous, formed through inter-objective relations and massively distributed in time and space (ibid). Thus, brought together, what this "hyperobject" approach may promote is the decentralization of human beings in the constellation of heritage processes.

To some degree, this is of course an impossibility; heritage is in essence a human-centred phenomenon. But, as so simply and beautifully framed by Stein Farstadvoll (chapter 18),

"Just as we cannot always control what we inherit … it also follows that we do not always know who or what is the 'heir' of a place". And this particular nuance to the well-established discussion of ownership is, I argue, much needed in heritage understandings at present. One may even frame it as an aspect of the wanted decolonization of heritage, activated through the challenges of the Anthropocene (see Harrison and Sterling 2020).

Ecotone

To state the already obvious, I should clarify that more than a conclusion of any sort, this closing text is a personal reflection and afterthought, and, while deeply inspired by the authors in this volume, it is not an attempt to conclude our arguments. In fact, as established through the variety of the preceding contributions, heritage is not a phenomenon that can or even should be concluded and settled. Much rather, it may be claimed to thrive only on and through movement, or in passing. It is passed down from generation to generation, passed through, inhabited and abandoned by constantly new ideas, entities and organisms and is continuously affected by processes of becoming and erasure. In passing, objects of heritage, material and immaterial, also affect and move their surroundings, including our knowledge of pasts and expectations for futures. Thus, while much of heritage definitions and management may be claimed to work towards increasing stability and security, I would argue that it is still, in the first place, this element of uncertainty, transformation and drift that directs what we recognize and care for as heritage. Rightly, therefore, heritage may be defined as a process (Harvey 2001; Smith 2006), as nomadic (Appelgren 2014) or, as shown in many of the preceding chapters, as a strain and tension between rivalling currents – between past and future, permanence and entropy, nature and culture, tangible and intangible, us and other, the known and unknown.

For this reason, not only some but most heritage phenomena could be further described as *difficult* and, so to speak, trapped in conflict and divergence. Not necessarily or solely in a negative or cynical understanding but equally in a generative and fruitful sense. Drawing on that dynamic in-betweenness, and on Morton's (2013) definition of the hyperobject, heritage can therefore even be imagined as an ecotone: a creative and potentially transgressive borderland between diverging domains.

Originating in biology and braiding the notions of *eco*logy and *tone, tonos* (from Greek) or *tension*, the concept of *ecotone* refers to a region, sharp or diffuse, where different ecologies, biomes and communities converge (Naiman and Décamps 1997), such as land and sea, forest and grassland or saline and fresh water. The ecotone, therefore, is a zone that both combines and divides and which is always, and by definition, in transition. A dynamic expanse where movement, flow and drift are the most liable and unwavering constants. At the same time, however, and because of this relentless internal drama, the ecotone is also very much *a place of its own*. While its porousness renders it receptive to influence from both or all flanking biomes, it may itself also host other organisms and greater population densities than found in any of the neighbouring communes (Attrill and Rundle 2002; Walker et al. 2003). Interestingly, the ecotone's complexion, variety and wealth are therefore greater and other than the sum of its parts.

I like to think that the same holds true for many heritage places and phenomena. While one may rightfully view heritage as that which unites groups and generations, grounds a sense of belonging and anchors identities, it does so more by means of enduring plasticity and mobility than though any form of stable or static permanence. Like the ecotone, heritage combines and divides and by so doing constantly generates something new.

This element of drift, mobility and becoming is also what makes heritage attractive and alluring, since ideally, and possibly idyllically, it casts heritage as something open and inviting rather than predefined and closed. At the same time, this element of openness and becoming is also that which reinforces the somewhat opposing idea of heritage as fragile and infinitely subject to threat and loss (DeSilvey and Harrison 2020; Brown, Liuzza and Meskell 2019). While to some degree the very constitutive ingredient of heritage discourse and management, agitation for the future of heritage is a topic that has climbed quite explicitly on current agendas (see e.g. Harrison et al. 2020). With climate change, melting ice caps, rising sea levels, more turbulent weather forecasts, increasing erosion, air pollution and so on, the anticipated loss of both tangible and intangible heritage appears more immediate now than before. Hence, and ironic as it may seem, it appears that in the Anthropocene, "our" heritage – the heritage of the Anthropos – is concurrently that which is *threatened* and that which is *threatening* futures, for humans and many others.

While care for the fragility and transience of heritage is a more or less habitual part of heritage management, fragility itself – as a natural characteristic of heritage phenomena – is mostly not the matter of that concern. Quite the contrary, the elimination of uncertainty is often what lies at the heart of heritage projects (Appelgren 2014; Pétursdóttir 2020).

FIGURE 22.1 Writer Andri Snær Magnason talks to media by the foot of Ok, SW Iceland

Photo: Lára Björg Björnsdóttir, courtesy of Katrín Jakobsdóttir.

Heritage should be safeguarded and its fruitful management viewed as integral to sustainable development, natural and social (cfr. Figenschau 2019). However, and as many of the contributions to *Heritage Ecologies* illustrate (see e.g. Storm, DeSilvey, Bryant, Farstadvoll, Harrison Baker and Bangstad), thinking heritage in the Anthropocene may also require alternative perspectives. While preventing loss is inevitably an important undertaking, taking seriously fragility, instability, transience and drift – the specific characteristics of the ecotone – is, I would argue, just as important. Not merely as something to be restrained but equally as features that should be embraced (Pétursdóttir 2020). In other words, fragility itself may also require concern, care and commemoration (see also Figenschau, chapter 15).

Two tales of heritage becoming

In the summer of 2019, while this book was in the making, two parallel occurrences prompted my reflection on just this: on instability, transience and drift as integral to heritage making. Both events unfolded in the landscapes of Iceland, one I experienced first hand, the other through news, social media and witnesses' retelling. While both establish the centrality of instability and becoming in constitutions of heritage, they do so in rather different ways. One is disquieting, the other more hopeful.

In August 2019, Icelandic novelist Andri Snær Magnason and a team of geologists from Rice University in Texas commanded an expedition that immediately travelled the world through news and social media. Together, the author and the team of geologists escorted a small crowd of people, including the Icelandic prime minister and the minister of environment, to the top of the mountain Ok in southwest Iceland. Having reached the top and admired the view, Magnason directed a ceremony commemorating the glacier that once covered the very peak the troop was gathered on. The glacier, bearing the same name as the mountain – Ok-jökull – had earlier the same year been discredited its global status among glaciers by the team of geologists who for years had been following its constant and escalating retreat. As of 2019, the ice-cap that still covers the summit of Ok no longer met the criteria required; it was too thin and the ratio between summer loss and winter gain tilted the wrong way. Ok*jökull* was, according to definition, no longer a glacier – no longer *jökull*.

To acknowledge this unprecedented loss, and to complete the ceremony, Magnason and the two Icelandic ministers unveiled a memorial on the mountain summit. A brass plaque on a basalt boulder inscribed with, as it says, "A letter to the future". The letter, given first in Icelandic and then in English, reads:

> Ok is the first Icelandic glacier to lose its status as a glacier. In the next 200 years all our glaciers are expected to follow the same path. This monument is to acknowledge that we know what is happening and what needs to be done. Only you know if we did it.

A postscript below the letter dates the memorial to August 2019, followed by the words "415ppm CO_2", referring to a record-breaking level of carbon dioxide recorded in the atmosphere in May of that same year.

Following the ceremony, the small crowd descended the mountain, leaving the memorial on the summit as a new site of heritage and commemoration in the Anthropocene. Bearing in mind how the news story of the event and memorial immediately went viral and the boundless increase in tourism in Iceland in the last few years, I am confident it will become just that. A "destination". A place to visit. A place of heritage and commemoration. A sort of lieu de mémoire (Nora 1996), where something past, gone and lost is recalled – and mourned. And not least this last point seems telling for this particular case: Ok has become a place of mourning. A place of heritage *lost*, more than anything else.

This is interesting not least when considering that Okjökull was never really in any spotlight before its loss. In the context of Iceland's many and much larger glaciers, it was probably a sort of minor, sidelined and overlooked. Thus, Ok enters the heritage arena more as "a-glacier-lost", so to speak, than as the phenomenon as such. Its loss, in other words, is constitutive of its heritage status.

Ironically, while the brass plaque was unveiled at mount Ok, commemorating a glacier lost to climate change, I took land on Surtsey – an island born into the Anthropocene era. Together with a group of biologists and a colleague from archaeology, we were there to register the island's development, an annual expedition aimed at monitoring and documenting the colonisation of new plants, insects and animals, soil formation, erosion and so on. The becoming of land and nature.

Surtsey is part of the Westman islands archipelago, a group of 18 islands off of the south coast of Iceland. Together the islands form a single volcanic system, which during the last 10,000 years has seen at least 24 eruptions – two of which have happened after the settlement of Iceland. Surtsey was formed in one of the two.

The birth of Surtsey was first noticed by crewmembers on a fishing vessel who reported to land about strange sights, smoke and explosions above the sea surface in November 1963.

FIGURE 22.2 Trampled pathway leading between the "science station" and the seagull colony. The abandoned lighthouse on the peak in the background

Photo: Þ. Pétursdóttir

The eruption lasted continuously for almost four years, until the beginning of June 1967, and resulted in a barren island, 2.6 square kilometres in size, and the consequent southward extension of Icelandic terrain and jurisdiction. As phrased by one of the geologists observing the eruption:

> It is just as if one of the blissful islands ("insulae fortunatae"), which the cartographers of the Middle Ages placed in the sea to the south of Iceland but which no one ever had an opportunity to see, had become reality.
>
> *(Þórarinsson 1964, 54)*

Due to erosion, however, and the only gradual solidification of her young bedrock, Surtsey has changed in form and shrunk considerably in size since her birth and is today only half the size she was in 1967.

The island has been closely monitored from the very moment the eruption reached the sea surface and to the present day. Scientists first set foot on the island in May 1964, and it was declared a nature reserve already in 1965, while the eruption was still ongoing and Surtsey yet in the making. From that time, the island has been strictly and completely closed off from unlegislated *human* visitors and closely managed by the Icelandic environmental agency and the Surtsey scientific association. It is imagined as an utterly *natural* terrain and was as such listed as a UNESCO World Heritage site in 2008. Here, Surtsey's outstanding significance is grounded in its natural virginity, which offers the world "a pristine natural laboratory, free from human interference" (UNESCO), where undisturbed biological processes and the formation of an ecosystem can be directly monitored.

"A pristine natural laboratory, free from human interference", and yet, there we were. While the first steps onto land did indeed resemble an imagined moon landing, our arrival was quite rapidly and inevitably registered in the beach sand. And already on the eve of our first day, the whole landscape was rippled with our presence. Possibly because of the creeping feeling of trespass, we followed the same routes, tracing each other's footsteps over and over again, creating a network of pathways like ripples in the land. Or possibly it was because, as Robert Macfarlane (2013) states in the *Old Ways*, that "paths are the habits of a landscape", a sort of "consensual making" and route-finding between us and land.

Of course, alongside stated *natural* processes, bacterial growth, birds nesting and plants rooting, the island's cultural legacy has developed and flourished throughout the years. This both intentionally, through a considerate yet unambiguous scientific colonization, the naming of land, installation of infrastructure, studying, sampling and listing, and unintentionally through human terrestrial, airborne and seaborne influence and effluence. Ironically, thus, despite the very strict regime of stewardship, isolation and purity, aimed at keeping nature and culture separate, Surtsey also sustains an interesting uncensored, though mostly overlooked (and even silenced), ecology that conspicuously defies any binary distinction between human and non-human.

For one, the scientific presence on the island is in fact quite distinct, registered in various structures and installations: a science station or barrack, a helicopter platform, an abandoned lighthouse, a weather station, several reference points for aerial photography and GIS mapping, standing and fallen cairns, bore holes, label posts marking the location of species and spread of vegetation, sampling posts and so on – not to mention all the gear that is shipped onto and off the island in relation to each expedition. While it

FIGURE 22.3 Drift matter, wood and plastic on Surtsey's shore

Photo: Þ. Pétursdóttir.

is possible to view all these structures and instruments as alien intrusions – as a necessary infrastructure of scientific observation but *external* to the *natural laboratory* of Surtsey – it soon becomes evident that things are more entangled. That is, that these *unnatural* installations are in fact constitutive organs of the claimed *natural* processes monitored and measured (cf. Barad 2012).

For example, there are several registrations of an evident tangling of human relics and nonhuman becomings on the island: The most lush vegetation and intense soil formation are found in and around the colonies of seagulls and humans, respectively; Some vegetation has been shown to initially follow the pathways of scientists, mapping their repetitive movements around the island; lichen settle the leaching and nutritious concrete surfaces of the lighthouse and the perimeters of the helicopter platform even before they make an attempt with the rough lava; birds may braid their nests with materials snatched from scientific installations; the ravens especially, famous for their aesthetic eye for colour and glitter, harvest around the human colony picking up pegs, rope, plastic and more to construct their eccentric habitats inside the now cool volcanic crater; many of the first winged insects have been caught on the insides of windows in the science station and lighthouse; and a popular destination in the search for new creeping colonizers is the drift beach, turning around wooden and plastic objects washed in from near and far.

These are not unexpected findings. But they are rarely named (see, however, Ólafsson and Ásbjörnsdóttir 2014; Lárusdóttir 2017). The long list of scientific publications assembled in celebration of Surtsey's 50th anniversary in 2017 did not embrace this form of human-nonhuman world-making, and the "natural laboratory" designation by UNESCO

FIGURE 22.4 Lush grasses enclose the helicopter platform in Surtsey

Photo: Þ. Pétursdóttir.

ignores it altogether. Also telling of this "insular" perspective is the fact that during 50 years, research in Surtsey has been driven exclusively by *natural* scientists. My colleague and I were possibly the first humanities scholars (with explicit research aims) to set foot on the island. Pointing this out is not to downgrade the scientific work done on Surtsey, which is no less than groundbreaking, but to stress that it does make a difference. As Karen Barad argues, it matters who and what is making the measuring and monitoring. Measurements, as she says, "are not simply revelatory but performative: they help constitute and are a constitutive part of what is being measured" (Barad 2012, 6). For this reason, because "measurements are world-making", it matters greatly how something is observed and explored. And it is not simply a matter of an "object *reacting* differently to different probings but *being* differently" and hence, what is at stake "is the very nature of nature" (Barad 2012, 6).

Insular versus ecological thinking

Islands, like Surtsey, have for centuries been constitutive of Western notions of nature and environment. Insulated, remote and far away from the disturbing noise of urban centres, islands have been viewed as natural extensions of controllable laboratory environs and thus ideally suited for scientific work (Hennessey 2018). This understanding was the basis for the early protection and UNESCO listing of the Galápagos archipelago, which Darwin tellingly described as "a little world within itself" (Darwin 1845). In 1961, just before the birth of

Surtsey, Jean Dorst, deputy director of the natural history museum in Paris, described the significance of the islands so:

> Owing to the remoteness of the archipelago, the number of ancestors is of course very limited. Hence a simplification in the fauna which makes the laws of evolution much easier to distinguish than in the rest of the world, where the complexity of natural phenomena and the multiplicity of ancestors complicate inextricably the tracing of relationships. The Galápagos Islands thus stand out as Nature's experimental station.
>
> *(Dorst 1961, 30)*

This in many ways resembles the articulation of Surtsey's uniqueness and scientific merit in relation to its world heritage listing half a century later. And there is definitely much truth to this designation. At the same time, however, there is reason to ask whether such metaphors of isolated and insular Natures (cfr. Morton 2007, 2010) really serve sustainable goals at present – in the Anthropocene era. Is it as feasible now to seek simplifications and more easily traceable trajectories? Or is it, as revealed in the becoming of Surtsey and unbecoming of Okjökull, turning ever more evident that it is complexity – mobility, transience and drift – that need to be confronted and embraced? And thus, that "staying with the trouble" (Haraway 2016), with entangled nature-cultures, is in fact more *representative* of "the nature of nature" than seeking ways to simplify it. I would argue so, and this is also what I see as a common and very important thread running through *Heritage Ecologies*. A wish to think heritage through complex ecological relations, through ecotones, and to embrace how different stories, chronologies, species and kinds come together and contribute to its constitution and becoming.

References

Appelgren, S. 2014. "Heritage, Territory and Nomadism – Theoretical Reflections." In *Vägskälens kulturarv – Kulturarv vid vägskäl: Om at skapa plats för Romer och resande i kulturarvet*, edited by I. M. Holmberg, 243–256. Gothenburg: Makadam Förlag.

Attrill, M. J. and Rundle, S. 2002. "Ecotone or Ecocline: Ecological Boundaries in Estuaries." *Estuarine, Coastal and Shelf Science* 55(6): 929–936.

Bærenholdt, J., Haldrup, M., Larsen, J. and Urry, J., eds. 2004. *Performing Tourist Places*. Aldershot: Ashgate.

Barad, K. 2012. "What Is the Measure of Nothingness? Infinity, Virtuality, Justice." *Documenta* 13: 4–17.

Brown, N. E., Liuzza, C. and Meskell, L. 2019. "The Politics of Peril: UNESCO's List of World Heritage in Danger." *Journal of Field Archaeology* 44(5): 287–303.

Bryant, L. and Joy, E. A. 2014. "Preface: Object/Ecology." *O-Zone: A Journal of Object Oriented Studies* 1(1): i–xiv.

Darwin, C. 1845. *Journal of Research Into the Natural History and Geology of the Countries Visited During the Voyage of H.M.S. Beagle Round the World* (2nd edn). London: John Murray.

DeSilvey, C. and Harrison, R. 2020. "Anticipating Loss: Rethinking Endangerment in Heritage Futures." *International Journal of Heritage Studies* 26: 1, 1–7.

Domínguez Rubio, F. 2014. "Preserving the Unpreservable: Docile and Unruly Objects at MoMA." *Theory and Society* 43(6): 617–645.

Domínguez Rubio, F. 2016. "On the Discrepancy Between Objects and Things: An Ecological Approach." *Journal of Material Culture* 21(1): 59–86.

Dorst, J. 1961. "Where Time Stood Still: The Galápagos Islands and Their Prehistoric Creatures." *UNESCO Courier* 9(September): 28–32.

Figenschau, I. 2019. "The Heritage of War and the Discourse of Sustainability." *Norwegian Archaeological Review* 52(2): 89–108.

Finn, C. 2020. *Ice Without, Fire Within*. London: Archaeopress.

Haraway, D. J. 2016. *Staying With the Trouble: Making Kin in the Chthulucene*. Durham, NC: Duke University Press.

Harrison, R., Desilvey, C., Holtorf, C. and Macdonald, S. 2020. *Herritage Futures: Comparative Approaches to Natural and Cultural Heritage Practices*. London: UCL Press.

Harrison, R. and Sterling, C., eds. 2020. *Deterritorializing the Future: Heritage in, of and after the Anthropocene*. London: Open Humanities Press.

Harvey, D. C. 2001. "Heritage Pasts and Heritage Presents: Temporality, Meaning and the Scope of Heritage Studies." *International Journal of Heritage Studies* 7(4): 319–338.

Hawkes, J. 1951 [1978]. *A Land*. London: David & Charles.

Hennessey, E. 2018. "The Politics of a Natural Laboratory: Claiming Territory and Governing Life in the Galápagos Islands." *Social Studies of Science* 48(4): 483–506.

Lárusdóttir, B. 2017. "Að nema land með nafni: sagan af Surtsey." *Skírnir* 191(Spring): 166–185.

Macfarlane, R. 2013. *The Old Ways: A Journey on Foot*. London: Penguin Books.

Massey, D. 2005. *For Space*. London: Sage.

Morton, T. 2007. *Ecology Without Nature: Rethinking Environmental Aesthetics*. Cambridge, MA: Harvard University Press.

Morton, T. 2010. *The Ecological Thought*. Cambridge, MA: Harvard University Press.

Morton, T. 2013. *Hyperobjects: Philosophy and Ecology after the End of the World*. Minneapolis, MN: University of Minnesota Press.

Naiman, R. J. and Décamps, H. 1997. "The Ecology of Interfaces: Riparian Zones." *Annual Review of Ecology and Systematics* 28(1): 621–658.

Nora, P. 1996. *Realms of Memory: Rethinking the French Past*. New York: Columbia University Press.

Ólafsson, E. and Ásbjörnsdóttir, L. 2014. *Surtsey in Focus*. Reykjavík: Edda Publishing.

Pétursdóttir, Þ. 2020. "Anticipated Futures? Knowing the Heritage of Drift Matter." *International Journal of Heritage Studies* 26(1): 87–103.

Þórarinsson, S. 1964. *Surtsey: Eyjan nýja í Atlantshafi – The New Island in the North Atlantic*. Reykjavík: Almenna Bókafélagið.

Smith, L. 2006. *Uses of Heritage*. London: Routledge.

Staiff, R., Watson, S. and Bushell, R. 2013. "Introduction – Place, Encounter, Engagement: Context and Themes." In *Heritage and Tourism: Place, Encounter, Engagement*, edited by R. Staiff, R. Bushell and S. Watson, 1–24. Abingdon: Routledge.

Stewart, K. 2008. "Weak Theory in an Unfinished World." *Journal of Folklore Research* 45(1): 71–82.

Walker, S., Bastow Wilson, J., Steel, J. B., Rapson, G. L., Smith, B., Warren, McG. K. and Cottam, Y. H. 2003. "Properties of Ecotones: Evidence from Five Ecotones Objectively Determined from a Coastal Vegetation Gradient." *Journal of Vegetation Science* 14(4): 579–590.

INDEX

22 July terrorist attacks 185–191, 193–195, 197, 199, 201–204

Achuar, the 139
Adam, Barbra 189
Adorno, Theodor 258
affect 12–14, 24, 39, 44, 108–110, 114, 126, 131–132, 134, 136–137, 139–142, 165–167, 169, 171, 173, 178, 180; affective museums 135–137, 141–142; objects of affect 108, 117, 136, 147, 158, 174; politics of affect 109
Age of Devastation, the 83–84, 86–89
age-value 229–230
Albrecht, Glenn 51
Alexander, Jeffrey 188
Anaconda Smoke Stack, Montana 215, 217–218, 220–221
Ancient Egypt 129, 137
Anderson, Benedict 185, 195
Anthropocene, the 3, 7–12, 15, 22, 24, 29–102; Anthropolithic age, the 4; Anthropozoic age, the 4, 15; heritage of 9–10, 24; and the nature/culture–divide 5, 7, 32, 44, 53–54, 61, 65–70, 82, 93, 167, 310, 342
anthropocentrism 5, 139, 141–142, 208–209, 211–212, 216–220, 246, 306, 308, 315
Appadurai, Arjun 40
Appelgren, Staffan 34
Aquilombamento 351, 361, 366
Architectural Institute of Scotland 117
archives 10, 16, 19, 35, 43, 120–121, 126, 241, 357, 362
Arctic, the 88, 307, 311, 316, 318–320
Aristotle 74, 77
Åsberg, Cecilia 321–322
Ash (*Fraxinus excelsior*) 299
Assmann, Aleida 188, 239
asters 292
Auerbach, Erich 49
Australia 176; Burrup Peninsula 165–166, 169–170, 175, 179; Dennawan 44; Great Barrier Reef 68–69, 78; Great Sandy Desert 167; Karratha 176–177; Murujuga 166–167, 169, 171–175, 177, 179–180; Pilbara 14, 165–169, 171–173, 175–179; Roebourne 173, 176–177; Western Australia 166–167, 170, 178
Australia ICOMOS 12

Baker, Janice 14, 392
Bal, Mieke 188
Bangstad, Torgeir Rinke 178, 295, 392
Barad, Karen 19, 33, 96, 114, 141, 308–311, 319–320
Barker, June 44
Barndt, Kerstin 233, 288, 295
Barthes, Roland 149, 153–155, 157–158
Bartolini, Nadia 149
Bastian, Michelle 318, 323
Bastille, Paris 81
Bateson, Gregory 387
Bauer, Andrew M. 247
Bayon image, the 113, 121–122
Bayon, the 108
Benjamin, Walter 136–137, 158, 201, 258–259
Bennet, Jane 17, 74
Bennett, Wayne 174
Berkeley Pit, Montana 213–216, 219–221
Berque, Augustin 84–85
Bhopal disaster, the 319
Bibliothèque Nationale de France, Paris 122, 124
biodiversity conservation 22, 33, 42
birch tree (*Betula pubescens*) 289–291, 299, 302, 330–331
biting stonecrop (*Sedum acre*) 293

black mold (*Stachybotrys*) 333–334, 336, 344–345
Blue Latitudes Initiative 96
Bogost, Ian 396
Bohlin, Anna 34
Boko Haram 73
Bonnichsen, Rob 85
Bowker, Geoffrey 43
Boxwood (*Buxus sempervirens*) 330
BP oil spill *see* Deepwater Horizon oil spill
Braidotti, Rosi 308, 319, 322
Bramble (plant) 289, 292, 299, 303
Braudel, Fernand 202
Brazil 350–351, 353–355, 359; Boa Esperança, Minas Geiras 22, 351–362, 366; Massambará 362–365; Paraíba Valley 351, 354, 359, 362, 364; Porto das Flores 357, 361; Rio de Janeiro 136
Brennan, Teresa 173
British Association for the Advancement of Science 117
British Library, London 116, 121–122
British Museum, London 14, 129
brownfields 17, 222–223, 227–228, 236, 241
Bryant, Levi 5, 225, 247, 385–386
Buen Retiro Park, Madrid 328
Burdock (plant) 292
Burra Charter 12
Burtynski, Edward 87
Buser, Marcos 34, 39
Byrne, Denis 13

California redwood (*Sequoia sempervirens*) 300
Cambodia 107–108, 110–114, 117; Angkor 13, 107–109, 111, 113, 115, 117
Canvey Wick, Essex 228
Carson, Rachel 5
Castor bean (*Ricinus communis*) 330
Central Interim Storage Facility for Spent Nuclear Fuel, Oskarshamn 37
Chaco War, the 93–94
Chakrabarty, Dipesh 32, 306, 318
Chernobyl disaster 319
chestnut (*Castanea sativa*) 250
China: Three Gorges Dam 68
Clark, Andy 75, 210
Coelho, Rui Gomes 22, 392
Cold War, the 268
Colebrook, Claire 15
Coleman, Claire 173
colonialism 142, 166, 169, 173, 177, 255–256, 350–354, 366, 386
connectivity ontology 59–61
correlationism 66, 70, 79
creeping jenny (*Lysimachia nummularia*) 338
Czikszentmihalyi, Mihaly 210–212

Daisy 292
Dandelion (*Taraxacum officinale*) 299
Darwin, Charles 3, 399
Das, Sanatu 95
Day, Mick 51
de Bolla, Peter 158–159
Deepwater Horizon oil spill 95
Delanty, Gerard 188
Demos, T.J. 166
Denmark 159; Copenhagen 146, 148; Rubjerg Knude, Jutland 8; Zealand 151, 159
Derrida, Jacques 7, 38, 43, 173, 204
DeSilvey, Caitlin 16, 20–21, 55–56, 59–60, 62, 121, 170
Dias, Nelia 35
Divine Cat, the 129–130, 132–133, 137–138, 141
Dock (plant) 292, 299, 302
Domínguez Rubio, Fernando 19, 241, 252, 254, 392
Donaldson, Mike 169
Dooren, Thom van 323
Dorst, Jean 400
Douglas fir (*Pseudotsuga menziesii*) 300
Douglas, Mary 34
Driessen, Clemens 285
Driscoll, Kári 16

earthhuman 175–176, 180
ecology 3–5, 95, 226, 246–247, 259, 292–293, 328–334, 336, 340; deep ecology 5; definition 3, 331, 392; non-equilibrium models 7, 292
ecology of practices 19–21, 33
ecumene 84–85, 87, 88
Edinburgh Photographic Society 117
Edwards, Elizabeth 108, 117–118, 125
Egypt: pyramids 68
Ellam, Bradon 168
Ellis, Erle C. 247
Emma Withnell Heritage Trail, Pilbara 176
Emmet, Robert S. 50, 58
endangered language preservation 33–35
endangerment 9, 33
entanglements 17, 21–24, 289–380
entropy 16, 251, 254–255, 257, 285–286, 297–298, 302, 393
environmental humanities 5, 49, 126, 306
Erll, Astrid 226
Ernst, Wolfgang 16
Ethnological Society of London 117
Eucalyptus (*Eucalyptus globulus*) 253, 260

Fagerland, Tor Einar 197
Faro Convention 7
Farstadvoll, Stein 9, 21–22, 234, 308, 320–321, 392–393
Felman, Shoshana 188
feminism 306–309, 317
Ferrez, Marc 354–355, 359, 366

Final Repository for Short-Lived Radioactive Waste, Forsmark 35, 37
Finland: Inari 370–372, 376–379; Lapland 22, 370, 374, 376, 392
Finn, Christine 391
fireweed (*chamaenerion angustifolium*) 331
Fiskartorpet, Stockholm 313–315
Florence, Hercule 353
Flying Foam Massacre, 1868 177
forget-me-not 299
Foucault, Michel 70
Fredengren, Christina 320–322, 392
Fuji-Hakone-Izu National Park, Japan 58

Gabon 310; Oklo 310
Galison, Peter 40
Gastein prison camp, Norddalen 267–268
Gell, Alfred 134
Germany 52–53; Emscher region 222, 229; Landschafstpark Duisburg Nord, Duisburg 20, 285–288, 293–294, 296; Nürnberg 187; Ruhr region 222, 227–231, 236, 242, 285, 292, 294–295
Gervais, Paul 319
Ghosh, Amitav 31, 318, 320
giant sequoia (*Sequoiadendron giganteum*) 300–301
Glasgow Photographic Association 117
Goat's beard (*Aruncus dioicus*) 340
González-Ruibal, Alfredo 17, 87–88, 93–94, 161, 276, 319
González Zarandona, José 169–170
Gorse (*Ulex europaeus*) 248, 252–254, 256, 260
Government Quarter, Oslo 185, 191, 193, 197, 199
Grafton, Anthony 190
Graham, Brian J. 58
Great Acceleration, the 83–89
Gregory, Kate 177
Guattari, Felix 5, 126, 168, 171

Haeckel, Ernst 3–4, 15
Hällström, Raoul af 374
Halperin, Ilina 174
Hamersley Ranges, Celia Sandy 175
Hancock, Nuala 136
Haraway, Donna 125, 131, 308, 310–311, 317, 321–324, 350
Harrison, Rodney 19, 53, 59, 62, 120, 161–162, 395
Hartog, François 85, 88, 186
Hart's tongue fern (*Asplenium scolopendrium*) 299
Harvey, David 60
Haukkapesäjoki prison camp, Inari 371–373, 375, 377–379
Hawkes, Jacquetta 391
Hawkweed 299
Hazel 299
Heather (*Ericaceae*) 248
Heidegger, Martin 70–71, 78
Herder, Johann Gottfried 202–204
heritage: difficult heritage 50, 60, 165–166, 177, 187; ecology 3, 5, 14, 24, 147, 178, 309; heritage ontologies 33–34, 59–61, 125; heritage photography 107, 125–126; heritagization 199, 307, 320, 322; and humanism 6–7; industrial heritage 49–52, 54–55, 60, 170, 178, 208–209, 212, 216, 222–223, 229–242, 285–296; intangible heritage 12–13, 35, 40, 58, 81, 307, 392–394; more-than-human heritage 10–11, 110, 114, 121–125, 187, 223–225, 227, 242, 307–309, 319–323; national 135, 200, 315; natural heritage 32–33, 35, 68–69, 228, 306, 309, 396–399; as process 12; unruly heritage 11, 69, 73, 76, 227–228; world heritage 43, 170, 230, 397, 400
Hesjedal, Anders 75
Hetherington, Kevin 31, 35, 38
Hevia, James 108, 117, 125
Heyd, Thomas 174
Himalayan balsam (*Impatiens glandulifera*) 298
Himalayan cedar (*Cedrus deodara*) 300
Hirschberger, Gilead 187
Hitchings, Russel 288
Hodder, Ian 75
Högberg, Anders 34, 313
Holifield, Ryan 51
Holling, Crawford Stanley 57
Holocene, the 82–83, 171, 247, 313, 316, 318
Holtorf, Cornelius 34, 45, 313
Hong Kong 136
Hooper, Arthur 44
Horkheimer, Mark 258
Houston, Donna 51
Howard, Peter 58
Humboldt, Alexander von 3
Hume, David 383–384
Huyssen, Andreas 43, 258
hyperobject 31–32, 34, 92–93, 171, 225, 392–393

IBA Emscher Park 229–231, 234
Ice Age, the 83
Iceland: Okjökull 395–396, 400; Surtsey 396–400
image-object 110, 120
Industrial Revolution, the 83
Industrienatur 222–225, 228, 231, 234–235, 241–242, 292–294
Ingold, Tim 17, 138, 149, 158, 190
International Centre for Agricultural Research in Dry Areas (ICARDA) 10
International Committee for the Conservation of the Industrial Heritage, the (TICCIH) 311
Iovino, Serenella 321
Iron Age, the 247

Japan 338; Fukushima 86
Japanese knotweed (*Fallopia japonica*) 234, 298, 337–338
Jørgensen, Dolly 53–54, 231
Joy, Eileen A. 5, 225, 392

ka nimbia 133–134
Kankiniemi prison camp, Inari 370–371, 373
Kant, Immanuel 70, 74, 77
Kilmahew/St. Peter's Seminary, Cardross 20, 286, 296, 298–299, 302, 303–304
Kitzbühel prison camp, Norddalen 264–266, 268, 271–274, 276
Knittel, Susanne 16
Knox, Malcolm 168
Kokerei Hansa, Dortmund 223–225, 230–237, 240–242, 295
Koloniehaus Museum, Filadelfia 93, 98
Koselleck, Reinhart 185, 202
Köttet av en kvinnlig gud (2019) Sara Söderberg 317, 321
Kowarik, Ingo 54, 231–232
Krämer, Sybille 189
Krupar, Shiloh R. 61, 293

Lacan, Jacques 69, 247, 387, 390
Lake Chad, Chad 73
Laos 94, 99, 102; Long Tieng 94, 99
Lapland War, the 374
Larsen, Bjørn Tore 203
Latour, Bruno 8, 72, 81–83, 117, 134, 178, 317
Latz, Peter 287–288
Latz, Tilman 286, 292
Laub, Dori 188
Lawson cypress (*Chamaecyparis lawsoniana*) 299–300, 330
Lebanon: Bekaa 42
LeCain, Timothy 72, 167, 392
LeMenager, Stephanie 93
Leopold, Aldo 49
Lévi-Strauss, Claude 67
Library of Congress, Washington D.C. 120
lifetimes 15, 186–189
Linnaeus, Carl 3
Lipp, Wilfried 230
Lippard, Lucy 166, 173
Lorimer, Jamie 228–229, 285
Louvre, Abu-Dhabi 136
Luhmann, Niklas 71–72, 189–190
Lyell, Charles 3, 315, 318–319, 321
Lyell's Oak 307, 312, 315–316, 319–321

M+, Hong Kong 136
Macdonald, Sharon 52, 165, 187
Macfarlane, Robert 397
Maidenhair fern 299
Malanggan 139
Mallnitz prison camp, Norddalen 267–268, 273–274
Malpas, Jeff 156
Margulis, Lynn 71
Marila, Marko 392
Marris, Emma 294
Masco, Joseph 54
Massey, Doreen 392
McKibben, Bill 171
McLuhan, Marshall 72, 226
McQuire, Scott 121
media ecology 225–227, 309
Meillassoux, Quentin 66
memory 6, 14–17, 183–280; cultural memory 223, 225–227, 242; ecology 3, 208–210, 219–222, 224, 227, 234, 241–242, 264, 277; material memory 16–17, 84–86, 212, 215–220, 223, 231–232, 238, 275; media 223, 225–227, 238–239, 242; memorialization 199, 202; post-humanist understanding of 16; re-membering 208, 213, 216, 219–221; studies 16, 121, 225–227, 239, 277–278
Memory Wound (2014) Jonas Dahlberg 198–203
Mennonites 93–94
Merchant, Carolyn 317
Merton, Robert K. 189
Meskell, Lynn 187, 199
Mirzoeff, Nicholas 359, 361
Monk's House, Rodmell 136
Moore, Jason 350
Morris, David J. 188
Morton, Timothy 34, 92, 95, 171–172, 225, 392–393
Mowaljarlai, David 174
Muaze, Mariana 354
Mullein 292
Mulvaney, Ken 169, 172
Museum of Science and Art, Edinburgh 117
Museum of Tomorrow, Rio de Janeiro 136

Næss, Arne 5
narrow-leaved ragwort (*Senecio inaequidens*) 293
Nativity with Donors and Saints Jerome and Leonard (1510–1515) Gerard David 237
Natural History Museum, Oslo 342
naturecultures 21, 54, 61, 93, 224
Navaro-Yashin, Yael 133, 136
Negarestani, Reza 171
Neolithic, the 247
neophyte goldenrod (*Solidago canadiensis*) 293
Nettle 299
New Climate Regime 81, 83
new materialism 5–6, 59, 74, 147, 214–218, 223, 246, 306, 308, 310
Nixon, Rob 319
Nordblad, Julia 313
Norra Djurgårdstaden, Stockholm 307, 312, 316, 318–320, 322

Norsk Oljemuseum, Stavanger 94–95, 100
Norway 94, 100, 185, 189, 195, 197, 203–204, 308, 328, 330, 338, 340; Fannefjorden 328; Lyngenfjorden 367; Molde 21, 308, 328, 340; Norddalen 17, 265, 267–268, 272; Oslo 15, 185, 187, 191–193, 199; Sørbråten 199–203; Spitsbergen 35; Stavanger 100; Svalbard 10, 35–37, 40, 42–43; Utøya 15, 49, 185, 187, 190–195, 197–200
Norwegian Broadcasting Company *see* NRK
Norwegian Petroleum Museum *see* Norsk Oljemuseum, Stavanger
NRK 191, 193, 195, 197–199
Nye, David E. 50, 58

Oak (*Quercus robur*) 250, 255
object-oriented ontology 147, 308–309, 320, 389
oikos 3, 6, 9
Olivier, Laurent 12
Olsen, Bjørnar 34, 68, 75, 148–149, 155, 158–159, 170, 259, 308
ontic wilderness 67, 71
ontological turn, the 13, 33, 82
Opperman, Serpil 320
orephosis 165, 167–168
Ötzi 88
Ovenden, Richard 116, 118

Pale chanterelle *(Cantharellus pallens)* 313
palliative curation 170, 174
Papua New Guinea 133, 139
Paraguay 98, 99; Chaco region 94, 102; Filadelfia 93; Loma Plata 94, 99
Parikka, Jussi 227, 318
Paris Agreement 92
Paterson, Alistair 177
Pearce, Susan 7
Pearse, Guy 168
Periwinkle (*Vinca minor*) 328, 338
Peters, John Durham 226
petrocultures 92–93, 96
petroglyphs 165–167, 169, 172, 175, 179–180
Pétursdóttir, Þóra 34, 68, 78, 148–149, 156, 158–159, 166, 170–171, 259, 307–308, 320–321
phenomenology 79, 174, 186, 386–388
Pitt Rivers Museum, Oxford 139
plantain (plant) 299
Plantationocene, the 350–351, 366
Pleistocene, the 83, 172
Pollan, Michael 287
Pomian, Krzysztof 202
Portugal 350; Alentejo region 363
post-humanism 5, 13, 16, 19, 59, 76–78, 147, 166, 171, 223–225, 306–307, 310
Postman, Neil 226
post-processual archaeology 83
Probyn, Elspeth 173
processual archaeology 83
Prosper-Haniel colliery, Bottrop 223
Proust, Marcel 210
Prøvestensfortet, Copenhagen 146, 148, 150–151, 154–156
punctum 149, 155–158
Purple loosestrife (*Lythrum salicaria*) 234

Raspberry (*Rubus idaeus*) 331
Rathje, William 85
Reaganism 388
Reflecting Absence (2004) Michael Arad and Peter Walker 202
Renaissance, the 133, 135, 237
Retiro landscape garden, Molde 328, 330–331, 333–334, 336, 338, 340, 342, 344–345
Rhododendron (*rhododendron ponticum*) 297–299, 302
Roberts geranium (*Geranium robertianum*) 234
Robin, Libby 57
Rochberg-Halton, Eugene 210–212
Romantic Period, the 109
Roscoe, Paul 133–134
Rose, Deborah B. 168, 173, 311, 318, 322–323
Rosenberg, Daniel 190
Rosenberg, Elissa 292–293
ruderal species 10, 16, 54, 223–224, 228–229, 231, 241, 299, 338

Samuel, Raphael 120
Saxifrage 299
Schaefer, William 108, 112, 114, 124
Schjeldal, Peter 204
Scotland: Glasgow 286
Scots pine (*Pinus sylvestris*) 299
Secret War, the 94
Sernander, Rutger 307
Serres, Michel 69, 76
SFR *see* Final Repository for Short-Lived Radioactive Waste, Forsmark
SGSV *see* Svalbard Global Seed Vault
shattered time 185, 188, 193, 199
Shuar, the 139
Silver birch (*Betula Pendula*) 236
Silver fir (*Abies alba*) 328, 333, 338
Simmel, Georg 285, 287–288, 296, 302
SKB *see* Swedish Nuclear Waste Management Company
slavery 350–366
Sloterdijk, Peter 33
Smith, Laurajane 12, 58, 177, 307, 310
Soapwort (*Saponaria officinalis*) 234
Sorokin, Pitirim A. 189
sow thistle (*Sonchus oleraceus*) 299
Spain 246, 248; Arén 250; Galicia 17, 247–251, 255, 257; Meilide 250
Spelman, Elizabeth V. 55–56

Spittal prison camp, Norddalen 266–271, 277–279
Spruce (*Picea abies*) 342
Stengers, Isabelle 19–20, 33, 223, 317, 372
Sterling, Colin 13–14, 19, 392
Stewart, Kathleen 131–133
Stiegler, Bernhard 226
Storm, Anna 34, 233, 392
Striphas, Ted 122
surface ecology 147, 149, 155–158, 161, 163
Svalbard Global Seed Vault 35–37, 40, 42–43
Swanton, Dan 294
Sweden: Forsmark 35–37; Husarviken 306–308, 312, 316–317, 319, 321–322; Oskarshamn 37
Swedish Nuclear Waste Management Company 34–35
Sycamore (*Acer pseudoplatanus*) 333, 338
Syria: Aleppo 10, 42

terrorism 15, 185–191, 193–195, 197, 199, 201–204
Thompson, Michael 34
Thomson, John 107–108, 110–124
Tilley, Chris 174
time-shattering event 185–187, 195
timelines 186–201, 306
toxic waste 9, 16, 34, 38–43, 50, 54, 58, 95–96, 214–220, 225, 239, 312, 319, 385
Truelshøj 150–156
Tsing, Anna 43–44, 308, 312, 322, 350
Tuominen, Suvi 392
Turkey: Çatalhöyük 75
Tutsan 299

Ukraine: Chernobyl 86
UNESCO 12, 68, 170, 230, 399
United Arab Emirates: Abu-Dhabi 136
United Kingdom, the 385; London 120
United States 359, 383, 385; Anaconda 213, 215–217, 219; Ashio 219; Butte 213–214, 216–220; New York 359, 365; Redwood Forests 68
Unruly Heritage: An Archaeology of the Anthropocene 34
USSR 388

Valley of the Sun, Burrup Peninsula 173
Varutti, Marzia 14
VG (*Verdens Gang*) 194, 197–198
Vidal, Fernando 35
Vietnam Veterans Memorial (1982) Maya Lin 202
Vogel, Susan 136

wall-rue (*Asplenium ruta-muraria*) 234, 236–239, 242
Wellcome Collection, London 107, 111, 120, 122
wet collodion process 110, 112, 114, 120
Wild carrot (*Daucus carota*) 234
Willow 289
Willowherb 292, 299, 303
Witmore, Christopher 76, 81
Wolin, Richard 258
Wood fern (*Dryopteris filix-mas*) 333
Woods, Lebbeus 55–56, 62
Woods, Whitney E. 58
World War II 17, 52, 83, 227, 233, 264–279, 369–371, 374, 376, 379

Yaburura 177
Yaburara Heritage Trail, Pilbara 177
Yangoru Boiken 133
yellow brain fungus (*Tremella mesenterica*) 333, 336
Yew (*Taxus baccata*) 330
Yindibarndi 175

Zeche Zollverein, Essen 230
Zerubavel, Eviatar 185
Zylinska, Joanna 108, 113, 121